HI-FI IN THE HOME

HI-FI
IN THE HOME

John Crabbe

Editor, *Hi-Fi News & Record Review*,
Studio Sound and *Audio Annual*

BLANDFORD

First published in 1968
Second Edition 1970

ISBN 0 7137 0494 2

Printed in Great Britain by
Fletcher & Son Ltd, Norwich
Bound by Richard Clay (The Chaucer Press) Ltd,
Bungay, Suffolk

ACKNOWLEDGEMENTS (1st Edition)

Thanks are due to numerous individuals and manufacturers in the audio industry for much valuable information and many photographs, and particularly to the staff of *Hi-Fi News* and *Tape Recorder*, with David Kirk and Frank Jones clarifying my thoughts through helpful discussions, and Anne French coping with a multitude of hand-written pages full of complex amendments. I much appreciate the work of Rex Baldock, the audio consultant, who read through the typescript looking for technical errors, and Stanley Whitehead who gave me the reactions of a non-technical layman. John Williams managed to find time between guitar recitals to read through the script with a musician's eye, and Geoffrey Horn, who runs an excellent hi-fi shop in Oxford, gave me the benefit of a retailer's comments. Of course, none of these kind people must be held responsible for any mistakes that might have crept in or for my views on controversial matters. My wife, Gwen, must receive the final accolade for her heroic tolerance – for many months – of an unhusbandly husband, for keeping the children out of my hi-fi hair, and for much valuable help on points of style and readability. But again, if the book is unreadable that is my fault, not hers!

J.C.
Norwood, 1968

NOTE FOR 2nd EDITION

Recent developments in tape cassettes and four-channel stereo have been covered, prices of equipment and possible budgets amended, and the list of suggested recordings updated. Also, chapters dealing with equipment now carry itemised page headings to facilitate quick reference to the various sections. Other minor corrections, additions and changes of emphasis have been made, including an expanded and rationalised bibliography, a few more terms in the glossary, and some fresh items in the collections of equipment illustrated.

J.C. 1970

CONTENTS

Page

ACKNOWLEDGEMENTS 5

INTRODUCTION: THE OBJECT OF THE EXERCISE 8

Chapter

1. WHAT ARE MUSICAL SOUNDS? 14

   The nature of musical sounds and how our ears respond to them.

2. SORTING OUT THE STRANDS 36

   Which features of these sounds determine the technical requirements for recording and reproduction.

3. BASIC COMPONENTS FOR MUSIC REPRODUCTION 45

   A look at the basic elements common to all music reproduction, as a prelude to defining hi-fi.

4. HIGH FIDELITY I: TAPPING THE SOUND SOURCES 66

   Pickups, turntables and tape recorders.

5. HIGH FIDELITY II: RECEIVING, AMPLIFYING AND REPRODUCING SOUNDS 98

   Tuners, control units, power amplifiers, loudspeakers.

6. MUSIC TO LISTENER OR LISTENER TO MUSIC? 135

   Space, ambience, stereophony and music's environment.

7. HOW TO CHOOSE EQUIPMENT 154

   Practical guidance, with hints and tips on judging quality.

8. INSTALLATION 192

How to set-up your equipment, various problems, making the best of room acoustics, etc.

9. ADVENTURES IN LISTENING 233

Choice and care of records, notes on planning and giving recorded concerts – making music in the home a community activity.

10. MUSIC IN THE HOME 252

Speculation on present and future, psychology and aesthetics of listening, musical values at home and in the concert hall.

GLOSSARY OF AUDIO TERMS AND ABBREVIATIONS 272

BIBLIOGRAPHY 317

INDEX 324

INTRODUCTION

THE OBJECT OF THE EXERCISE

In the last forty years a vast new public has come to appreciate good music through the technical media of radio and gramophone. This audience starts by listening to broadcast Promenade Concerts or buying occasional records, and eventually its more enthusiastic members go along to hear the real thing in concert hall or opera house. The experience of musical reality often makes a deep impression, underlining the fact that music reproduced domestically via loudspeakers is, for most people, very much a second-best business. Nevertheless, as techniques have improved and the age of high fidelity has arrived, so the unavoidable gap between reality and its reproduced illusion has narrowed. The quality of musical experience possible in the home is now very high and can offer, in some cases, serious competition to the concert hall itself. This is in terms of technical sound quality alone, and when other factors are taken into account, such as the standard of performance, avoidance of audience noise, and the social and geographical complexity of visits to live performances, the appeal of good reproduction can be very strong.

However, for many the very term 'hi-fi' has unfortunate connotations, being used in a gimmicky mod-pop way to advertise cosmetics, appearing in gilt lettering on some tasteless highly-polished radiograms, or evoking visions of technical fanatics making life intolerable for their families with constant playing of percussion records at high volume and masses of wire and electrical instruments strewn across the floor. Apart from those for whom the subject has such off-putting associations, there are many thousands of ordinary concert-going music lovers who have come across hi-fi only via advertisements. The majority of these have probably not heard a really satisfactory demonstration of reproduced music, and consequently tend to be sceptical about the ability of mere equipment to re-create the invigorating experiences known in concert hall or theatre, recital room or jazz-cellar. Others, perhaps attending a properly presented domestic gramophone concert, are first taken aback by the discovery that listening to reproduced music can be a very vital activity, and then brought to the verge of disbelief when told that the sounds heard come from ordinary commercial discs and could be simulated in their own homes via standard hi-fi equipment.

This book is written for all three groups of people in the hope that those who have not previously done so may be persuaded to seek out and hear for themselves the high level of musical quality now attainable in a domestic setting. Those who have toyed with the idea of 'going hi-fi' but who are deterred from actual purchase by the incomprehensible technicality of many advertisements, leaflets and demonstrations, should also find the following chapters useful in their efforts to distinguish between ends and means in a field where music and engineering tend to become confused.

The expression hi-fi means quite simply high fidelity, or a high degree of truthfulness to the original sounds in the reproduction of music, and those whose acquaintance with hi-fi and stereo has inclined them to believe that the business is a commercial racket have been the unfortunate victims of technique in the hands of the non-musical. When technicians with some musical sensibility are really given a chance the results can be most satisfying, and certainly should not exhibit the strident 'edgy' reproduction still assumed by many to characterise hi-fi but which is avoided when good equipment is used correctly. Unfortunately there is a strong tradition among manufacturers tending to emphasise the technical rather than the musical merits of reproducing equipment, and this bolsters the conviction of the unconverted that the whole business is a mere hobbyists' playground. But the music lover's increasingly important role as a customer is beginning to influence Sales Managers, so we can, perhaps, expect a gradual shift of emphasis away from engineering and towards music.

Music heard in the flesh is a unique experience. Much of the uniqueness is due to the precise and detailed character of sounds we hear and the nature of their relationship to the hall or room in which they originated. With adequate technical resources to hand most of the atmosphere of an original performance may be caught and recorded to be reproduced later with – in some cases at least – a quite surprising degree of realism. But the 'realism' is, of course, a subjective illusion helped along by one's involvement in the music itself, and for this reason judgements are bound to vary with personal taste and with the type of music. But at its best the technique of sound reproduction has now reached the point where listeners can accept and then ignore the obvious and irreducible differences between reality and illusion, and as a visit to the concert hall is not always convenient or possible it is now feasible to provide a very satisfying substitute at home.

However, there are many music lovers who remain quite content with their conventional radiograms, record players and transistor radios, accepting a total distinction between real music and its domestic duplication. Others have developed a liking for music based entirely on mediocre reproduction of radio and gramophone records, and have therefore never experienced the many subtleties of tone-colour on which so much orchestral music depends for its full impact.

Although faithful regeneration of a wide range of instrumental timbres is by no means the sole object of hi-fi, it is certainly its starting point. As the value of this very first step towards better music in the home is sometimes challenged, it will be worth digressing for a few paragraphs to examine tone-colour as an element in musical structure and appreciation.

The first point is that if music merely comprised successions of notes arranged in varying degrees of contrapuntal complexity, composers would not bother to specify different instruments for the various parts – apart, that is, from the need for a wide pitch range. Throughout the history of music, evolution of instruments and the development of composition have gone hand-in-hand. In the Baroque period, for instance, composers were acutely aware of the need for contrasting tone-colours to add variety and interest to concerted music; concertos by Bach and Vivaldi often progress, movement by movement, through differing groups of instruments, and much of the beauty is lost if the changes are removed. Later, with the emergence of sonata-form, the constant reappearance and transmutation of phrases and melodies is aided enormously by a corresponding instrumental diversity.

Each instrument also demands its own style of writing, and if its correct tonal character is not produced (or reproduced) the intentions of composers are not fully realised. Mozart saw the emergence and acceptance of the clarinet as a serious instrument, and his wonderful Quintet and Concerto (K. 581 and 622) are the first major works to exploit fully this instrument's lovely rippling quality. Fortunately, Mozart never had to tolerate poor reproduction of clarinet tone, but had his patrons always listened to music via radiograms or record players he may not have felt so inspired by its instrumental possibilities. Similarly, when Beethoven introduced the trombone to the symphony for the first time with that thrilling outburst in the finale of the fifth symphony, the impact was unique and entirely dependent on the instrument's particular character. For that matter, similar arguments apply to the saxophone in the hands of Ravel, to Wagner's tubas, and to the bass clarinet in Stravinsky's 'Rite of Spring'. Jazz also has much to do with instrumental qualities, a vital link with the performer's feelings often depending on just so much huskiness from a saxophone or on an exact balance between bite and mellifluence in expressive trumpet playing.

In fact, the characteristic sounds of various instruments are an essential part of much Western music, for, in addition to clarifying the texture of complex scores, they each have qualities appropriate to different musical moods. The lonely sadness of the cor anglais used by Sibelius for the 'Swan of Tuonela', the cascading horns in the finale of Dvorak's G-major symphony, the ecstatic beauty of quiet string playing in Vaughan Williams' 'Tallis Fantasia' – in these and a multitude of other works the total musical effect depends as much on the quality of the sound itself as on a succession of notes in the score. If this were not so we would be content with the piano transcriptions of classics enjoyed by our great-grandfathers before radio and the gramophone had opened up a new path to music in the home.

Faithful representation of instrumental qualities is, then, an essential part of any plan to recapture the rich musical experience of an actual performance. Without the full panoply of tone-colours clearly displayed we have, in one sense, a mere silhouette of the real thing, with the texture obscured and many subtleties lost.

It is sometimes argued that fidelity in reproduction need be no higher than is necessary simply to permit differentiation of the various instruments. Imagine what would happen to the reputation of a conductor who applied such a modest standard to orchestral players at rehearsal! In practice, good string players try to avoid a rough and edgy tone, the brass section aims for power and penetration without ugly blasting, the oboist hopes to maintain a sweet and delicate tone without going wheezy or sour, and so on. If such things matter in the concert hall, why should they be less important at home?

Anything that degrades musical quality, whether it be human or technical, is worthy of attention, and the following chapters are intended partly for those whose temperament puts them in sympathy with this rather uncompromising viewpoint. Less perfectionist readers will not, I hope, give up before concluding that this business of high fidelity is at least worth investigating. Whether we like it or not, an increasing proportion of music listening will take place via loudspeakers in the home; if those who care for music apply the right standards and make the right demands, this listening revolution need not be a disaster and could be very fruitful. The technical and musical background to this process of changing listening habits is related in the following pages, with the object of introducing ordinary music lovers to what may well be a large part of their own future.

The book proceeds roughly as follows. First, starting at grass roots, an attempt is made to describe the nature of real musical sounds and how our ears respond to them; then we see which features of these sounds determine the technical requirements of a recording and reproducing chain. Next comes some consideration of the basic elements needed for music reproduction, as a prelude to two sections describing and explaining the requirements for very high quality sound. These two chapters are the technical core of the book, and if not absorbed fully at a first reading will form a convenient base for reference back as practical points are brought up in later sections. Then comes a fundamental question: is the object of music reproduction, figuratively speaking, to bring performers to listeners' homes or transport listeners to the concert hall? This raises the whole matter of music's acoustic environment, and the subjects of space, ambience and stereophony. The latter receives fairly full consideration, as stereo has proved to be quite the most important single technical advance in sound reproduction since the coming of the long-playing record, with more purely musical implications than is commonly realised. Then some guidance is offered on the choice of equipment, with hints and tips on judging quality and value-for-money. Installation in the home follows, with notes on some common pitfalls and problems, and a special concern for provision of maximum musical enjoyment despite variations in room acoustics. Such matters as the fitting and interconnecting of equipment and positioning of loudspeakers are covered here, making this section the nearest we shall get to hi-fi as a constructional hobby.

As the main point of high-quality domestic sound is (or should be) to serve the art of music, the next section deals with choice and care of records and the use of equipment controls to give pleasing musical results, with some suggestions for those who would like to give planned concerts for friends who may not otherwise hear good music with anything like its real-life impact. The final chapter attempts to assess the present state of the art, examine some basic limitations, take a look at possible future developments, and then delve a little into the psychology and aesthetics of listening. Consideration of remaining differences between reality and its reproduction here provides the starting point for a discussion of various psycho-acoustic problems, leading to some speculation on ways in which this twentieth century music-in-the-home may actually provide new types of musical enjoyment and create fresh aesthetic values.

Despite my intention to write a book which could be understood by previously non-technical music lovers, some specialised words have been used, and each such term is defined in a Glossary at the end. I suggest that the reader refers to this section just as soon as an unfamiliar word raises a barrier to proper understanding. Where first introduced or when first used in a specialist context, the more important or obscure technical expressions will usually appear in italics. Also, there is a steady introduction throughout the book of symbols and abbreviations, which are explained or made obvious when first used and are thereafter employed whenever the text might otherwise become cluttered with oft-repeated technical phrases. However, to help those reading isolated chapters or using the book simply for reference, definitions of these and other abbreviations commonly found in audio literature are included in the Glossary. No mathematics is used and there are no theoretical circuit diagrams, though graphs abound as these are able to convey otherwise difficult relationships in an easily assimilated fashion. The reader unfamiliar with the graphical approach will, I am sure, find it well worthwhile when tackling a previously mysterious subject. Magazines or books referred to in the text and a selection of other books for further reading are given in a Bibliography, and manufacturers whose products are mentioned are also listed.

Being published in Britain, an assumption is made at one or two points in the book that the reader is resident in the U.K. However, all the basic information and descriptive material is relevant to high-quality domestic music reproduction anywhere in the world, and readers in North America, Australia, New Zealand and English-speaking Africa will find 95 per cent of this course in hi-fi equally applicable in their own countries.

1. WHAT ARE MUSICAL SOUNDS?

MUSIC REPRODUCTION may be likened to a journey from reality to illusion via a number of technical stages. The reality consists of actual sounds emitted by musical instruments blown, plucked, scraped or struck by hard-working musicians in a hall or studio. The illusion exists in the minds of listeners and, if perfect, comprises subjective aural impressions identical to those obtained by sitting-in at an actual performance. In so far as the impressions *are* identical, the path taken may be considered circular, with the traveller arriving back at the starting point. In practice, of course, start and finish are never exactly coincident: the route is not a true circle and there is some blurring of the reproduced image – rather as in a poorly focused photograph.

Deviations from perfection arise because of the immense range and complexity of real musical sounds, and as these sounds are the raw material of domestic high fidelity, this chapter is concerned with their fundamental properties and how the ear responds to them. It is hoped seldom to lose sight of this elemental musical foundation stone, for though music is an abstract, non-representational art, its subtleties depend entirely on minute sonic details. A journey beginning and ending with music should not at any point become so abstracted that all musical significance is lost in a maze of technical specifications.

However, before tackling musical sounds as such it will be as well to clarify the simple word 'sound' itself, as it has two basically different meanings, one objective and the other subjective. The first is concerned with physical vibrations which may be classified and explained in scientific language; the second usage refers to what we hear – the actual aural experience corresponding to, but not the same as, the measurable external disturbances. We shall be jumping from one to the other very frequently in this book, so the reader should be alert to the distinction between acoustic sound (vibrations) and mental sound (sensations). A deaf person might understand and remember an exhaustive mathematical analysis of a C-major chord, but this would bring him no nearer to the subjective experience of hearing one.

Sound (objective) consists of physical palpitations – commonly in the air, but not necessarily so. These vibrations are very rapid by everyday standards. For the lowest note on a piano the string oscillates back and forth $27\frac{1}{2}$ times in one second, while the highest note represents over 4,000 such motions. The number of complete to-and-fro movements, or *cycles*, per second is called the *frequency*, one cycle per second being represented by the unit called a *Hertz*

(Hz). Whatever the instrument and however high or low the frequency, once the surrounding air has been set in motion the vibrations travel outwards as *pressure-changes*. It may be imagined that a vibrating object alternately pushes against and moves away from the surrounding air molecules, thus compressing or rarefying the air according to its own pattern of motion. This is represented in **fig 1,** which shows how the air might look if we could see the pressure differences at a given instant. Once generated in air, these pressure-changes constitute *sound-waves* that move away at a speed of about 1,130 feet per second. The word 'waves' arises because of an analogy with ripples on the surface of a pond, where it can be seen that successive peaks or troughs are more closely spaced for rapid undulations than for slow ones. With sound-waves the corresponding points of high or low air pressure (acoustic peaks or troughs) are also closer together for high frequencies than for low, the actual distance for one cycle being known as the *wavelength* (see **fig. 1**). It is convenient to remember that the wavelength in air of the standard A (440 Hz) is just over $2\frac{1}{2}$ feet, a distance that doubles for each octave down and halves for each octave up.

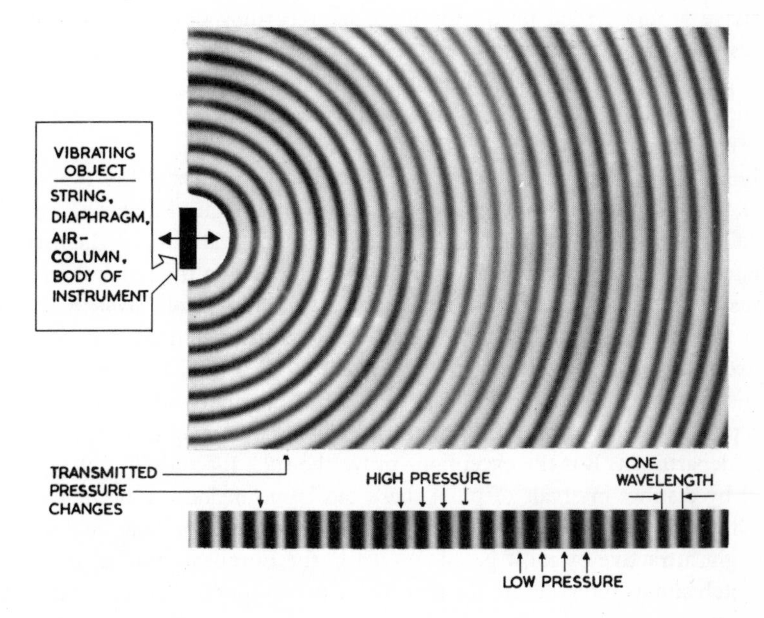

Fig. 1 Sound-waves travel away from a vibrating object as a pattern of pressure differences in the air.

But the analogy with waves on water must not be overstretched, for whereas these carry energy outwards as expanding rings on a plane surface, sound-waves move out in all directions more in the manner of expanding spheres – apart from some special cases where the sound tends to 'beam'. When musical instruments are played, an incredibly complex and constantly changing pattern of ever-expanding sound-waves is generated, to pass, perhaps, a pair of ears or a microphone which responds in a manner not unlike the floating cork on a pond that bobs up and down in sympathy with sundry ripples. These elusive and transitory tremors in the air are not themselves music – they are subtle sonic messengers created by brass, wood, steel, gut and skin from the muscular, nervous and mental efforts of performing musicians. The message is conveyed to the listeners' ear drums, where by a complex acoustic and nervous process not yet fully understood the sound-waves create nervous pulses which enter the brain. Then mind comes into the picture and we hear the sound (subjective) as music.

What we hear depends on the precise nature of the sound patterns travelling through the air, and as this is the link in the music-making chain where the message is intercepted by microphones for recording or broadcasting purposes, we shall take this as our reference point for a look at musical sounds. The pure sound-waves themselves may be related to known and remembered sound qualities on the one hand and to their method of generation in musical instruments on the other.

Beneath the formal superstructure of music we perceive the four basic elements: rhythm, melody, harmony and counterpoint. Similarly, we can usually ascribe four characteristics to the instrumental 'voices' themselves, whatever parts they may be allotted by the composer in the musical scheme. Thus a violin produces notes of recognisable *pitch* at a certain *loudness* in relation to what has gone before; its *tone-colour* is distinct from that of other instruments, and it is played in a particular room or hall that adds its own acoustic colour or *ambience*. These four features apply to all musical sounds except a few without precisely identifiable pitch emanating from the percussion department. But the exceptions prove the rule, for without notes separated by distinct intervals of pitch there can be no melody or harmony, and pitchless rhythm, however colourful, cannot continue for long and still be thought attractive by most people reared in the European musical tradition. So pitch and its intervals are the first things to consider in a survey of musical ingredients.

Pitch is what we hear and describe loosely as 'low', 'medium' or 'high' – vague language representing fairly accurate impressions enabling us to place

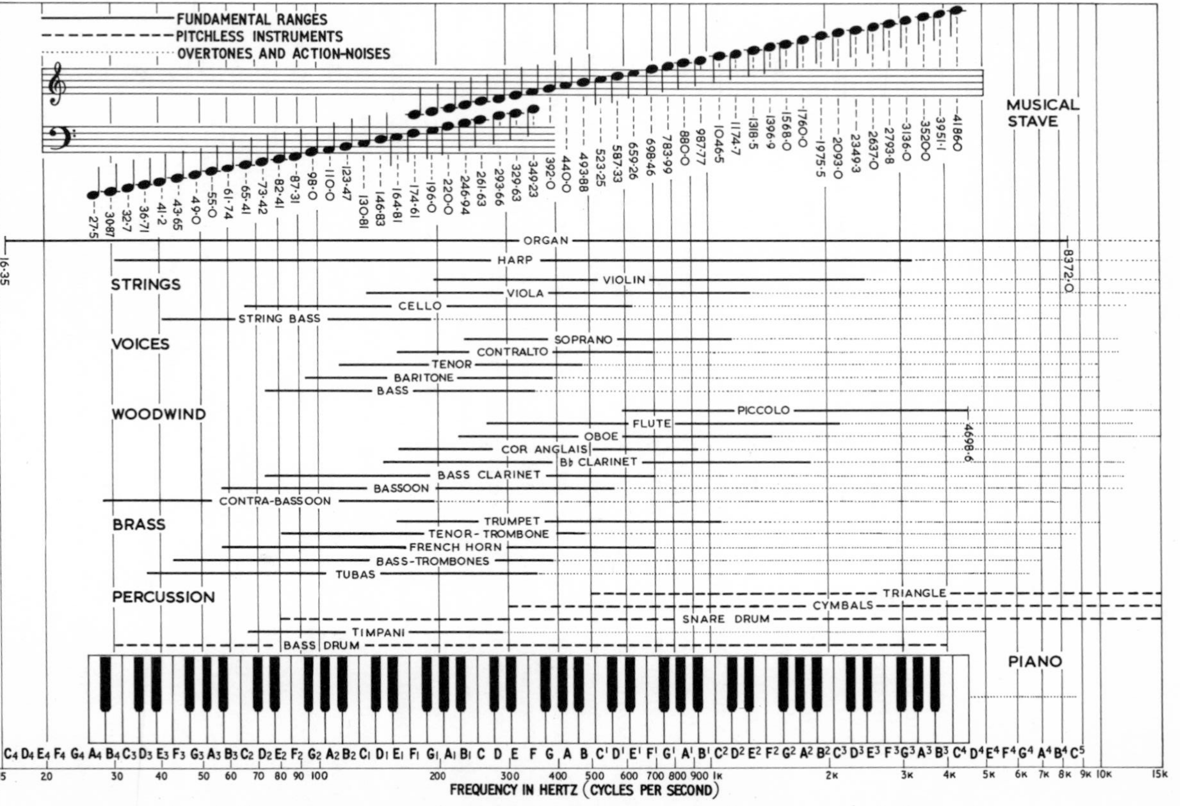

Fig. 2 Frequency compass of orchestral instruments and voices shown in relation to the musical stave.

notes in the musical scale. Some instruments cover limited ranges of notes only, while others – such as the piano or harp – are more versatile. **Fig. 2** illustrates the pitch coverage (in some cases only approximate) of the principle orchestral instruments and some voices, shown in relation to the piano keyboard and printed stave. We have already noted the wide frequency span of musical sounds, and the horizontal scale along the bottom of **fig. 2** with lines extending vertically through the various instrumental groups gives this in Hertz and thousands of Hertz. In addition, the individual frequencies of all the white piano notes are given beneath the stave.

It is obvious from this diagram that what we hear as pitch is related very definitely to frequency, deep sounds going with low frequencies and high ones with high. Further study of the drawing shows also that the band of frequencies covered by an instrument is related very much to its physical size. The string bass, for example, extends from 41·2 Hz to 196 Hz, where its much smaller but similarly shaped relative the violin takes over and goes up to 2,637 Hz. For readers to whom notes come more easily than frequencies, these spans correspond to E_3 – G_1 and G_1 – E^3 in the equally tempered scale, using the rationalised notation marked beneath the keyboard in **fig. 2.**

The question of size and frequency takes us a step further in our enquiry into musical sound, for now we touch on the basic principles underlying all instrumental behaviour. We have seen that sound consists of rapid fluctuation of pressure in the air, but these oscillations originate either from vibrating objects 'coupled' to the air in some way or from enclosed volumes of the air itself. The original vibrations, whether of air-columns, stretched strings, reeds, diaphragms, metal tubes or wood blocks, are generated by *resonance*, a characteristic whereby objects satisfying certain mechanical requirements tend to vibrate or oscillate at specific frequencies if 'excited' from outside. In most cases, because of the underlying physical properties of resonance, the lower the desired frequency the greater the necessary size of an instrument.

Stringed instruments are the most important group employing 'vibrating objects', and as for a practical range of string tensions and suitable loudness of tone the strings need to be longer and thicker for lower frequencies, this inevitably makes the double bass larger than the violin. Similarly, the bass drum has a larger skin than the timpani, and the relative massiveness of harp and piano is determined by the lowness of their bottom octaves.

In wind instruments the frequency of a note depends mainly upon the effective length of a resonant air-column, for as the speed of sound in air is fixed (more or less) and the duration of each cycle of vibration or pressure-change must become greater as the frequency is lowered, it follows that the air-columns – and with them the sizes of instruments – become larger as one probes more deeply into the bass. To take two woodwind extremes, piccolo and contra-bassoon have approximate total air-column lengths of one foot and 16 feet respectively, for bottom notes of 587·3 and 29·1 Hz. This is another way of looking at the wavelength idea mentioned earlier, for in fact the effective air-column length in woodwind instruments often corresponds to a half-wavelength for the note being played, and a simple inverse length/frequency relationship must hold.

Returning to **fig 2**, it is evident that most musical instruments with definite pitch come well within the range of the piano keyboard. Indeed, apart from the piccolo, which transgresses upwards by a mere two semitones, only the organ explores the extremes. At 16·35 Hz, the note C_4 is not only the lowest normally found in music (a few freak organs go down to C_5 and up to C^6), but is very near the practical limit of human hearing in the downward direction. Below this one begins to sense the separate cycles or pressure changes, rather as in cinematography the individual pictures cease to merge into a moving pattern when projected at a similarly low repetition rate. At the other end, C^5 (8,372 Hz) is heard as a distinct but extremely high note by most people, though in practice organists would normally use the top octave for harmonic enrichment only, not for melodic lines.

The frequency ranges of some pitchless percussion instruments are indicated by thick dotted lines in **fig. 2.** The triangle, cymbals and snare drum are shown as extending to 15,000 Hz – the approximate upper limit of human hearing for most people – but they are not unique in this. The solid lines show only the range of *fundamentals* on other instruments, with finer broken lines representing overtones or *harmonics* that extend well beyond the upper notes themselves.

In the limit, then, musical pitch can extend over a frequency ratio of 500 : 1, corresponding to the full chromatic range of 109 semitones in the span of nine octaves between C_4 and C^5. With the commonly used equal temperament scale, the frequency difference between each successive semitone is just under 6 per cent, and when allowance is made for inevitable mis-tuning in practical music-making, and for the fact that many instrumentalists drift away from tempered intervals when playing alone, it is apparent that actual musical notes may crop up *anywhere* in this enormous range. In practice we can be a little more conservative, especially if the very largest organs are excluded. In orchestral music the range of fundamental notes is confined to 29·14 Hz (A_4 – contra-bassoon and Wagnerian contra-bass tuba) at the bass end, and 4,698·6 Hz (D^4 – piccolo) at the top.

But so far we have investigated only one of the four aural features that characterise musical sounds, and as we examine tone-colour it will be evident that a simple statement about pitch or frequency – particularly in the deep bass – is not adequate for an understanding of how or what we hear. Similarly, we shall see that both pitch and tone-colour depend upon loudness, which in turn becomes involved with ambience.

Apart from pitch, every musical sound has an individual character loosely categorised as tone-colour or timbre. Instruments playing the same note produce different sounds – if they didn't composers would not be bothered with orchestration and music would be a less important art. Reference to **fig. 2** shows that all the major tuned instruments except the contra-bassoon, piccolo and double bass may be used to play middle-C (261·63 Hz), yet if each instrument in turn produced this note before an audience of blindfolded concert-goers it is likely that most listeners would at least attribute the sounds to their correct instrumental families. The viola might be confused with violin or cello, oboe with cor anglais, and French horn with trombone or tuba, but use of an alternative common note or a slight change of loudness would soon sort things out.

In addition to the basic tonal differences between the various orchestral sections, each instrument has its individual range of tone-colours varying with pitch, loudness, style of playing, or all three. To take an extreme case, the French horn may produce a round, mellow sound or – the player pushing his fist firmly into the bell and blowing hard – a strident, brassy, attention-attracting noise. Two horns playing the same notes in these different styles have little in common but pitch, a fact noted and used by many composers in the search for colourful musical expression.

Less dramatic but very important variations may be found in the sound-qualities of nearly all instruments. A violin, for instance, may be bowed near the bridge or near the finger board; it may be muted, plucked, or struck with the wood of the bow; the bow may be bounced rather than drawn across the strings; the player may sound harmonics or – by quivering his bow – create tremolo effects; and by double or triple stopping it is possible to produce chords. All these modes of playing result in separate sound qualities distinct from that associated with simple bowing of a single string, and all are used by composers for various musical effects. When, too, we remember that for much of the time string players waver their pitch by means of vibrato, and that in a symphony orchestra there may be up to thirty violins playing at once, it is obvious, before we even begin to think about the acoustic facts underlying the aural impressions, that musical sounds are enormously rich and complex.

What produces this richness, even in individual instruments? Why, when a trumpet, violin, soprano voice, flute and harp all sound the same note, do they sound it so differently? The answer has two parts. Firstly, an apparently simple, single note is made up from various constituents – usually a vibration corresponding to the pitch of the note itself (the fundamental) plus sundry overtones and other colorations. Secondly, a note must start, continue and finish, and the rapidity of beginning and end, steadiness or otherwise of the middle portion, and change of overtone content while the note is sounding – all these features affect tonal quality.

Let us start with overtones. We have conceived sound as vibrations in the air at particular frequencies, created in many musical instruments by mechanical devices such as strings or reeds buzzing back and forth. Imagine this oscillatory process slowed down to the tick-tock rate of a pendulum clock, and picture the pendulum itself – with a pen attached at the bottom – tracing a line on a moving band of paper as in **fig. 3.** The simple, steady motion produces a pattern of the sort illustrated, known technically as a *sine-wave* because of the mathematical law that it follows. If the pendulum swings at twice the fre-

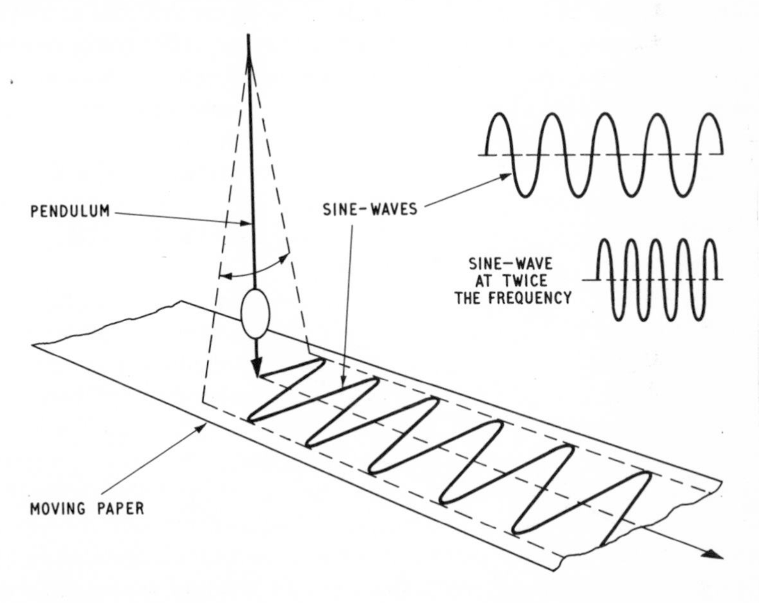

Fig. 3 Motion of an oscillating pendulum recorded on moving paper reveals a sine-wave pattern. Any sound or vibration, however complex, has a corresponding graphic waveform.

quency, the vibration-curve or *waveform* will be more cramped as shown, but otherwise the basic shape is retained. The moving paper represents passing time, and each complete motion of the pen from the central position out to one extreme, across to the other side and then back again to the centre-line, constitutes one cycle. The prongs of a tuning fork move in a similar fashion, but at an audible frequency, transmitting sound to the surrounding air. As the sound-waves pass any particular point the air pressure moves up and down at the same frequency, and if a microphone is placed near a tuning fork and the resulting signals are displayed on an appropriate electronic device, a sine-wave pattern like those in **fig. 3** will appear. There is a connection here with the sound-waves depicted in **fig. 1**, for the points of high and low pressure correspond to the upward and downward peaks respectively, while undisturbed air would be the equivalent of the dotted straight centre-lines in **fig. 3.**

In musical reality sounds seldom have this pure uncomplicated quality, as vibrating objects tend to move at several frequencies at once. A simple string, for instance, may oscillate as a whole or in parts which are simple fractions of the whole. **Fig. 4** shows several such modes, which could all be operating together, and some of which a skilled violinist could use in isolation, playing harmonics on their own to extend the effective range of his instrument well beyond the modest upper limit of E^3 shown in **fig. 2**. If we consider air pressure in a tube instead of string movements, the situation depicted in **fig. 4**

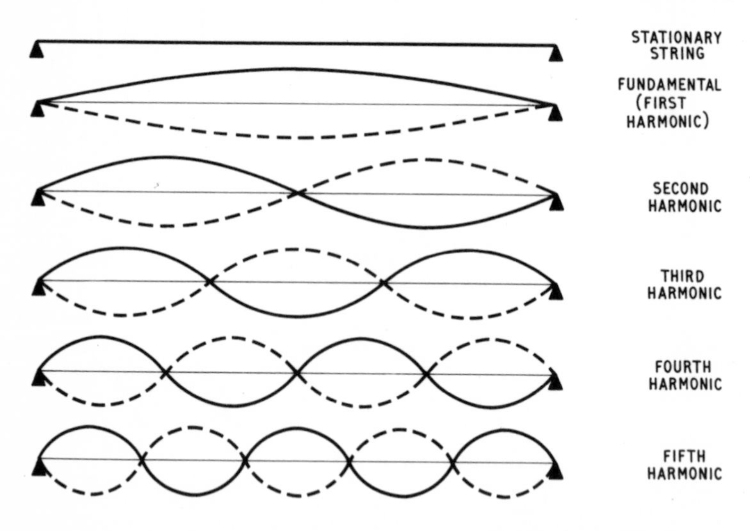

Fig. 4 Some modes of string vibration, exaggerated for clarity.

could apply equally well to wind instruments. With uniform media such as strings or air columns, these additional vibrations – or partials – tend to involve fairly exact fractions of the total length, and in consequence their frequencies are usually simple multiples or harmonics of the fundamental note. It is partly because of this harmonic relationship that a note rich in overtones is still heard as a single sound – the character of a note may change from one instrument to another, but the partials do not fall apart. Practical strings rarely adopt in isolation any of the modes shown in **fig. 4**: many harmonics will usually be generated together, the proportions changing from instrument to instrument, from note to note, from one loudness to another, and frequently from moment to moment within one note. The actual waveforms of superficially simple musical sounds are therefore rather complex and 'spiky', the precise formation providing some of the answer to our question about the origin of tone-colour.

Fig. 5 shows a typical example, together with an harmonic – or *Fourier* – analysis revealing the contribution made by each partial to the overall waveform. This one breakdown involves fairly elaborate graphical or electronic procedures to reveal the structure, or acoustic spectrum, of a single cycle from one note played in one of many styles by one of many possible musicians on one instrument. With percussive sounds of indefinite pitch the vibration-curves are even more involved, as the cyclic pattern corresponding to a specific frequency is missing. However, although the variations and complexities might daunt a scientific investigator, there are some useful general rules which help understanding of instrumental sound qualities in terms of overtone structure.

Bowed strings are generally very rich in harmonics – rich in the sense that nearly all the possible multiples of the fundamental note are present in diminishing amplitude right up to or beyond the twelfth in the harmonic series. The precise proportions depend on the manner of bowing from instant to instant, and when the various special techniques mentioned earlier are taken into account we can understand why the ear generally finds the string family less tiring than the woodwind or brass. The woodwinds themselves work by excitation of the natural resonances of air columns in tubes, the important variations being in the method of setting the sound-waves in motion. In the flute and piccolo, simplest of woodwind instruments apart from their forerunners the recorders, the player's breath is directed in oblique fashion at the edge of a wind-hole in such a manner that changes of pressure cause the air column to vibrate. The flute is unusual in having relatively weak overtones, only the second harmonic and sometimes the third contributing appreciably to the tone

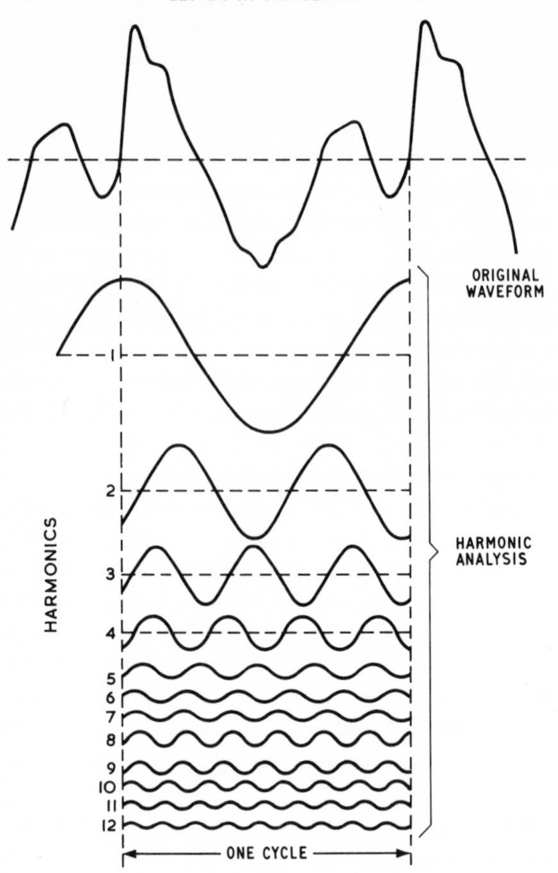

Fig. 5 Complex waveform and its harmonic analysis, showing how an apparently irregular pattern is really the sum of a series of sine-waves.

quality – apart, that is, from frequently very audible breath noises which the ear cannot always separate from the tone itself. The first harmonic, incidentally, is not normally named as such in acoustics because it is simply the fundamental itself.

Other woodwinds employ cane reeds to excite the air columns, the clarinet's combination of single reed and cylindrical bore giving an easy, bubbling sound characterised acoustically by exceptionally weak lower even-numbered harmonics, a strong third, and significant energy between the eighth and eleventh partials. The oboe and cor anglais, double-reed instruments with a tapered air column, have a bitter-sweet tone difficult to pin down in terms of harmonics and apparently dependent on a natural emphasis of certain frequency bands regardless of the actual notes played. This means that whenever a note's harmonics fall within these bands these are the overtones that are most important. This characteristic is known as a *formant,* also noticeable in the bassoon, whose folds and lower rate of taper combine with a double-reed to produce yet another set of tone qualities.

Whereas woodwind instruments use the natural fundamental resonances of air columns to produce their notes, with the tone-colour enriched by whichever harmonics happen to turn up, the brass family leans heavily on the harmonic modes themselves for production of the wanted notes. We have seen that strings and air columns will resonate at a number of harmonically related frequencies, and in the case of trumpets, trombones, horns and tubas the player uses his lips as 'reeds' to excite harmonics corresponding to the required notes, adding extra lengths of tubing as needed by means of valves or slides. Thus the character of the resulting sound depends somewhat upon the player's technique, the more permanent features of brass tone being related to the various proportions of bore and bell, and to different shapes of mouthpiece. The two types of French horn tone already mentioned are paralleled in the trombone by the overpowering majesty of brazen fortissimos on the one hand, and the solemnity of quite chords on the other. Even the trumpet can be strident or smooth to order, though the tuba is generally more consistent except when used to poke musical fun. As with other instruments, much (although not all) of the variety in brass tone-colour can be traced to the balance of overtones in the acoustic waveforms; but in this case the partials are usually 'harmonics of harmonics' instead of harmonics of the fundamental air-column note itself.

Some interesting points concerning the upper and lower frequency extremes arise from the harmonic nature of musical waveforms. We saw from **fig. 2** that nearly all actual notes occur at frequencies below about 4,000 Hz, but the dependence of tone-colour on overtones above the fundamental means that components contributing character to musical sound exist at much higher frequencies. Consequently the bulk of instruments produce significant sound elements up to 10,000 Hz, and many go beyond this to the limits of human hearing, as shown by the broken lines in **fig. 2.** Also contributing sound at these very high frequencies are the inevitable buzzes and creaks, hisses and squeaks which accompany the intended notes, while percussion instruments such as cymbals, snare drums and triangle produce complex sounds without definite pitch but with marked concentrations of acoustic energy in a frequency region that is otherwise almost devoid of fundamentals.

At low frequencies the rich harmonic make-up of musical sounds combines with a certain characteristic of human hearing to produce a curious phenomenon. The bottom note on the piano has a frequency of 27·5 Hz and when it is sounded we certainly hear A_4; but harmonic analysis of this note as produced by typical pianos reveals an almost non-existent fundamental tone, nearly all the energy appearing in the harmonics. The note A_4 is therefore created in the listener's mind because the ear tends to fabricate a note when a number of the tones in its associated harmonic series are present; in other words, the ear somehow picks out a note corresponding to the *difference* between successive harmonics, and this, of course, is the same as the fundamental itself. It happens that all the lower harmonics are very powerful in the piano, thus helping the ear to compensate for the weak fundamental. The actual tonal quality would be quite different if the 27·5 Hz component *were* relatively strong, but only deep organ pedals can do this at such low frequencies, and many orchestral notes falling within the piano's bottom octave depend to some extent on the ear's capacity for filling in the pitch of the fundamental. We shall return to this feature of low frequency sounds later, as it naturally has a bearing on technical requirements for bass reproduction.

So far, this examination of tone-colour has concentrated on overtones and their proportions; but this is only part of the picture, for we judge instrumental quality by listening to successions of notes which have beginnings, middles, ends and transitions from one to another. That the *shape* of a note influences its subjective quality is easily proved by an experiment with a piano and two tape recorders. If piano notes are played and recorded in the normal fashion and the tape is then replayed *backwards* (achieved by recording on a half-track machine, inverting the tape, and playing back via the lower track on a quarter-track model), the resulting sound is totally unlike anything ever heard from a piano. In a piano the strings are struck suddenly and the tone dies away gradually, but when the pattern is reversed the sound builds up from nothing to a sudden cut-off in a most weird fashion. The effect has been described by G. A. Briggs* as sounding 'like a home-made harmonium suffering from anaemia and groaning in agony'. So here is a case where overtone structure is subsidiary to *attack*, as it it is in all percussive devices and in plucked instruments like the harp, guitar and harpsichord. The characters of other instrumental sounds are not so drastically dependent on the mode of starting, though if the beginnings of notes are removed by cutting tape-recordings it is not always then possible to recognise otherwise easily distinguished instruments.

* *Musical Instruments and Audio*, page 47.

But whether or not the beginnings are crucial, their shapes are still important, a multitude of different tonguing, lipping, plucking, hitting and bowing techniques all contributing to tonal diversity. This is emphasised by the relative failure to fabricate really convincing instrumental sounds by purely harmonic means. The various organ stops labelled with the names of other instruments sometimes offer fair imitations – but a listener would rarely be fooled. In other words, the harmonic series provides only a framework or scaffolding for instrumental tone-colour, and a number of other complex factors play important parts.

In fact, quite simple sustained notes on an instrument like the flute may vary enormously in both amplitude and harmonic content from moment to moment or even from one cycle to the next. Dr. W. H. George, one-time head of the department of physics at the Chelsea College of Science and Technology, and an authority on musical sounds, has conducted many investigations into the actual behaviour of instruments, and in **fig. 6** we see two of his pictures showing low and high flute notes. The individual cycles are just perceptible, the changing darker bands on the low note indicating variations of harmonics. The length of time for each of these examples is under half a second, yet the fluctuations are considerable. Pictures of this sort could be produced for most instruments, the changes in shape along the length of notes – and in the rise and fall at beginning and end – throwing yet more light on the not-so-simple subject of musical timbre.

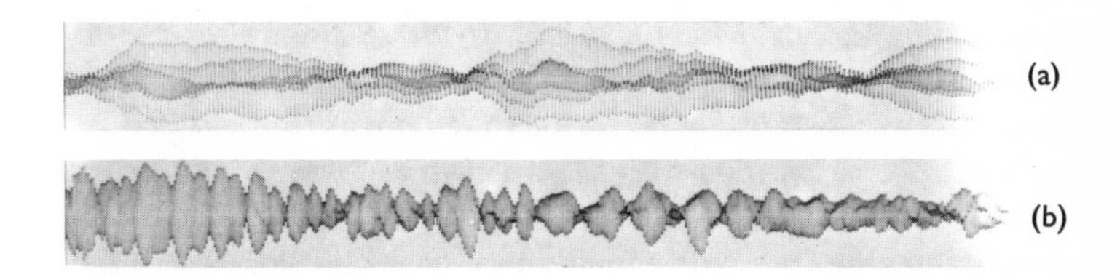

Fig. 6 Two photographed vibration-curves of flute notes. On (a), a low note, we see variations of harmonic content in addition to changes of amplitude, while the higher note (b) exhibits severe fluctuations of level (*photo courtesy of Dr. W. H. George*).

A major element influencing the tone-colour of stringed instruments is the coupling provided to transfer sound energy from the strings to the surrounding air. A thin string vibrating on its own between absolutely rigid supports is unable to displace much air, because as it oscillates the air particles slip past without appreciable disturbance and very little change of pressure occurs to be radiated away as sound. This limitation is overcome in the violin family by using a bridge to transfer the vibrations to the soft-wood belly and thence via a sound-post to the hard-wood back of the instrument. The relatively large area of the instrument's body acts as a more efficient sound radiator than a lone string, and the enclosed volume of air and various wooden parts have their own natural resonances which add colour to the final sound. It is well known that some types of violin are desired for their tonal qualities, which depend on quite minute differences of shape, finish, maturity of wood, etc.

Acoustically, the varieties of string-tone between various instruments played by the same performer arise from the many ways in which everything that is not just strings and bow adds its own little series of resonances or formants to the already complex string vibrations. It seems that violins have an important group of formants between about 3,000 and 6,000 Hz, but that good old Italian instruments are characterised by a fairly even distribution of resonances within that band, whereas inferior violins tend to have a few rather strong isolated regions of tonal accentuation at high frequencies.

The sounds of the 47 harp strings are similarly enhanced and modified by a large resonant cavity in the underside of the triangular frame that rests against the player. The piano also depends for the fullness of its tone upon a large sound-board, and notwithstanding the markedly non-resonant character of this in comparison with, say, the body of a cello, it is still notable that no two pianos sound the same.

Similar considerations apply to many wind instruments, as the actual conduits containing the air columns may also vibrate despite their rigidity and massiveness in comparison with the enclosed air. It is interesting, for instance, that players can distinguish between the sound qualities of wooden and metal flutes. Another complication with the woodwinds is that vented air columns do not resonate in an exact theoretical fashion, the 'unused' parts of the columns beyond the operated vents adding a quote of coloration to the sound. Thus still more formants join the queue of tone-colour contributors.

The question of formants brings us naturally to the human voice. Evolved to communicate meaning rather than pitch, this instrument leans very heavily on tone-colour as determined by formants in the production of vowels. The larynx-tone itself, generated by vocal folds (commonly misnamed 'cords') interrupting the air flow from the lungs, is extremely rich in harmonics. But this rather 'spiky' sound has to pass up the pharynx and via the mouth and nasal cavities before reaching the outside air, and these various air volumes act as resonators to favour certain frequency bands. The actual formant regions vary with the shape of the mouth and positions of tongue and epiglottis according to the sounds being uttered or sung. Because no two humans are the same shape and size, each voice has its own unique tonal quality quite apart from questions of pronunciation.

Apart from the significance of shape and attack in individual notes, we have not yet considered the vital matter of loudness. Practical music may be played at any level between *ppp* and *fff*, but we must try to pin down these musical terms with more precise language in readiness for the sorting-out process in the next chapter. According to some classic investigations conducted by the Bell Telephone Laboratories (U.S.A.) many years ago, sound-power or intensity produced by musical instruments varies between extraordinarily wide limits. A 75-piece orchestra at its very loudest created the acoustical equivalent of 70 electrical watts, while a solo violin playing *very* quietly produced only 0·0000038 watt – an intensity ratio of 18 million to one, corresponding to a sound-pressure ratio of 4250:1 (for the technically minded, one is the square root of the other). Differences of this order are somewhat difficult to manage in calculations, but fortunately our ears oblige by sensing the *proportional* changes of sound level rather than the absolute levels; in technical language, this means that they respond *logarithmically*. The outcome is that each time the sound intensity is doubled we hear an equal change of loudness, even though one such change might be from one to two violins and another from one to two similar brass bands. This is analogous to our hearing of pitch intervals, as, for instance, $G_3 - G_2$ gives the same subjective change as $G^2 - G^3$ despite the fact that one gap contains only 49 Hz and the other 1,568 Hz. It is for this reason that the approximately linear arrangement of semitones on the piano keyboard is directly equivalent to a logarithmic frequency scale (**fig. 2**), the latter serving to compress a wide frequency range into something corresponding to our musical impressions.

Returning now to sound levels, a convenient unit called the *decibel* (dB) performs a similar compression task by translating the orchestral sound intensity ratio of 18 million into 72½ dB. Other measurements and calculations in recent years have tended to confirm this extreme dynamic range of around 70 dB, though in practice it appears that a range of 60 dB is rarely exceeded except on the most massive musical occasions involving large choirs and much brass and percussion. It is important to remember that this 60 dB is a *ratio* of loudest to softest and not a measure of actual sound levels. However, to confuse matters the decibel is also used to indicate sound pressure above a reference point corresponding to the threshold of hearing, though this threshold changes with frequency, hearing generally being most sensitive in the region

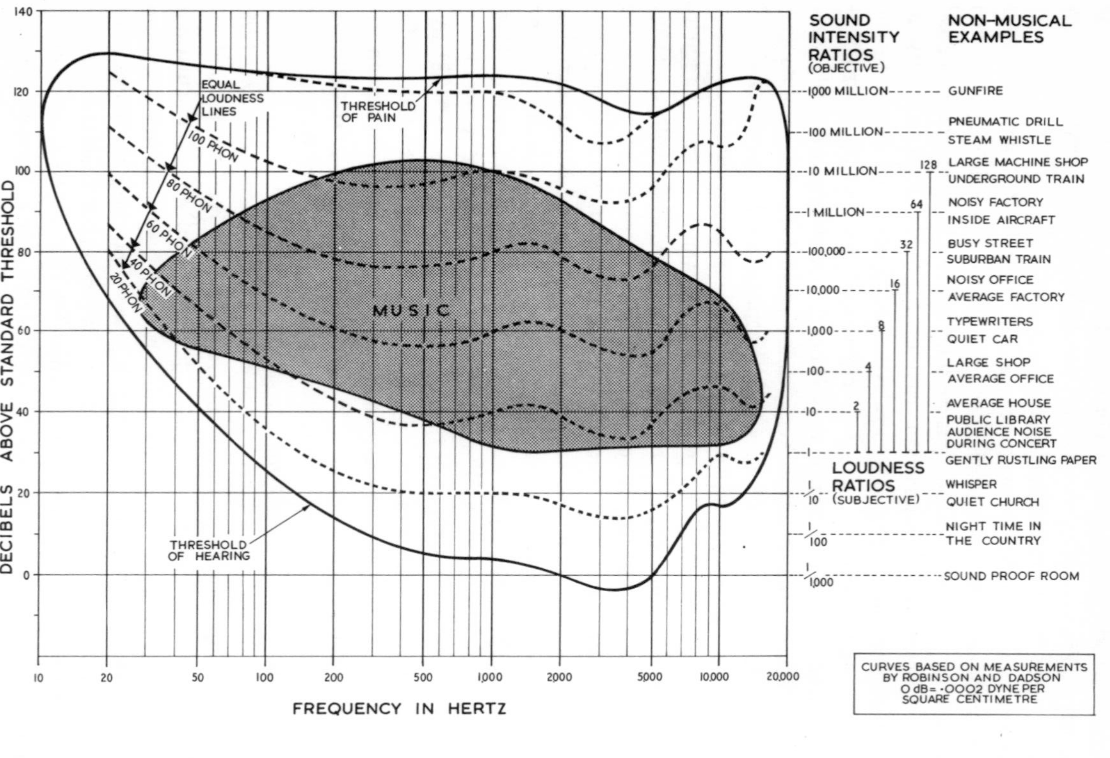

Fig. 7 Place of music within the sensitivity range of human hearing. All limits are approximate and statistically determined.

of 3000 Hz. At very high sound pressures we reach the threshold of pain, where discomfort or feeling take over.

The place of music within these boundaries is shown in **fig. 7**, where the 72 dB dynamic range is accommodated between 30 and 102 dB on a vertical scale with zero corresponding to the standard nominal lower threshold. Use of the horizontal scale gives the distribution of sound energy according to frequency within the shaded area, the boundaries of this being approximate only and representing the significant limits for most purposes. Also included for reference are the corresponding ratios of sound intensity and average subjective loudness, both referred to the lower music level of 30 dB. Note that each increase of 10 dB (at, 1,000 Hz) doubles the loudness. The lower musical limit of 30 dB above threshold refers to what may be attained in exceptionally quiet surroundings with intimate chamber music; but in the concert hall the irreducible audience and other background noise is usually at a higher level, tending to mask quiet music, so that performers respond accordingly by playing a little louder. These factors, together with the position of the average concert hall seat, mean that the musician's *ppp* often corresponds to approximately 40 dB (above lower threshold) at the listener's ears. The reader should not worry about how the decibel is derived – a mathematical operation beyond the scope of this book – but it is important to grasp that any information given in decibels necessarily uses some reference level, either directly or by implication. In audio, the standard lower hearing threshold (corresponding to an agreed physical sound pressure) is commonly used for 0 dB, and if any sound is said to have an *intensity* of 60 dB, this usually means 60 dB above the standard zero; whereas music with a *range* of 60 dB may well have its lowest point at 40 dB intensity and would therefore reach a maximum level of 100 dB. Be alert to these superficially different uses of the decibel, which is essentially a device for indicating the ratio of one quantity to another and is not a tangible measure like the watt or an ounce weight.

Nearly all music uses contrasts of loudness as part of its stock-in-trade, and as the orchestra grew and the Romantic period followed the Classical the dynamic range available to and demanded by composers expanded towards its present-day magnificence. Up to the time of Mozart the maximum range was around 40 dB, but as the trombones, bass drum, cymbals and tubas crept in the level went up, until today's expanded post-Wagnerian orchestras – with much additional percussion – can produce quite awe-inspiring sounds at the flick of a baton. This does not mean that twentieth century music is all loudness and noise, for though Mahler's 'Symphony of a Thousand' and, say, Prokofiev's fifth symphony might raise the roof by momentarily reaching levels of 100 dB or – for listeners near the front – perhaps 105 dB, one can find many passages of quiet, limpid beauty. The hush as the dying chorus fades into nothingness at the end of Holst's 'The Planets' – the audience almost holding its breath – is probably the most effective example of large-scale concert hall music descending to a sound level of 40 dB or even lower. But one can find endless illustrations of dynamic range in music, all of which make their point in practice without any thought of decibels or acoustics. It is as well to remember, however, that musical sounds are complex enough from considerations of pitch and tone-colour alone, so that adding a possible intensity range of nearly 20-million-to-one (albeit reduced to 4000:1 in terms of sound pressure and then simplified by a 72 dB label) makes the eventual task of reproduction no easier.

Another important aspect of loudness is its frequency-dependence. The change of hearing threshold with frequency has already been mentioned and appears as the curved bottom boundary in **fig. 7.** From this it is apparent that the ear needs higher sound levels at the extremes, particularly at low frequencies, to produce a just audible sensation; conversely, if we move far enough back from the concert platform or operatic stage for the frequency extremes of some musical sounds to fall below the threshold line – or to a level where they become masked by the background noise – we begin to receive weaker impressions of the extreme bass and treble. The dotted equal-loudness lines in **fig. 7** show how the ear's sensitivity varies with sound level above the lower threshold, and from this we can deduce that whenever music is louder or softer than intended, either by virtue of a poor seat or poor reproduction, the aural balance is modified and tonal qualities are changed. In other words, if the shaded area in **fig. 7** is moved downwards, not only will the extremes cross the hearing threshold, but the loudness relationship between low and middle frequencies will also shift, thus altering the apparent balance of instrumental overtones and therefore the tone-colour. This is why the fullness of double bass tone sometimes seems to decline in relation to the other strings as one moves further back in the concert hall. At the high frequency end any additional level needed for a given impression generally becomes greater as people grow older – a polite way of saying that we all suffer from creeping high frequency deafness; but many children and even a few adults can hear up to and above 20,000 Hz. However, it is easy to misinterpret curves such as those in **fig. 7,** for they are obtained with pure single-frequency tones, and what the ear does dynamically with musical sounds is not fully understood.

It will be noticed that the loudness lines in **fig. 7** are equally spaced at 1,000 Hz, this being the accepted reference frequency in this and other audio matters.

There is also a unit of loudness level called the *phon* which is the same as the decibel (above zero) at 1,000 Hz. This conveniently takes account of the ear's changing sensitivity with frequency by specifying, for instance, that any sound producing a sensation of loudness equivalent to 60 dB at 1,000 Hz has a level of 60 phons, even though it may be at some other point in frequency on the relevant loudness line or comprise a complex sound with many frequency components.

As a little light relief after these few rather over-technical sentences, it may be worth pondering on the fact that our ears are so sensitive that the 3,000 Hz tones used for establishing the lowest threshold level involve ear drum movements of about one ten-thousand-millionth of an inch. The corresponding sound pressure fluctuations are equivalent to the change of atmospheric pressure resulting from a vertical movement of one thirty thousandth of an inch. If we could nod our heads at audio frequencies it would be rather too easy to create some very loud noises. In fact, if our ears were just a little more sensitive we would be conscious of the minute but ever present random movements of the air molecules themselves. Even at the threshold of pain (around 120 dB) the changes amount to less than one tenth of one per cent of normal atmospheric pressure.

Before leaving loudness, there is one peculiarity of hearing that has a bearing on pitch. We assumed earlier that the objective characteristic called frequency had a direct and simple relationship to the subjective sensation called pitch. This is a common-sense assumption, but unfortunately it is not quite true. As the sound level is raised above 40 dB, frequency being held constant, low notes seem to go down in pitch and high notes go up. This is only a slight effect mainly operative around 100 Hz (G_2), the apparent pitch at this frequency falling by about 10 per cent as the level is raised from 40 to 100 dB. Here again, what one hears will depend on one's nearness to the instruments and on the part played by possible offending notes in the musical texture. It might be thought from this that an instrument's fundamental (as heard) could go out of tune with its harmonics, but in practice the ear obliges by keeping the components of complex waveforms locked together, thus avoiding yet another tangle in the web of musical sound.

Of the four primary aural characteristics of this nexus listed earlier, only ambience now remains. Ambience may be defined as the acoustic colour added to music by the space in which it is performed. Different types of music need, ideally, different acoustic settings to make their full impact. Gregorian chant and much early organ music demand a cathedral-like acoustic which would obscure much of the detail in, say, a Mozart piano concerto. Similarly, the

Lutheran churches whose less reverberant acoustics permitted Bach to compose relatively brisk and complex choral music with string parts – as in the 'St. Matthew Passion' and 'B Minor Mass' – would be unsuitable for Gabrieli's slow-moving antiphonal music for choirs and brass. Music of the Classical period generally benefits most from performance in halls of modest size, giving a fresh, intimate sound; but Beethoven foreshadowed the Romantic age, with its demand for a more opulent and full-bodied sound. Opera generally needs a theatre with acoustics permitting easy understanding of fast librettos, though Wagner is a special case demanding a 'Romantic' sound and many people would sacrifice the odd word or two in a Verdi opera for the sake of added opulence in that composer's great climaxes.

There are many factors involved in the acoustical behaviour of halls, theatres and studios, but one which is generally agreed to be paramount is the *reverberation time* (r.t.). When sound is emitted in a hall it is reflected back and forth between walls, ceiling and floor until it eventually decays to an insignificant level, the rate of decay depending on the size of the enclosure, the absorbency of materials on the boundary surfaces, and other factors. The time taken for a sound to decay by 60 dB is called the reverberation time, or period, the generally accepted desirable figure for orchestral music being in the region of 1¼–2 seconds. A cathedral might have a period of five or more seconds, while a small hall with much absorbent material can be as low as one second. At one extreme music will sound grand and sonorous but somewhat muddled (depending on the score), while a very short period will give great precision and clarity but a rather hard and assertive tone quality. Another complication is that the decay time usually changes somewhat with frequency, so that two halls with similar general characteristics may give, in one case, a full-bodied bass, and in the other a rather thin sound, because of reverberation periods that rise and fall respectively at low frequencies.

Thus a hall or studio adds its own particular character to the music performed within it, and as most music depends on the enhancement provided by its acoustic setting, we must accept the fact that sonic ambience is a necessary ingredient of the art. When listening to music in a hall, most people are not consciously aware of the space around them as a separate acoustic entity, but the music is certainly heard within that particular framework, the reverberation arriving at listener's ears from various directions and thus providing a certain atmosphere or sense of 'being there' which is a major factor in the enjoyment of live performances. Besides providing a setting for the music, a good hall adds fullness of tone, rounding off some rough edges; it blends the string sections of the orchestra without obscuring their separation, and provides the musicians themselves with something more substantial than a dead open space into which to play, so that they can assess more accurately the quality of their efforts. In addition, the apparent loudness and dynamic range of music is affected by the characteristics of the hall in which it is performed, possibly due to the manner in which our ears respond to reverberation. This may be related to the fact that reverberant sound from quiet music must disappear into the ambient background noise more rapidly than that following loud passages, though these very subtle aspects of hearing are not yet properly understood.

What are Musical Sounds? was the title at the head of this chapter, and the reader will see that the answer is not simple when we probe beneath superficial aural impressions. Still only the fringe of the subject has been touched: I have not even mentioned the totally inharmonic overtones of most tuned percussion instruments, or the doubtful accuracy and instability of many supposedly harmonic partials in others. At every step we see complexity behind apparent simplicity (there are 243 strings for the 88 notes on a grand piano, for instance), though perhaps we should not be surprised at richness and diversity in the ingredients of a great art.

I have attempted to summarise the more important factors in the make-up of musical sounds, as a sort of launching platform for the rest of this book. But even this modest isolation of the more obvious features is a step away from real musical experience, for our minds do not apprehend tone-colour or pitch as isolated qualities unless we are listening as musicologists or critics. We hear collections of notes or rhythms played by instruments, and we must hope that when hi-fi in the home has been perfected there will be no technical distractions to prevent the plain, natural enjoyment known in real life, whether it be from a Haydn quartet or Berlioz mass, a Schubert song or Shostakovich symphony, a Dixieland Jazz session or a Rodgers and Hammerstein show. Then we shall have come full circle and reality and illusion will be one.

2. SORTING OUT THE STRANDS

HOW ON EARTH can the complex array of strands that make up musical sounds be sorted and simplified for the purposes of recording and reproduction? After Chapter 1 this may seem a challenging question, but the matter appears in a slightly different light if we look at the simple sorting that has already taken place in the two narrow channels leading to our ear drums. The ear is sensitive to changes of air pressure only within the frequency and intensity boundaries shown in **fig. 7**. This means that everything we hear is due to the tiny volumes of air next to our ear drums undergoing fluctuations at rates between 15 and 20,000 times per second, by amounts corresponding to quite extraordinarily minute fractions of the prevailing atmospheric pressure. Thus the most involved admixture of sounds created by the mightiest gathering of musicians is reduced to two similar composite vibrations.

To understand this apparent miracle we must return for a moment to **fig. 5**, which shows how one cycle of a complex vibration may be analysed into its constituent harmonics. While containing all the elements shown beneath it, the original waveform is a single continuous curve representing the way in which air pressure moves up and down when all the harmonics are added together. Although such a vibration-curve comes, in this case, from the sound of a single instrument, it is easy to imagine a dozen separate instruments each producing a tone equivalent in frequency and intensity to one of the harmonics shown. At some point in space where an ear or microphone might be placed, the changing pattern of air pressure resulting from the various sources would be just the same as that obtained from the original integrated sound represented in **fig. 5**. If the frequencies and amplitudes of the twelve sound-sources were not quite correct a somewhat different total waveform would be created, but at any particular place the air pressure would still fluctuate in a manner that could be drawn as one unbroken line.

In practice, of course, each individual instrumental waveform will be rich and complex within itself, and of differing frequency, overtone structure, amplitude, growth and decay. But still the individual vibrations all add together from instant to instant, for any particular group of air particles carrying the sound cannot be both compressed and rarefied at the same time. So, if we forget for the moment the slight differences between the sounds at our two ears, it can be seen that any music is finally simplified and transformed into one vibration pattern in the air. **Fig. 8** shows such a pattern as picked up by a microphone during the playing of an orchestral chord, and if we could expand this picture several hundred times we would see a single continuous line flit-

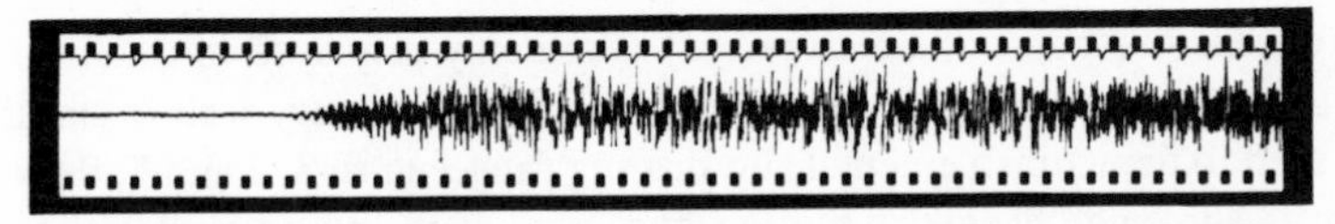

Fig. 8 Photographed vibration-curve of a chord made up from many notes played simultaneously. Recording made at an orchestral rehearsal (*photo courtesy of Dr. W. H. George*).

ting up and down and carrying a multitude of smaller undulations representing every overtone and every loudness of all the many instruments playing.

Our task in this chapter is to look more closely at this flitting line. We must abstract from it those features carrying the vital information needed to regain music from a mere waveform, which is all we have – in electrical form – once music has entered the recording/reproducing chain via a microphone. Very conveniently, the four characteristics of musical sounds examined at such length in the last chapter all play a fairly simple part in moulding the pattern of vibrations to be picked up by ears or microphones.

Starting with pitch and frequency, we saw that sound energy might be present at any point in a spectrum extending from under 20 Hz right up to the limits of human hearing approaching 20,000 Hz, and that although actual musical notes themselves seldom reach beyond 4,000 Hz, the complex structure of overtones requires a more extended range if the full detail and sparkle of instrumental tone-colour is to be captured. The numerous elements in musical sound add together, as we have seen, to produce a single ever-changing vibration pattern; but this integrated signal still carries all the frequencies present in the unamalgamated sounds radiated by separate instruments and voices. A microphone used for recording must therefore respond evenly over the full range of frequencies. Similarly, the whole recording and reproducing chain must give equal treatment to all frequencies if what we hear from our loudspeakers at home is to be a repetition of the original sound.

Thus the first abstraction on our journey from music to its reproduction is the concept of *frequency response*. We saw in connection with **fig. 7** that our ears respond unevenly as the frequency is changed, and that if music is heard an an unnaturally low or high loudness its tonal balance may be upset. Conversely, if the overall level is correct but the reproduced intensity varies with frequency, the balance of overtones within instrumental waveforms (tone-colour) will be changed. In extreme cases the balance between instruments will be upset, especially when widely separated pitch ranges are involved – violin and double bass or piccolo and bassoon, for instance. Also, a very irregular response might exaggerate or stifle particular groups of notes regardless of the instruments producing them.

Frequency response is a feature very commonly displayed in the form of a graph such as that in **fig. 9**. Here, the horizontal frequency scale is similar to that used to depict instrumental and hearing ranges in **figs. 2** and **7**, the vertical scale showing changes of audio signal level as measured electrically from devices such as amplifiers, tape recorders or gramophone pickups, or perhaps as measured acoustically from a loudspeaker. We have seen that constancy of response over the musical frequency range is desirable if the sounds reaching a microphone are eventually to be recreated with realism at the listener's ears. For this reason the solid line in **fig. 9** is much better than the dotted curve, one indicating a 'flat' response from 30 to 14,000 Hz, and the other falling away badly at the frequency extremes in addition to being rather 'bumpy' in between.

It is important to accustom oneself to illustrations of the **fig. 9** type as they show at a glance what would otherwise be obscured in lists of figures. Such a list for the dotted curve might be presented as in the table, and it can be seen that some considerable exercise of graphic imagination is called for if the shape of the response is to be pictured. As we shall see in later chapters, such shapes may, with care, be interpreted in terms of actual musical sound quality, an equation very difficult with mere figures. Be wary, however, when comparing apparently similar graphs of this sort, as a set of figures can be made to give a relatively flat-looking response by choice of a suitably cramped vertical scale for the decibels.

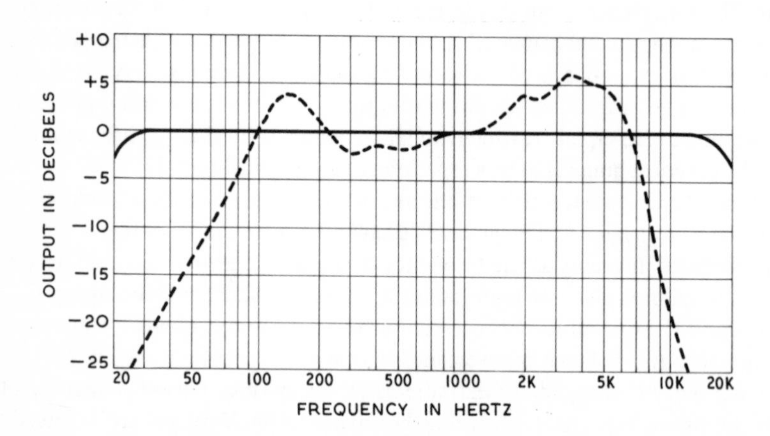

Fig. 9 Response curves of the type that may be found in audio literature. In a hi-fi system the flat extended response (solid line) would be preferred to the peaky but limited one (dotted line).

FREQUENCY	OUTPUT	FREQUENCY	OUTPUT
30 Hz	−23 dB	1000 Hz	−0 dB
50 Hz	−13 dB	2 kHz	+4 dB
100 Hz	0 dB	3 kHz	+5 dB
150 Hz	+4 db	3½ kHz	+6 dB
200 Hz	+1 dB	5 kHz	+5 dB
300 Hz	−2 dB	7 kHz	−3 dB
500 Hz	−1 dB	10 kHz	−19 dB

One small difference between the symbols here and those employed earlier is use of the letter 'k' before Hz to represent kilohertz or thousands of cycles per second; thus 3,000 Hz becomes 3 kHz or, in published literature employing earlier symbols, one may find 3,000 c/s or 3 Kc/s. One cycle per second (c/s) is now generally represented by the Hertz, and assuming the reader to have absorbed this, one thousand cycles per second (Kc/s) will from now on be kHz.

A smooth and extended frequency response, then, is the technical requirement arising from abstraction number one. Some of the practical problems created by this requirement will be examined, together with the question of *how* smooth and *how* extended, in the high fidelity context of later chapters. If the reader studies advertisements and technical reports in the hi-fi and gramophone magazines he will find that frequency response is one of the most commonly mentioned features of reproducing equipment, the sort of information contained in graphs like **fig. 9** often being abstracted and condensed still further to give expressions such as '23 Hz − 17 kHz ± 1 dB' or '75 Hz − 7½ kHz ± 6 dB'. These figures actually represent the responses in **fig. 9**, the plain and dotted curves deviating to limits of one decibel and six decibels respectively within the stated frequency bands. Taking 0 dB (left-hand vertical scale) as a reference level, the dotted curve moves upwards to a high point of +6 dB at about 3½ kHz and is down to −6 dB at 75 Hz and 7½ kHz. Thus between these limits the response is said to be within plus and minus six decibels (±6 dB), usually taking the signal level measured at 1 kHz as a reference for 0 dB.

The next musical characteristic to be identified in that flitting line of air pressure or electrical waveform is tone-colour. This, it will be remembered, depends upon the maintenance of certain proportions between overtones and fundamentals, on subtle changes in amplitude and overtone structure as notes progress, and on percussive qualities in the starting of many instrumental sounds. Fortunately a lot of these points are covered automatically by a good frequency response, as any natural pattern in the distribution of acoustic energy over the musical spectrum – and any changes in that pattern – will be reflected faithfully if the total recording/reproducing chain has no bias for or against particular frequency bands. Once again, a 'flat' frequency response is desirable.

The percussive element, however, poses additional problems. In the last chapter we saw that musical instruments make use of resonance for the generation of specific notes, though resonance is in principle a characteristic of any mechanical object. In the creation of music, strings, skins or air-columns are adjusted to vibrate efficiently at desired frequencies, though for the reproduction of music there are, as we have seen, no specific 'desired frequencies'; indeed, any favouritism is highly undesirable. But however smooth the response in a sound reproducing system, it is difficult to eliminate entirely all resonant characteristics – especially from loudspeakers. In practice this often means that while most music may sound well balanced for most of the time, percussive sounds will be either dulled or superficially exaggerated, with a corresponding but more subtle degradation of those instrumental qualities dependent to some extent on initial attack. The sharp impact of such sounds, known in audio jargon as *transients,* tends to 'excite' such natural resonances as may still be lurking in the reproducing system, just as a sudden rut in the road will draw one's attention to particular vibrations in a vehicle. Thus we come to the need for good *transient response* as the servant of instrumental tone-colour, the preserver of inner clarity in complex music, and for the avoidance of unwanted coloration or aural 'character'.

Transients, with their occasional very high amplitudes, bring us to the loudness or dynamic range aspect of the single complex waveform chosen as the basis of our sorting process. We learnt from **fig. 7** and associated comments in the last chapter that the maximum dynamic range of music is about 72 dB, corresponding to a sound intensity ratio of 18 million to one. This is the acoustic measure of music's enormous range of power, and this same range must be handled domestically via loudpseakers. Imagine a car required to perform with equal felicity, but without gear changes or slipping of clutch, at any speed between 30 m.p.h. and one tenth of an inch per hour. At one speed it would take an hour to travel from London to Guildford, while at the other speed the car would just be arriving at the Guildhall if it had set out when Sulla was leading Rome's victorious armies into Greece nearly a century before Christ. This, by analogy, is what audio equipment must do if a hushed solo violin and full orchestra and chorus are to be reproduced in natural acoustic proportions.

Fortunately the laws of nature ease the arithmetic of such huge ratios – and place less strain on the imagination – by permitting us to deal, in practice, with sound pressures rather than powers or intensities. As a change in one equals only the square root of a change in the other, the rather impracticable ratio of 18 million is reduced to something over 4,000. Likewise, in the electrical circuits handling audio signals all the way from microphone to magnetic tape or disc groove, and thence again from gramophone pickup or tape-head to loudspeaker, it is usually convenient to think and work in terms of voltages and currents rather than powers. This means that if we display our single complex musical waveform diagrammatically, the positive and negative peaks corresponding to the very loudest music will be some 4,000 times higher than the changes of sound pressure (acoustic) or voltage (electrical) produced by the quietest solo note. Some idea of scale may be obtained from **fig. 10**, where the base line apparently representing 'silence' in places is actually something like six times the thickness needed for a very quiet music signal.

Thus it will be seen that even in terms of voltage changes in audio equipment a musical dynamic range of over 70 dB is not without its difficulties. The main problem is of the 'devil and the deep blue sea' type, the devil looking in at the loudest passages and swishing the tips off with his sword if they are of magnitudes that would drive amplifiers or pickups beyond their physical limits, and the deep blue sea being an irreducible background of *noise* generated at a low level in all audio equipment and waiting to swallow up very quiet music or the minute signals representing reverberant ambience. The 'zero' line in **fig. 10** may be likened to this background noise, which must be quieter than the very quietest music if it is not to obtrude. As we have seen, the line

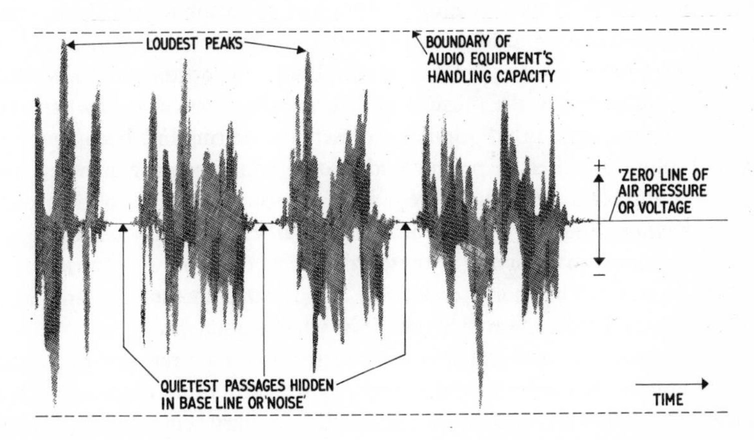

Fig. 10 Diagrammatic representation of complex musical waveform. Shaded area actually comprises a single continuous line oscillating up and down in a very involved fashion as in fig. 8.

as drawn is already thick enough to swamp very quiet passages, and if we regard this thickness as irreducible it is clear that the only way to drag very low music levels out from the brine is to raise the amplitude of everything. Alas, as soon as we start doing this we reach the dotted lines symbolising the physical limits set by equipment design. The ratio of peak signal handling capacity at the one extreme to background noise at the other may be specified in decibels and is known – appropriately enough – as the *signal-to-noise ratio*. This should, ideally, be at least equal to the dynamic range of the type of music to be recorded and reproduced.

Not only must this ratio be adequate, but for realistic reproduction equipment must be capable of creating at the listener's ears sounds of the same maximum loudness as those heard in the concert hall. Reference to **fig. 7** shows this to be around 100 phons for the very loudest music, and calculation reveals that the acoustic power needed to produce such a loudness in a domestic room of some 2,000 cubic feet capacity is about one third of a watt. Taking into account the rather low electro-acoustic conversion efficiency of the average loudspeaker, this gives an electrical power requirement in the region of 5–30 watts from the associated amplifier. Thus we arrive at the concept of maximum *power handling capacity*.

There is yet another technical point arising from this business of dynamic range and peak powers. Even when the inherent noise level is low enough and the maximum power capacity high enough to accommodate 70 dB of music, there is the possibility that an audio system will not respond with quite equal efficacy at low, middle and high signals levels. Such uneven or non-linear behaviour introduces what is known as amplitude distortion, whereby extra harmonics are added to the musical waveforms (*harmonic distortion*) and the various elements within that original endlessly undulating line become mixed and multiplied together to produce additional components which are harmonically unrelated and therefore add a rough harshness to the sound (*intermodulation distortion*). The word 'distortion' is frequently used in audio literature to apply to all or any of these misdemeanours, but it is fairly widely accepted that total harmonic distortion is a reliable measure of amplitude non-linearity and more often than not this is what is quoted.

Only one group of musical sound characteristics now remains to be considered in our sorting of technical strands, namely space and ambience. There are two aspects to this, each with a technical corollary, one regarding signal level and the other concerned with spatial distribution. By the very nature of its subtle contribution to music, with reverberation decaying to a point 60 dB below the initiating sound before it is regarded as 'finished', ambience requires room to express itself at very low signal levels. Technically, therefore, it is satisfied if the dynamic range criteria already mentioned have been met. The spatial element in music encompasses both ambience and the disposition of orchestra and/or singers, involving a technique for presenting the reproduced sound pattern over the sort of angle found in real life, without exaggerating the apparent sizes of individual instruments or singers, and within a natural acoustic setting. This technique is stereophony, to be discussed at length later but which, we may note for the present purpose, involves two basic points: (i) use of two identical recording and reproducing channels – once the initial signals from various microphones have been mixed and allocated into two groups; and (ii) care in the choice of characteristics, positions and surroundings of a pair of loudspeakers at the end of the reproducing chain.

Having considered the physical elements in musical sound as they bear on reproductive matters, including the need to minimise additives such as background noise and distortion products, there remains one other important requirement that is not immediately apparent from a superficial study of the basic musical material. Whenever sounds are recorded, that single endlessly fluctuating line (or two such lines for stereo) encompassing all the complexity of music within its agitations must somewhere be laid out, or wound in a spiral, *as a line*. And as a line it must be pulled or driven past a fixed point at some predetermined speed if the recorded material is to be reproduced correctly. Thus LP records must be rotated at $33\frac{1}{3}$ revolutions per minute and most tapes played at a linear velocity of $7\frac{1}{2}$, $3\frac{3}{4}$ or $1\frac{7}{8}$ inches per second. Any fixed deviation from the correct replay speed will not only affect total playing time, but will change also the number of cycles scanned per second in any given note, thus altering the frequency and therefore the pitch. A disc record of a piece of music in the key of E-flat will, if played on a turntable running 6 per cent slow, offer something in the key of D, a contingency clearly to be shunned if only to avoid confusing listeners with perfect pitch. More important is the prevention of short-term *changes* in speed, as the resultant wavering of pitch caused by slow changes (*wow*) or roughening of sound caused by more rapid changes (*flutter*) is easily noted and objected to by the majority of ordinary listeners.

This completes our initial sorting process, by means of which we have extracted seven strands from the musical complexity revealed in Chapter 1. As an exercise in the philosophy propounded at the beginning of that chapter, let us now see how these seven reproductive criteria would affect the sounds of a piece of music – taking, conveniently, the seven movements of Gustav Holst's 'The Planets'.

Mars, bringer of war, takes us into the tumult of a full orchestra playing noisy music, with much hard work from the brass departments. Here there is plenty of scope for *non-linear distortion* to rear its ugly head, and the ear will tire easily if the harshness of intermodulation is added to the massive but nevertheless not unpleasant sounds intended by the composer. Venus, bringer of peace, imposes no strain, but her repeated calls on the French horn certainly sound curious if their pitch is steered off course through *wow*; and the woodwind – particularly the oboe – has a roughened, bubbly sound when marred by *flutter*. Mercury, winged messenger, flits lightly from instrument to instrument; the pinpointing of sounds within the orchestra and the effect of a spacious acoustic as a setting for this delicate scherzo is aided, without doubt, by *stereophony*. Jolly Jupiter, with cymbal clashes and bouncing orchestral chords, demands good *transient response* if the music's zest is to be captured without coloration and if the strings are to retain their natural attack even when playing legato in the famous maestoso tune in the movement's middle. Sad and aged Saturn depends, despite his measured tread, on many subtleties of tone-colour, with a multitude of timbres from woodwind and strings, spiky brazen chords, bold trumpet calls, the metallic clatter of orchestral bells and, underlying his progression, a plaintive foundation played on the double basses. For untroubled realism in reproduction this all requires a *smooth frequency response*, without emphasis or absence at any point in the spectrum. Uranus, the magician, conjures his slap-stick way through a panoply of sonic effects until, at the height of involvement in his magic, a full organ shatters the spell with a massive glissando, demanding for an instant a *full reserve of power* to avoid overload and distortion in reproduction. Finally, Neptune spins his mystic way around the Solar System's periphery, with a wordless chorus eventually fading to inaudibility and demanding absolute silence from the audience and a corresponding *low noise level* in reproducing equipment.

For convenience in this little tour only one of the seven technical requirements has been applied to each of Holst's seven movements, though in fact of course most will apply to all the music for much of the time. Also applying to most music reproduction, at whatever level of fidelity, are certain basic techniques. The non-technical reader should have a look at these before attempting to use the concepts of transient response, noise-level, etc., in a hi-fi fashion, so we move on now to consider a simple transistor radio.

3. BASIC COMPONENTS FOR MUSIC REPRODUCTION

DOMESTIC RECREATION of music ranges from the squeaky efforts of a small transistor radio to the full-scale realism approached by the best high fidelity stereo installations; but at any point in this range we are still dealing with reproduction, and certain basic principles and components remain the same. What are the minimum essentials in this business? In fact, what does a small radio *do* to create the impression, if not the verisimilitude, of a real musical performance?

We have seen that musical sounds comprise certain types of vibration in the air, so the first requirement in reproduction is some method of recreating such vibrations 'to order'. The essential component for this is called a loudspeaker (commonly, just 'speaker') and the orders are in the form of electrical impulses or, to be more exact, a continuously alternating electrical current corresponding to the complex but unilinear waveform underlying our analysis in the last chapter. It so happens, as hinted there also, that for various technical reasons it is far more convenient to handle audio signals in electrical form than as mechanical vibrations. It is true that music is *stored* mechanically and magnetically on discs and tapes, but whenever recording or reproduction is actually taking place the sounds that enter microphones or emanate from loudspeakers are travelling through wires and various electrical components in the form of rapidly changing currents and voltages. Amplifiers or amplifier stages are used to make these currents or voltages larger up to the point where they are strong enough to actuate loudspeakers, which are essentially transducers for converting electrical energy into acoustic energy.

There are various types of speaker, but the most common is called the *moving-coil*, and in this the audio currents pass through a coil of wire suspended in a fixed magnetic field (see **fig. 11**), the laws of nature arranging that the varying additional magnetism created by these currents reacts against the fixed magnetism and causes the coil – known as a voice-coil or speech-coil – to move back and forth in sympathy with the applied electrical signal. The coil is attached to a conical diaphragm that moves with it, the whole assembly being free to flex by virtue of the suspension arrangements. As the cone oscillates to and fro it alternately compresses and rarefies the air in its immediate vicinity, thus producing sound which radiates outwards like that from musical instruments as discussed in connection with **fig. 1** (page 15). The waveform of this sound follows, ideally, the electrical input. When a high note is played the speaker cone vibrates rapidly and the air is set into motion at the same high

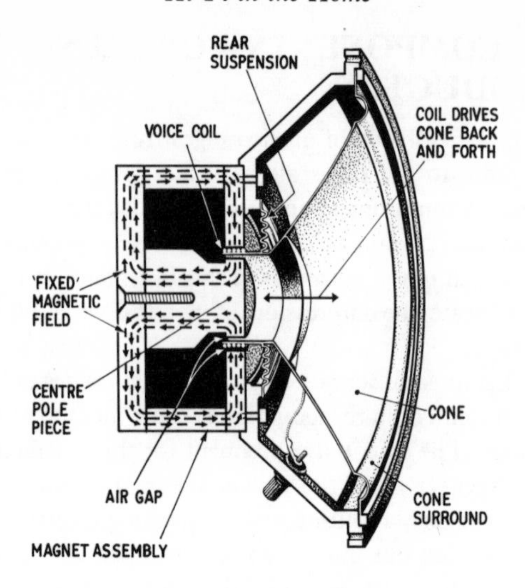

Fig. 11 Sectional view of a moving-coil loudspeaker unit. Signal currents in the voice-coil produce a varying magnetic field which reacts against a fixed field from the magnet system. This causes coil to move in the air gap.

frequency as that around the original musical instrument. Conversely, for low notes the cone and air vibrate more slowly and we hear a correspondingly deeper sound. With complex musical waveforms all the various components and multitudes of overtones are carried, as we have seen, on a single line of pressure, electrical current in the speaker voice-coil being the equivalent of this. At any instant that current must have a specific value and can only be travelling clockwise or anticlockwise around the coil; likewise, the speaker cone can only be moving in or out at a particular velocity. Yet a loudspeaker is clearly capable of sounding like a number of separate instruments and voices producing different notes and timbres at the same time, which perhaps underlines from another angle the idea and fact of a unilinear waveform as the basis of sound reproduction. Other types of speaker employ the same general principle of sound generation as the moving-coil unit, for with only one special exception all have diaphragms of some sort which are caused to vibrate by electrical means, thereby creating corresponding disturbances in the air.

There are endless difficulties in making a loudspeaker do its job properly, mainly because diaphragms are awkward mechanical things which tend to vibrate rather more energetically at some frequencies than at others, due to natural resonances. Also, even if the voice-coil vibrates exactly in accordance with the electrical signals and the cone follows the coil faithfully at all frequencies and at every amplitude (very rarely the case), it does not follow that the *air* vibrations will be equally correct, as these depend on all sorts of geometrical and dimensional factors around the speaker unit. At low frequencies, for instance, the repetition cycle takes so long that sound-waves have time to travel round the speaker from one side of the cone to the other before very much movement has taken place, and as one side of the cone is compressing air while the other is rarefying it, this results in a cancellation of air pressure and very little sound is radiated. To overcome this difficulty it is necessary either to seal the 'leak' between the two sides of the cone by putting the speaker in a closed cabinet, or to make the relevant air path very long. The first solution brings fresh problems, for if the enclosed space is small the cone movement is restricted at low frequencies and if it is large the cabinet may be cumbersome; the second solution in any case demands rather large dimensions if performance is to be maintained at bass frequencies. There are other approaches to this problem, some of which will be examined in more detail later, but it can be stated as a general rule that loudspeakers for reproduction at high quality of an extended range of frequencies in the bass are likely, with their enclosures, to be physically larger than those found in transistor radios. Apparent exceptions to this are usually only achieved at the expense of conversion efficiency, meaning that more electrical power in is needed for the same acoustic or sound power out.

But bass reproduction is only one part of the speaker story, for there are numerous other design problems, with as many difficulties at middle and high frequencies as at the bottom end. Whereas large speaker units and/or enclosures tend to be needed for bass, bigness generally prejudices good performance in the upper octaves. At middle frequencies these conflicting requirements overlap, with little help from the ear, which is very sensitive to colorations and minor errors of tonal balance in this region. But here we verge again on specifically hi-fi matters, which must be held at bay while we look at those other links in any ordinary music reproducing chain needed to generate the voice-coil currents. Let it suffice to note at this more primitive stage that an inexpensive moving-coil speaker unit of less that two inches diameter, mounted on the tiny baffle found in a pocket radio, may still – for all its manifest deficiencies – provide *some* music in the home, using essentially the same electro-acoustic principles as those employed in many of the most elaborate hi-fi systems.

Similarly, the other parts of such a radio contain the bare bones of much that is found in more sophisticated gear, so it will be useful to stay with this unpretentious device for a few paragraphs. We have seen that audio currents must be driven through a speaker voice-coil, and as the laws of electricity require that voltages be applied across the coil to achieve this end – much as a pressure or 'head' is needed to drive a flow of water through the pipes to a tap – this means that *power* must be available. Even with a massive aerial array only a minute power may be extracted from passing radio waves, as older readers may recall from crystal set days, so some means must be found of amplifying very small signals. As the power is not contained within the signals themselves it must come from a separate and easily tapped source: a battery in portable radios and the mains supply with larger equipment. But such sources are not, of course, themselves audio signals, so the flow of power from them must somehow be controlled by the wanted music waveforms. The transistor and its predecessor the valve (vacuum 'tube' for American readers) will do just this job, requiring, respectively, relatively small currents or voltages fed in which have the effect of modifying the flow of current through the device according to the desired changing audio pattern. The main current flow thus controlled comes from the external power source and may be at a quite high amperage with transistors or at a high voltage with valves, eventually being fed via transformers and/or other appropriate electrical components to the loudspeaker. This amplifying or power controlling process may be likened to powered steering in a car, where the wheels turn in the direction and to the extent designated by the driver without extracting appreciable effort from his muscles. In a small radio there will be one or two transistors with this power amplifying function, others being employed to bring the audio signal current or voltage up to a level suitable for controlling or 'driving' the output stage, and the remainder involved in selecting, amplifying and decoding the received radio signals prior to the audio part of the circuit. Somewhere in the latter will be found a volume control, whose function is to adjust the amplitude of signals reaching the output stage, and thus the power fed to the speaker.

The audio signals referred to so far have been simple electrical replicas of sound waveforms, with every cycle of every sound frequency registered as a kink on an undulating line of voltage or current, and in theory retrievable as sound at any point if extracted and fed to a loudspeaker via a suitable amplifier. With radio, things are not so simple, for while what travels through space as radio waves is electrical in the sense that it is a species of electromagnetic radiation (which happens also to include heat, light and X-rays within its gamut!), the tiny disturbances created by such waves in an aerial comprise very much higher frequencies than those found in audio, and if amplified and fed to a loudspeaker as they stand will produce nothing but golden silence.

The frequencies commonly used for radio and television transmissions lie between 150 kHz and 200 MHz (MHz = one million cycles per second), or from 10 to 13,000 times the normal upper limit of human hearing. This being so – and it has to be so for technical reasons connected with transmission and reception – how is musical information added to and retrieved from radio waves? It is done by a process of *modulation*, whereby either the amplitude or the frequency of a normally steady and continuous radio emission is wavered at a rate corresponding to the audio signal and by an amount related to the amplitude of that signal. Thus what is transmitted is still a high frequency radio wave, but this acts as a 'carrier' of musical and other audio information. At the receiver this carrier-wave might pass an external wire or rod aerial ('antenna' for American readers), its electrical component inducing therein small current agitations; or in a transistor radio the magnetic component of the radio wave will be 'sucked in' by a ferrite rod aerial to induce similar radio frequency (RF) currents in a coil wound around the rod. Most small radios of this type are designed to receive only the lower part of the RF spectrum covered by the medium and long wave bands (MW and LW). The radio signal having been captured, it is then convenient in nearly all receiver circuits, in some cases after a stage of simple voltage or current amplification, to change the RF signal to another and easily amplified intermediate frequency (IF). The particular advantage of this is that the IF circuits can be permanently set for optimum amplification at their one frequency, any 'tuning' to cover one or more reception wavebands being accomplished within and before the frequency-changer stage. This conversion process, used in *superhet* receivers, leaves the audio modulation intact.

Having brought the carrier at its new frequency up to a satisfactory level, probably from some small fraction of a millivolt to something approaching a volt, the point is reached for extracting the audio signal by a demodulation or 'detection' process. With amplitude modulated (AM) transmissions the essence of this is to clip off by means of a *diode rectifier*, all those half cycles of the alternating high frequency waveform falling on one side of the 'zero' line, thus leaving a series of unidirectional pulses which, smoothed by appropriate components, coalesce to form an audio waveform in its own right. This signal is then passed on to the strictly audio frequency (AF) stages of the receiver. Frequency modulated (FM) signals require rather more complex treatment for derivation of the audio 'envelope', and as we cannot assume prior knowledge of electronic theory on the reader's part it must be accepted that appropriate circuits will provide an audio output proportional to frequency deviation with this type of signal. We shall in any case return to FM radio, as it so happens that for hi-fi purposes this is the only type of reception worth considering.

However, this brief look at a small AM transistor radio – housing, incidentally, and as a measure of our brevity, about sixty components – has covered many basic essentials and for the most part is relevant to valve circuits also. The reader may know that valves have internal filaments or heaters which dissipate additional power, and for this reason they are only normally found in mains operated equipment; any transistorised devices more elaborate than portable radios or miniature tape recorders will also usually be powered from the mains. A problem with all such equipment is that special additional circuits are needed to produce suitable supplies from the standard 200–240 volt AC mains. The letters 'AC' here mean *alternating current,* a type of electricity where the voltage swings alternately positive and negative in just the same manner as the AF signals with which we are now acquainted. However, in the case of the mains there is a fixed frequency of 50 Hz (60 Hz at 110 volts in the U.S.A.) and a larger reserve of power than will be required by the most massive audio amplifier. The AC nature of the supply has the advantage that a transformer may be used to 'transform' the 240 volts up or down to any desired figure, but the disadvantage that the power supply needed by amplifying circuits is 'DC' or *direct current*, in which the voltage is of fixed polarity and the current always flows in one direction. From our 'powered steering' amplifier analogy it is clear that in the absence of steering column rotation the car should just go steadily ahead, but if the mechanism supplying power to turn the wheels were itself in continuous oscillation from left to right the poor driver would be tempted to switch off and use his own elbow grease. For similar reasons, the current fed into amplifying valves or transistors from what is commonly known as the 'HT' (high tension) supply is DC, its flow being wobbled up and down about a mean value in accordance with the applied AF signals. Extra components, then, are used to produce DC supplies from an AC input, and unless the job of conversion is done very well some of the original mains 50 Hz – or its harmonics – intrudes as a 'ripple' on the HT supply, this being one possible cause of *hum*, a hi-fi nightmare to be considered in more detail in later chapters. Another common source of hum is the use of AC for valve heaters, though the relatively low voltage requirements of these are more easily met with AC than with DC by use of a transformer as mentioned above.

We have skipped glibly through circuits, with passing reference to a few key components but no hint as to size or appearance, so the reader may now care to glance at **fig. 12**, where a number of the more common constituents in

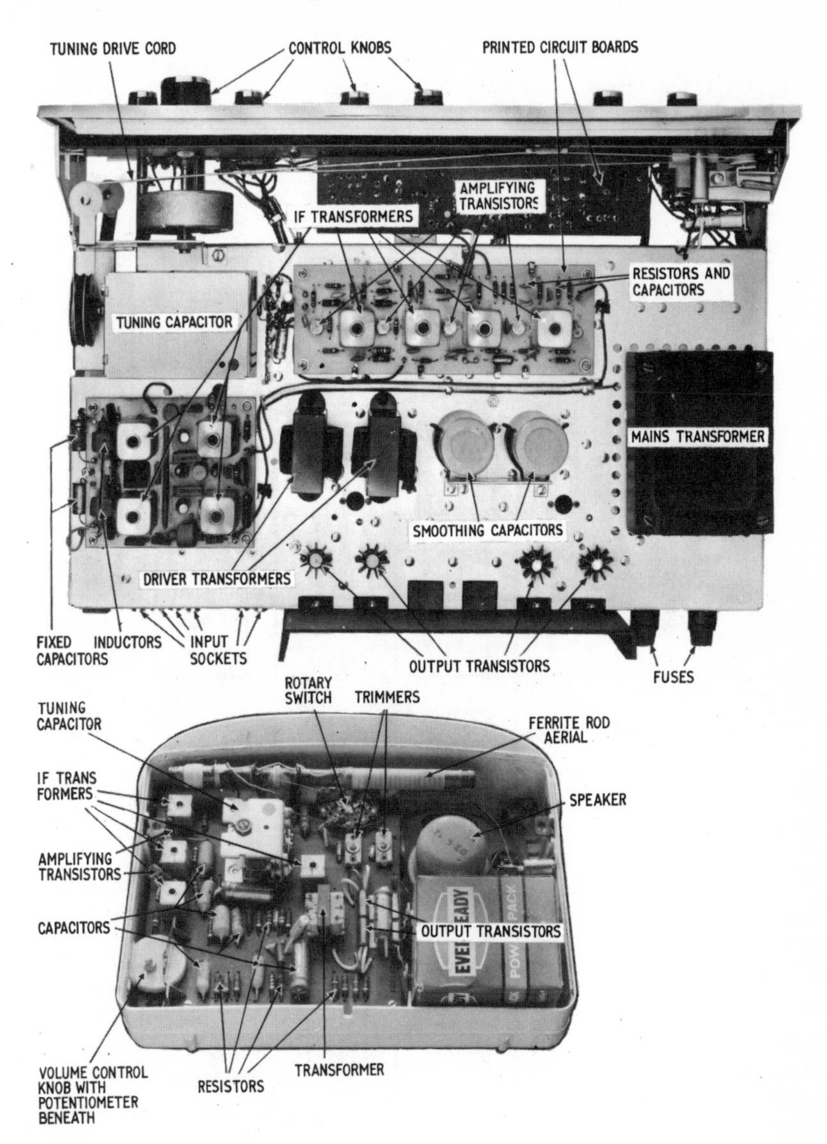

Fig. 12 A modest transistor radio (bottom) of the sort discussed in the text, with its sophisticated hi-fi equivalent (top), known as a tuner-amplifier and used with external loudspeakers.

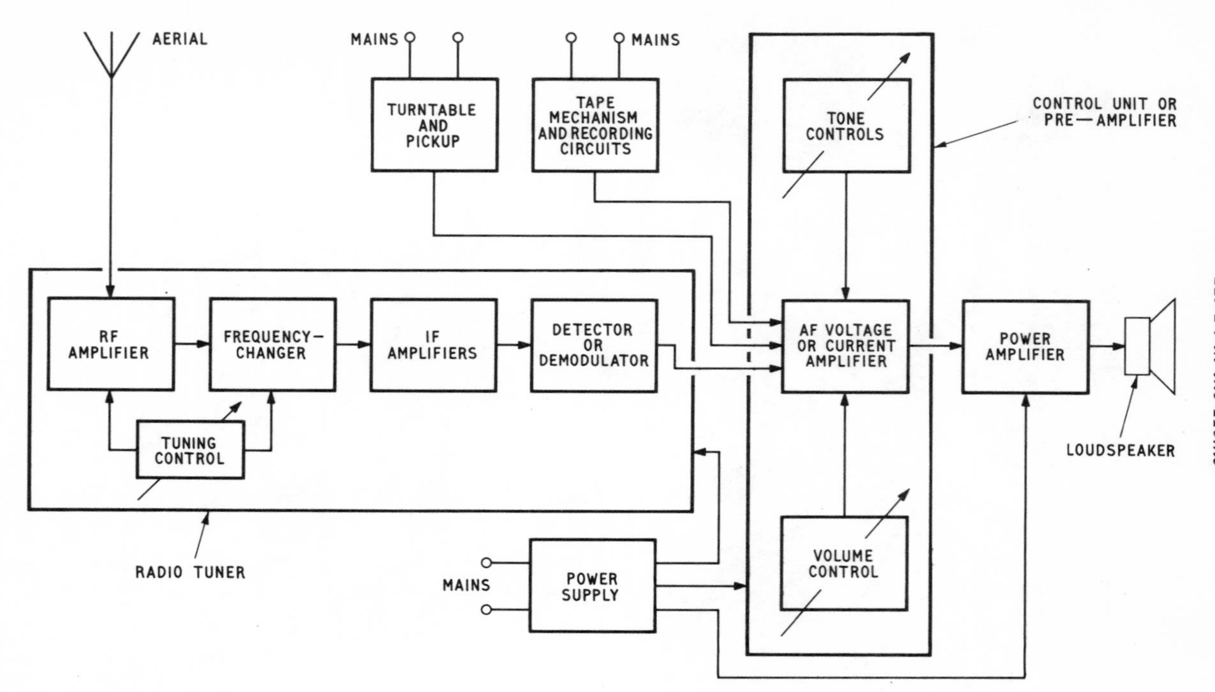

Fig. 13 Block diagram of basic elements for domestic music reproduction.

transistor circuits are labelled. Apart from the speaker, the functional parts so far considered in this chapter comprise the *electronic* elements of domestic sound equipment, and before moving on to pickups, turntables and tape recorders it will be useful to get these circuits into perspective by means of a further illustration, this time a 'block diagram' (**fig. 13**). Here we see all the parts mentioned in previous paragraphs arranged in functional order, with arrows showing the progression of signals through the system. Ignoring for the moment the two large enclosures labelled RADIO TUNER and CONTROL UNIT, the six boxes forming the 'spine' of this diagram are essentially the simple radio with which we set out, the power supply being a battery for a portable transistor model, or incorporating a transformer, rectifier and smoothing capacitors for AC mains use. Radio signals are picked up by the aerial, possibly a ferrite rod, and fed either directly to a frequency-changer stage or first amplified at RF; a variable tuning capacitor works in conjunction with both stages, each of which uses one transistor or valve and several inductors in addition to the usual small quota of resistors and capacitors found at each stage in nearly all electronic circuits. There may be one to three IF stages, each using a transistor or valve and a screened IF transformer or coil, followed by a detector circuit with a diode or diodes as its main components. From here, the AF signals are amplified by a further stage or stages, with volume and, possibly, tone controls intervening. Then there comes the power stage, sometimes (and with valve circuits nearly always) coupled to the speaker via an output transformer.

A radiogram simply adds a turntable and pickup unit to this assembly, audio signals from the pickup feeding in at the AF signal amplifying stage in place of those from the tuner section, selection being at the turn of a switch. Having added a gramophone signal source, removal of the tuner stages converts a radiogram into a record player. Likewise, replacing the gram unit with a tape transport mechanism, complete with some additional circuitry for recording purposes, converts a record player into a tape recorder. Finally, if the tuner circuits are designed for stereo reception, if the pickup is a twin-channel model, or if the tape-head and associated circuits are suitable for stereo, the AF signal amplifier, power amplifier and speaker may all be duplicated for stereophonic reproduction. Actually, apart from this last point the arrangement of units in **fig. 13** could be a full-scale hi-fi installation (**fig. 14**), in which case it is quite probable that the radio tuner grouping would in fact be a separate unit, as might also the stages linked here as a control unit. In hi-fi systems the box labelled POWER AMPLIFIER would normally comprise more than just the output stages themselves, and as everything must be doubled for

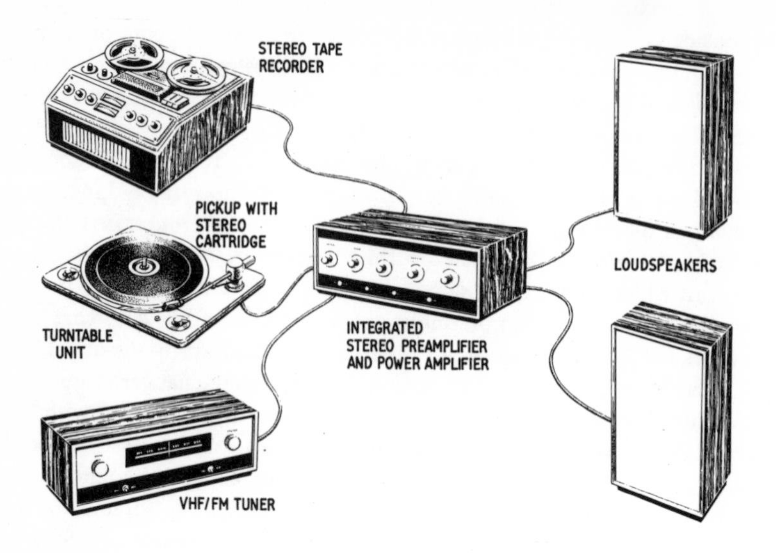

Fig. 14 Stereophonic high fidelity version of scheme shown in fig. 13. Usually, all the components depicted here except the loudspeakers – and possibly the tape recorder – would be housed in a single cabinet, though 'free-standing' amplifiers and plinth units for turntables may obviate this need in many homes. With most modern transistor systems the power amplifier and control unit are integrated as here, and in some cases the tuner will also be combined with the amplifier, then known as a tuner-amplifier.

stereo this part of the system can become rather bulky, especially when employing valve circuitry. Thus the reader will find some domestic installations with separate tuner, stereo control unit and stereo power amplifier, the latter sometimes even splitting into two separate single-channel amplifiers. However, transistors are changing this, their smaller bulk and lower heat generation permitting easy amalgamation of preamplifier and main amplifier on one chassis. In many cases even the tuner circuits are incorporated (**fig. 12**), leaving only the gram and tape signal sources to be accommodated separately. In due course the present change to smaller, cooler and lighter units associated with transistors will be followed by another revolution stemming from *integrated circuits* (I.C.). With these, most separate components of the sort seen in **fig. 12** disappear, together with the transistors, into small blocks of material. However, this is for a future to be explored more fully in Chapter 10.

We turn now to the remaining separate items, starting with the turntable, which is quite simply what its name implies: a table or platter that turns. Its purpose is to rotate disc records at the various standard speeds with sufficient constancy to avoid audible wow and flutter and with sufficient freedom from vibration and rumble to avoid adding extra noise to the reproduction. To meet strictly hi-fi requirements in these respects is not easy, but that is for the next chapter, and we shall simply describe here roughly how turntables are made to turn. **Fig. 15** shows a partially cut-away view of a typical turntable's underside, revealing that it is hollow, rotates on a central bearing, and is driven at its edge by an *idler wheel* which is in turn driven by a stepped pulley on the spindle of an electric motor. The motor cannot drive the turntable directly via its centre as the relatively slow rates of rotation required (78, 45 and $33\frac{1}{3}$ revolutions per minute – r.p.m.) are inappropriate to motor design. Neither is it practical to achieve the necessary 'gear ratio' by coupling the motor spindle directly to the turntable's periphery, as – apart from further technical objections – there are several speeds to be accommodated and this is most easily managed by introducing correctly proportioned steps either to the motor spindle or an extension thereof, coupling one of these steps, according to the speed required, to the turntable via a rubber-tyred idler wheel. The idler support bracket is pulled by a spring in a manner that wedges the wheel between pulley and turntable for most efficient transmission of drive. When the speed-change switch is operated the idler is retracted, moved up or down and re-engaged with a different step on the pulley, the idler normally remaining completely disengaged when the unit is switched off. Turntable drive systems of

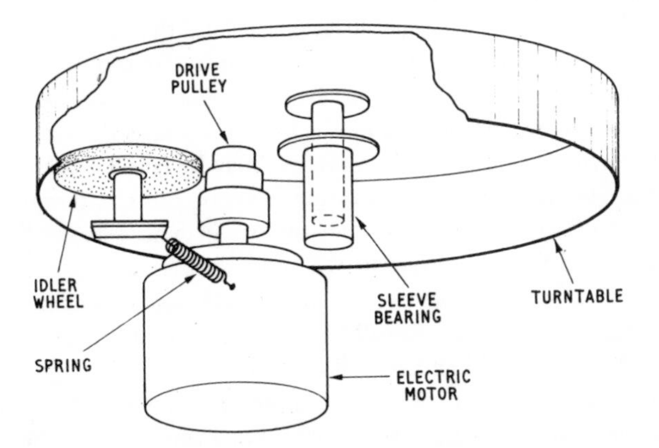

Fig. 15 A basic turntable drive system. Motor pulley drives inner rim of turntable via a rubber-tyred idler wheel, the latter being shifted by a speed-change mechanism to engage with various diameters of drive pulley.

this general type are very common, though there are considerable variations in detail and much additional complexity in most ordinary record players and radiograms through widespread employment of record changer mechanisms. Some turntables have an extra speed control giving a few per cent adjustment either side of the nominal switched speeds. This may be in the form of a variable electrical brake or a tapered spindle rather than a stepped pulley on the motor, the latter permitting fine speed control by shifting the motor or idler vertically to give a small change in drive spindle diameter at the operating point. But such refinements are normally confined to the more expensive units, to which we shall return later for a closer look at other hi-fi variations and precautions.

Closely associated with the turntable, and indeed often built on the same base-plate and, in changers, an integral part of the design is the pickup arm or (pointer to an acoustical past) tone-arm. The 'head' or front end of this houses a transducer whose function is to convert mechanical oscillations of the record groove, picked up by a *stylus*, into electrical signals for feeding into the pre-amplifying stages as indicated in **fig. 13**. This transducer, usually known as a pickup cartridge, has to be carried across the record by the groove with a minimum of hindrance and at a playing weight determined primarily by the cartridge characteristics and not by frictional and other limitations in the arm. In essence all pickup arms have pivots which allow easy movement both laterally and vertically, the electrical signals coming out via fine flexible leads near the pivot assembly at the arm's rear. Some adjustment of playing weight is often provided, either by movable counterweights behind the pivot or with variable tension on springs. For technical reasons related to arm geometry, the centre-line of the head or cartridge is off-set from the main arm axis, resulting in a disposition of forces tending to drag the pickup towards the record centre as soon as the stylus is on the disc. The reader may be familiar with this inward skating effect, which is overcome in various ways on some better quality arms – another matter for further consideration under a hi-fi heading.

The pickup cartridge has one of the most prodigious tasks in the whole sound recording/reproducing chain. The enormously complex musical waveforms having been cut mechanically as a succession of minute wiggles in a groove – to be duplicated in mass production by plastic moulding techniques – the groove is then dragged past a stylus which offers so little resistance to lateral and vertical movement that it follows every tiny kink and the most rapid changes of direction without demur. At one extreme, the very lowest recorded levels correspond to groove undulations of around two millionths of an inch in depth, or about one tenth of the wavelength of green light, while at the other end of the scale the stylus sometimes undergoes accelerations around a thousand times those experienced by a falling body under the action of gravity. Looked at another way, if an acceleration of this order (say 1300*g*) were maintained in one direction for one second, the stylus tip would then be four miles away and travelling at nearly 30,000 m.p.h.! When one considers that on stereo discs the two groove walls are independently modulated with slightly different signals, the whole business verges on the miraculous; and the mind is not helped into a more receptive mood on learning that in order to remain in contact with the groove walls under peak modulation conditions, and at a playing weight of a gram or so, the effective mass of the cartridge's moving parts as 'seen' by the groove at the stylus tip should not be greater than about one thousandth of a gram ($\frac{1}{30000}$ of an ounce or $\frac{1}{10}$ of the mass of an ordinary iron pinhead). Another fact providing food for thought is that with a playing weight of two grams the actual pressure at the two tiny areas where a normal size stylus makes contact with the groove is something like ten tons per square inch, yet the tip may show only slight wear after passing 2,000 miles of groove. To help the reader picture this microscopic world, **fig. 16** offers an artist's impression of a stereo groove magnified 500 times and being played by a stylus with a tip radius of under one thousandth of an inch (0·7 mil, one of the standard sizes), with some 10 kHz modulation on one groove wall not far from the record label. At this scale, the record's centre hole would be over 33 yards away. Thanks are due to Rex Baldock for this way of picturing the situation.*

We must now take a look inside the cartridge, leaving till later such matters as stylus tip shapes and sizes and the manner of resolving separate left and right components from the combination of lateral and vertical stylus motions caused by stereo grooves. The vibrations imparted to the stylus by the record groove are applied to a transducer element which generates their electrical equivalent, and in most cartridges the stylus movement is coupled to the generating element by means of a *cantilever*, known alternatively as a stylus bar or stylus arm. The stylus itself is generally no more than a tiny piece of suitably ground and polished sapphire or diamond, cemented to the end of a cantilever, the other end of which either couples directly to the generating element or is anchored to the underside of the cartridge by means of a small fixing screw. The latter system is employed in many of the cheap crystal cartridges commonly used in record players and radiograms, whereby stylus replacement is facilitated as the user simply replaces the complete stylus/cantilever assembly. This practice probably accounts for a common impression that it is the whole cantilever which comprises the stylus and not just the tiny

* 'Look at it This Way', *Hi-Fi News*, January 1965.

Fig. 16 Artist's impression of a stereo record groove being traced by a stylus tip. Note that each groove wall moves independently at 45 degrees to the disc surface. This is approximately 500 times life-size.

piece of pointed material at one end. With this type of cartridge the stylus bar usually rests across a block of plastic which transmits vibrations to the crystal element(s) inside, it being fairly common on older versions to fit a further cantilever on the reverse side of the assembly which couples to the same crystal, the whole cartridge being rotatable (see **fig. 17**) so as to make an alternative stylus size available for playing shellac 78 r.p.m. discs.

With crystal pickups electrical energy is generated by the *piezo-electric* effect, whereby when some types of crystal are stressed mechanically a voltage

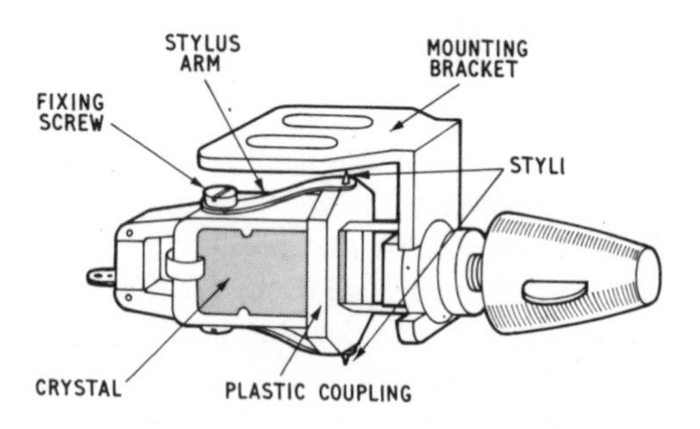

Fig. 17 Type of turnover crystal pickup cartridge found on older player systems, with alternative styli for 78 and LP records.

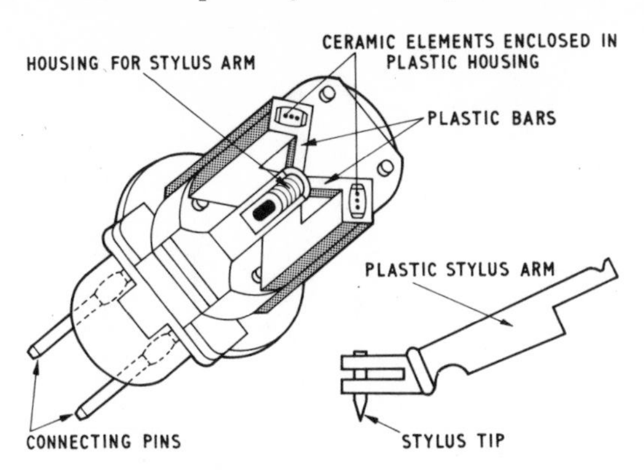

Fig. 18 Details of a high quality stereo ceramic pickup cartridge. The stylus bar is removed and shown separately; when in position it transmits motions from the two record groove walls via plastic bars to the left and right ceramic generating elements.

appears across the material. More sophisticated cartridges employing the same transduction principle are *ceramic* types, ceramic materials being man-made and therefore more amenable to accurate production techniques. An internal view of a high grade stereo ceramic cartridge is shown in **fig. 18**, with the stylus arm removed for clarity. The cantilever in this case is of plastic and is a push-fit into the body of the cartridge, where its motions actuate two ceramic elements placed at right-angles for generation of left- and right-hand signals.

Another type of transducer employed in stereo pickups is the *moving-magnet*, in which the cantilever terminates at the end remote from the stylus with a tiny bar magnet. As this magnet is moved by the stylus bar it induces a varying field in surrounding pole-pieces, this field passing through coils in which corresponding electric currents are induced. The functional elements in this type of cartridge are illustrated in **fig. 19(a)**, though such a view of the 'works' would never be possible in practice as the pole-pieces and coils are normally encapsulated in the cartridge moulding. However, a great convenience with this type of design is that the stylus/cantilever/magnet assembly may be removed intact simply by sliding out a small tube. Very similar to the moving-magnet in both principle and appearance is the *induced-magnet*.

Yet another pickup transducer type is the *moving-iron* or variable reluctance, (**fig. 19(b)**), in which the stylus can be in closer contact with the generating element, the frequency and transient response thus being less dependent on the mechanical behaviour of metal cantilevers (predictable), plastic cantilevers (less predictable) or plastic blocks (almost unpredictable).

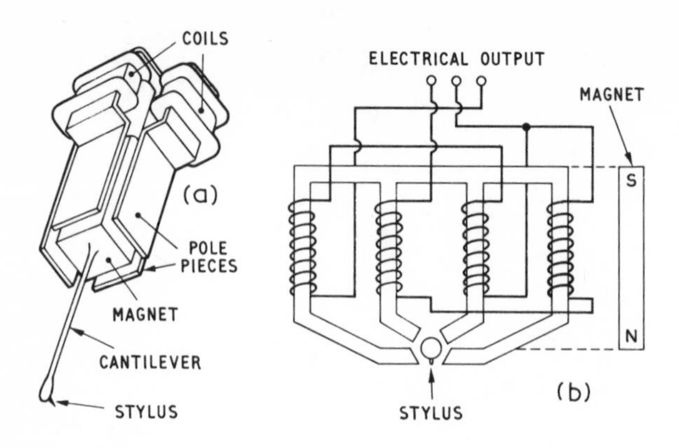

Fig. 19 Functional parts of a stereo moving-magnet pickup cartridge (a), and diagrammatic representation of a stereo moving-iron pickup (b).

Moving-coil cartridges will be found from time to time, as will sundry other devices employing strain-gauge principles or controllable transistor action. Whatever type of transducer is employed, there are a number of special technical considerations of particular concern for hi-fi use, things such as playing weight, tracking performance, tracing distortion, frequency response, the fundamental differences between piezo-electric and magnetic type, and several other factors. These will all be examined in the next chapter.

Thus we come to the final basic signal source for home music reproduction: the tape recorder. Tape recording is a technique for storing audio signals by creating a magnetic pattern in a thin layer of iron oxide particles deposited as a coating on a length of flexible plastic tape. Apart from that used in special closed cassettes, domestic magnetic recording tape is a quarter of an inch in width and is normally stored on spools, with the coated side facing inwards. The length of recording or playing time available with a given spool size varies according to: (a) the thickness of the tape used, there being five standard thicknesses; (b) the playing speed adopted, which will usually be one of three speeds although there are again five choices domestically; and (c) the number of *tracks* recorded on the tape width. In the early days of tape recording the full width of the tape was used for each recording, and when, later, two recordings were accommodated along the same length of tape they were called half-track recordings to distinguish them from the original full-track. Later still the tracks were further sub-divided and called quarter-tracks. This nomenclature has persisted, and the reader may consequently come across the terms half-track or two-track and quarter-track or four-track. The tape itself always looks the same, being brown, shiny on the uncoated side, and quite unaltered in appearance when a recording is present. Recording time, or rather a rough indication of how much tape has been used on a given occasion, is registered by some sort of tape position indicator (t.p.i.), usually a digital device rather like a mileometer.

Recordings are put on to, and replayed from tapes by means of a *tape-head*, a simple looking component across which the tape is pulled at constant velocity and which either generates a magnetic field from an audio input current, to modulate the tape coating (recording), or picks up such a field from the tape and generates corresponding electrical signals (replay). The important external feature of a tape-head is its magnetic *gap* (see **fig. 20** (**a**)), analogous to the air gap in a moving-coil speaker (**fig. 11**), though in this case carrying a field corresponding to the audio signal and filled with some non-magnetic hard-wearing material. In practice the gap may be barely visible as a fine vertical line in the tape-head's polished front, and a complete head will have a smooth, uncomplicated appearance (**fig. 20** (**b**)). Despite the gap's minuteness, the magnetic field that it interrupts reaches out slightly into the passing tape and induces some magnetism there to pass on as a recording.

To satisfy the various track position standards, and to avoid fluctuations of signal level and other technical difficulties, the tape must traverse the head at a constant height as well as at a fixed speed. This brings us to the mechanics of tape recording, which will be surveyed briefly before returning to the question of tracks and the electrical side of recording and replay. Tape transport

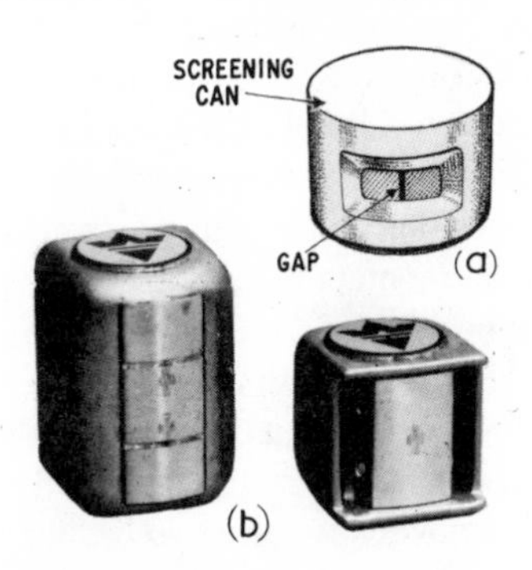

Fig. 20 External appearance (idealised) of a tape-head (a) showing magnetic gap; and photos (b) of some actual heads.

mechanisms vary more widely than do turntable drive systems, with a welter of belts, pulleys, levers, brackets, brakes, clutches and switches that will deter most users from tinkering; we shall therefore here examine only the essential elements.

Firstly, except in very cheap battery portable recorders the tape is *not* driven through the system by the spools. If it were, the tape speed would change continuously in accordance with the diameter of tape on the spools. Except when fast winding in one direction or the other, the feed spool and take-up spool have relatively static functions, one normally receiving a slight braking drag to prevent tape spillage and the other being rotated gently to take up the tape without excessive drag or a tendency to modify the tape speed, which, as we shall see, is controlled elsewhere. Separate electric motors are sometimes used for the two spools, with a large reserve of power for reverse or fast-forward winding. However, most machines employ one motor for all the tasks, with various coupling arrangements to transfer the drive to the feed-spool for reverse winding, to the take-up spool for fast-forward winding, or to the *capstan* for normal recording or replay, in the last case with some sort of slipping-clutch giving the take-up spool its necessary mild torque.

The capstan is an accurately ground spindle that protrudes through the tape deck top plate (see **fig. 21**), normally being coupled beneath to a flywheel which is in turn driven by the motor. The rotational velocity of the capstan sets the tape speed, and some mechanism analogous to that employed in gramophone turntables (**fig. 15**) is used to change the capstan speed in relation to the driving motor.

When the mechanism is operated for record or replay a rubber or plastic-composition *pinch-wheel* is pushed against the capstan, thus sandwiching or 'pinching' the tape and causing it to move from left to right at a speed set by the capstan's surface velocity. Suitably positioned guides (see **fig. 21**) control the tape's vertical position and stabilise its motion, and on many recorders *pressure-pads* move into position at the same time as the pinch-wheel to ensure that the tape makes intimate contact with the heads.

Two tape-heads are shown in **fig. 21**, one for tape *erasure* and one for record and replay. The erase head is fed with a continuous high-level signal in the region of 60 kHz when the machine is switched to the record mode, and an appropriate part of the tape width is thus scanned with rapidly oscillating magnetic energy which removes any earlier recording from the relevant track. Immediately after passing the erase head the tape traverses the record/replay head, which again scans the appropriate track and, depending on the mode of operation, either implants a fresh recording on the tape or extracts a signal

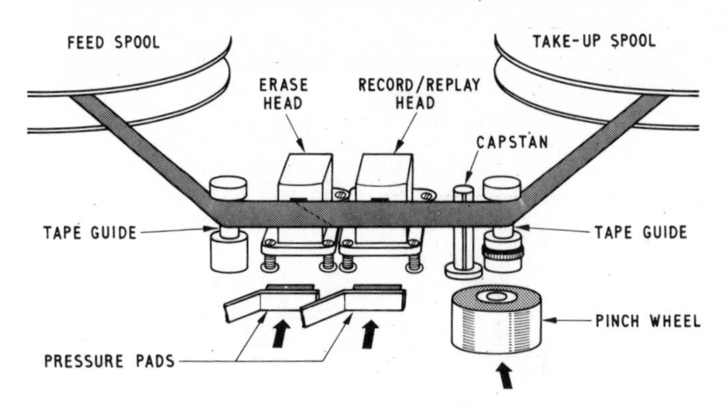

Fig. 21 Tape path in a recorder. When mechanism is switched on the pinch-wheel and pressure-pads move in as indicated, the tape then being gripped between capstan and pinch-wheel and forced into intimate engagement with the heads.

from it. In the latter case the recording remains on the tape, much as mechanical modulation in a disc groove manages to impart its 'information' to a pickup without being removed. In more expensive recorders there is sometimes a third tape-head; this is used for replay only, the second head then being confined to recording. This permits greater flexibility and refinement of design, a particular advantage being that when recording it is possible to listen to what is recorded on the tape an instant after it is recorded rather than much later, when one may discover faults that might have been avoided.

With two-track monophonic (single-channel) recording the magnetic gap in the head is so positioned in relation to the tape path that something a little less than the upper half of the tape width passes across it. This is shown in **fig. 22(a)** where, looking as it were through the tape from the uncoated side, it is apparent that as the tape moves from left to right a recording is implanted on the upper track. When the whole length of tape has been recorded in this manner the empty left-hand spool may be removed, the full take-up spool taken off the right-hand spool hub, inverted, and placed on the left-hand hub. If the tape is then rethreaded, the first track will be at the bottom and another may then be recorded from the same head gap (**fig. 22(b)**). For two-track stereo recording a two-in-one head is used, with two half-track gaps one above the other. In this case the recording goes once only along the tape length (**fig. 22(c)**), the top track carrying the left-hand channel and the bottom track the right-hand.

For quarter-track recording the tape-head is again a two-in-one device, but with appropriately smaller gaps arranged as shown in **fig. 23**. Here, for mono

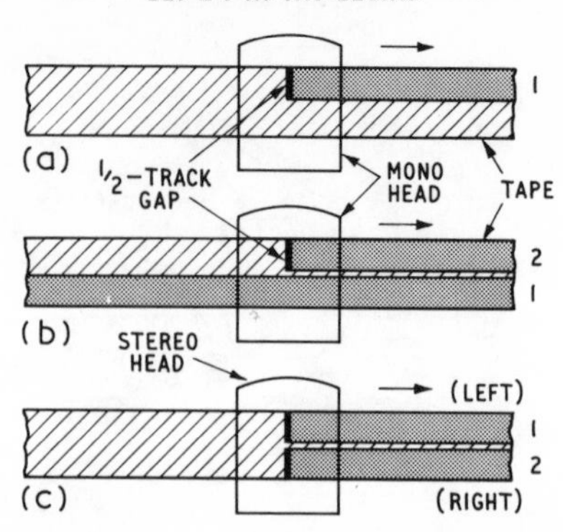

Fig. 22 Derivation of track positions on two-track (half-track) mono and stereo tape recordings.

recording one first uses Track 1, then the spools are reversed as for half-track work, bringing what was the bottom tape edge to the top for recording Track 4. The spools are again reversed, but for this next change a track selector switch has to be operated to bring the lower head gap into operation, thus recording on Track 3. Finally, with a last reversal, but leaving the lower track in operation, Track 2 is recorded. For stereo recording on four tracks both halves of the head are in operation, the left and right signals being implanted on Tracks 1 and 3 in one direction and on Tracks 4 and 2 on the other.

On the tape recorder's electronic side there are two special sets of requirements in addition to a continuous oscillation for erasure already mentioned. For recording purposes a high frequency *bias* waveform is added to the audio signal to overcome certain characteristics of the magnetic recording process which would otherwise introduce severe distortion. Provision of this bias and the erase power – often from the same oscillator circuit – and satisfaction of other technical points concerning the manner in which the signal is fed to the record head, normally means that several valve or transistor stages must be

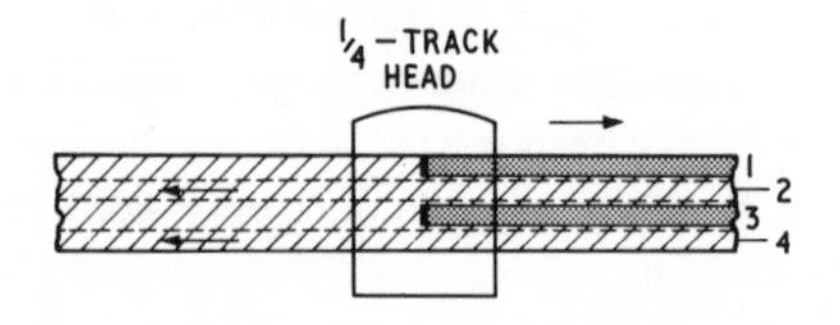

Fig. 23 Derivation of track positions on four-track (quarter-track) tape recordings. For stereo the L/R signals may be carried on 1/3 or 4/2.

employed specifically to make the recording process possible. Also, some means of indicating to the user the amplitude of signal fed to the head is incorporated, whether a meter or 'magic eye', so that overload distortion may be avoided. It is perhaps worth mentioning in the non-hi-fi context of this chapter that in some of the cheap battery-powered recorders mentioned earlier a permanent magnet is used for erasure and 'DC' bias is applied with the signal to the tape-head.

For replay purposes the output from the tape-head must be amplified up to a level suitable for ordinary handling in audio circuits, and, more important, its response must be corrected. As we shall see, gramophone records employ one agreed standard recording characteristic; but the frequency correction needed with magnetic tape changes with both tape speed and type of replay head, quite apart from the existence of various standards which might be applied to commercial tape records. Essentially, a tape-head responds to rate-of-change of magnetic flux as a recording passes along, and for a given recorded intensity it therefore follows that the higher the frequency the higher the output, and vice versa. Because of this, low frequencies have to be boosted very considerably, though in practice the output does not go on rising indefinitely at high frequencies as a point is eventually reached where a recorded wavelength on the tape is of the same order of size as the head-gap width. At such a frequency a recorded waveform would have both its positive and negative half cycles within the gap at one time; these would cancel each other magnetically and no output would be produced from the head. Clearly, high recording and replay speeds give greater wavelengths on the tape, and thus the cancellation point moves up in frequency. For this reason the optimum replay response curve changes with speed, as does the useful upper frequency limit, making high tape speeds preferable for hi-fi work. Detailed points related to this will be covered in the next chapter, though it may be noted that – other factors being optimised – a tape speed of 15 inches per second (i/s) is superlative, $7\frac{1}{2}$ i/s is excellent for most domestic musical purposes, $3\frac{3}{4}$ i/s can be very good, $1\frac{7}{8}$ i/s is suspect but improving, and $\frac{15}{16}$ i/s is almost useless for music.

Tape recording and recorders, like most of the other basic elements covered in this chapter, is a vast subject which can only be skimmed in a survey such as this. However, the information presented on the various links in the audio chain should provide at least a framework for the reader without prior electronic knowledge, and will help understanding when we move through hi-fi specifications on our way back to more musical matters.

4. HIGH FIDELITY I: TAPPING THE SOUND SOURCES

IN CHAPTER 2 we unravelled seven strands from the entangled acoustic texture of musical sounds, strands comprising essential guide lines to quality and realism in the reproduction of music. The more technical reader might reasonably have expected that analysis to be followed immediately by an extension and refinement of those seven pointers to cover specifications of high fidelity equipment. However, such specifications do, after all, arise from the limitations and possibilities of basic components, so another section intervened with a short course on the main links in the reproducing chain for the benefit of those – certainly a majority of music lovers – who have no notion of what items lie beyond the control knobs or how they work. Now, assuming the essence of the previous two chapters to have been absorbed, it will be useful to attempt some definitions, or at least descriptions, of hi-fi in terms of performance features for the separate components. This will be divided into two chapters, the first dealing with techniques used for converting the available sound sources into electrical signals, and the second covering reception, control and amplification of such signals and their conversion back to sound via loudspeakers. This amounts to chopping the component scheme laid out in **fig. 14** (page 54) into mechanical/magnetic and electrical/acoustic parts, with pickups, turntables and tape recorders in this chapter, and tuners, amplifiers and speakers in the next.

Firstly, partly as a reminder that most of the work has been done before domestic equipment even comes into the picture, and partly to warn that attempting to reproduce subtleties not included in the available signals is a waste of effort, it will be useful to examine the normally accessible sound sources themselves.

Radio is the most prolific source of music at low cost and good technical quality. But there are difficulties and limitations, for in the U.K. reception of AM transmissions on the MW and LW bands is subject to severe quality restrictions, and while the FM service is vastly better, the techniques used for 'piping' the audio signals around the country sometimes limit the upper end of the frequency scale. Another difficulty is that the dynamic range on music programmes is normally restricted to ease things for AM listeners, who inevitably suffer from a higher level of background noise and who still comprise (1968/69) a large proportion of any BBC sound radio audience. High background noise means that the very quietest music passages must be raised in level to avoid their loss through masking, and since there has to be finite peak modulation on transmitted radio signals, this means that the space available for manoeuvre between upper and lower limits is in practice often not much more than 30 dB. This is satisfactory for most music up to Haydn and Mozart but inadequate for realistic reproduction of nearly all later full orchestral works. But the degree of inadequacy depends on the skill used in compression, and as this is done manually there is some variation in the end result when judged subjectively. For instance, by easing down the volume for a bar or so immediately preceding a fortissimo tutti outburst rather than at the moment of impact, the studio operator can do much to retain an impression of the original dynamics. Also arising from concern for the average listener's eventual signal-to-noise ratio, broadcasting authorities often use something approaching the full modulation level for peaks in speech as well as music. In consequence, if music is reproduced at a natural loudness the announcements between items in broadcast concerts sometimes tend to leap out at the listener rather aggressively. Another limitation on speech quality is some exaggeration of low frequency chest tones due to microphone types and placings used by the BBC for news readers and many solo talkers.

These negative points, though serious, should not deter the reader from pursuit of music in the home via radio, for the BBC offers an unparalleled wealth of music, covers practically the whole population of the British Isles with a network of FM transmitters, and is slowly extending its stereo programme service to the major population centres via those stations. Sound radio listening via AM on the medium and long wavebands is probably going into decline in favour of FM at very high frequencies (VHF), and we may reasonably expect an eventual realignment of audio standards in accordance with the better system's capabilities. As this book goes to press a limited stereo service is available on VHF/FM from the BBC's transmitters at Wrotham (London and the South-East), Sutton Coldfield (Midlands) and Holme Moss (North), and it is clear that on those occasion when, say, a live musical event is relayed from a London concert hall without technical mishap under the control of an engineer with a perceptive approach to the music's dynamics, the result in the home is quite superb and enough to convert any visiting music lover to stereo/FM without more ado.

Gramophone disc records are probably the most important, if by no means the cheapest source of music for high-quality domestic reproduction. Indeed, it might fairly be claimed that practically the whole modern hobby of hi-fi is based on the LP record, which often kindles a first audio interest for technical and musical alike. As with radio, technical quality is variable; but the best discs are superbly good and it is generally possible to locate excellent recordings of most works in the standard concert and operatic repertoires.

There need be no serious degradation of musical dynamic range via the disc medium, as the practical extremes of groove modulation are equivalent to a span of 60 dB, the theoretical lower limit set by the vinyl material being some 80 dB below peak recording possibilities. In practice, some noise may be introduced at various points in the multitude of stages between studio microphones and commercial pressings – some discs even carrying recorded rumble and noise from passing traffic! – and during replay there are dust problems to contend with. But 55–60 dB in any case copes with all but the most exceptionally massive music, and if the listener still feels cheated of the full impact of molto fortissimo orchestral brass even when listening to the most dynamic of recordings via the best of equipment, this is probably due not so much to physical limitations of the disc medium as to various psycho-acoustic factors.

In a similar category are such matters as differences in instrumental and vocal/orchestral balance, reverberance or 'deadness', naturalness or artificiality of ambience, and other things dependent on microphone placings in the recording studio. Recording producers have various ideas on what is legitimate when music is to be consumed at home rather than in the concert hall, and such ideas are reflected as variations in the balance or character of sound on discs from different recording companies. This imposes not so much a limitation in quality as a variety of aural 'house styles', and as a considerable proportion of the record industry's vast annual non-'pop' output fully justifies the use of high quality replay equipment from all other points of view, we can award high marks to the gramophone record as a source of music for hi-fi in the home.

This leaves tape, regarded from time to time as a viable alternative to disc for commercial music recordings. There are many legitimate and interesting non-hi-fi uses of tape recorders which must remain outside the scope of this book, but the fortunate minority of enthusiasts able to record 'live' musical performances with good microphones and semi-professional machines will confirm that the *possible* standard of reproduction is very high. However, variations in quality due to deficiencies in mass-copying techniques, some restriction of dynamic and frequency ranges at the tape speeds commonly used for commercial recordings, the exasperating business of locating particular musical passages or movements and the corresponding waste of time running tape back and forth – these factors have until recently militated against tape as a substitute for the gramophone record.

Because disc recording and reproduction involves the seemingly crude mechanical business of cutting a pattern of wiggles and then using these wiggles – after transfer via 'Master', 'Mother' and stamper to commercial copies – to vibrate a pickup stylus, there is a widespread assumption that the disc medium must be inherently inferior to magnetic tape. It is often remarked that original recordings are always made on tape, and that the known and irreducible distortions involved in the disc cutting/processing/replay sequence are bound to degrade the sound we eventually hear. However, that the degradation can be minute indeed has been demonstrated publicly from time to time, as when Decca's recording chief Arthur Haddy has played master tapes and discs of the same musical passages to audiences who have been hard pressed to detect any difference. It is true that a carefully made 'one-off' copy, tape to tape, of an original recording can be extremely fine, but the practical problems of large-scale copying for commercial tape records to replay at speeds below $7\frac{1}{2}$ i/s have frequently involved inconsistency from sample to sample. Also, the potential signal-to-noise ratio of tape at low speeds is inferior to that obtainable from disc. However, the great success in recent years of tape cassette recordings for the popular market has prompted fresh research into this medium's hi-fi prospects, with promise of a brighter future. At present we must still regard commercial tape records with some doubt as a prime hi-fi signal source; but, as noted earlier, for private recordings tape can be superb.

Having glanced at the basic sound sources available to the home music enthusiast, we shall now take each of the main component links in a hi-fi reproducing chain concerned with converting the sources into electrical signals, and see what can be expected in terms of the seven technical strands.

First comes the pickup, which, as remarked before, has one of the most delicate and prodigious tasks in the whole gamut of sound reproduction. A tiny piece of sapphire or diamond is required to follow, mechanically and simultaneously, two recorded waveforms of almost unbelievable complexity, faithfully transmitting their vast range of amplitudes and accelerations to independent generators which provide exact electrical equivalents of the undulations implanted on the two groove walls. The word 'exact' is here used relatively, for no pickup is perfect and hi-fi models are simply somewhat less inexact than the types used in mass-produced record players and radiograms. How small that inexactitude can or should be is best considered, first, in relation to non-linear distortion, which we have seen leads to harmonic and intermodulation products generally regarded as objectionable. The term non-linear refers to the uneven response of a device – be it pickup, amplifier, microphone or speaker – to signals of differing amplitude, leading, for instance, to progressive 'clipping' of waveform tops as the reproduced material gets louder. Another way of looking at this is to regard any deviation from the recorded or transmitted waveform as a species of non-linearity in the sense that the original *line* is not being followed exactly. Waveforms will also be reshaped by a poor frequency response, but when this happens the original patterns may sometimes be recreated by adding suitable corrections. However, genuine amplitude non-linearities produce actual harshness which cannot be removed or corrected except in a limited and special way which we shall consider in the next chapter in relation to control unit facilities.

In gramophone pickups there is one definite and inevitable misdemeanour which causes trouble of this sort – it is called *tracing distortion* and arises because of a difference in shape between the cutter tip used to engrave the groove on the original lacquer disc and the stylus tip used to play the commercial pressing: one is a sharp-edged chisel and the other is approximately spherical. The problem is illustrated in **fig. 24,** showing a simple lateral waveform in relation to both cutter and pickup stylus tip. The cutter's sharp corners ensure that whatever motion pattern may be imparted to its centre (wavy dotted line) is accurately duplicated on both groove walls; but the spherical replay stylus cannot reciprocate, for its points of contact with the groove vary continuously as the modulation changes, resulting in a distorted motion pattern (angular dotted line) and consequent distorted electrical waveform out from the pickup. In extreme cases the relatively clumsy round tip cannot gain full access to sharp high frequency troughs, so it just bumps across as best it can, producing yet more distortion. This condition has just been reached at the two lowest points on the groove in **fig. 24**. Another complication is that a tip of finite radius will vary its depth of penetration according to the waveform

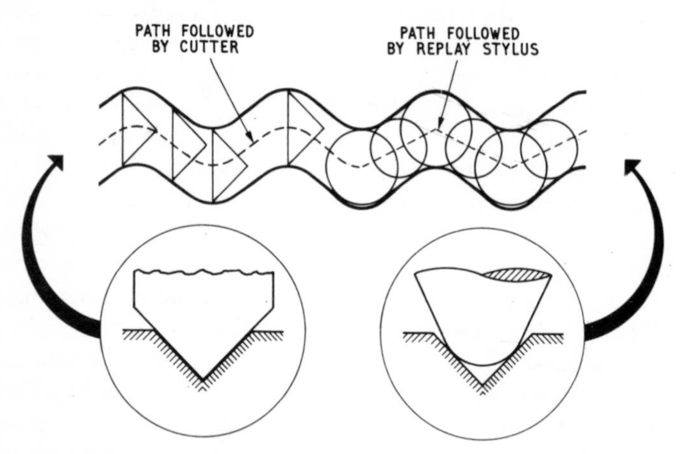

Fig. 24 Tracing distortion arises from differences in shape between the original cutter used by the recording company (a sharp-edged chisel) and the replay stylus – usually a cone with a spherical tip.

– hence the changing diameter for an exact fit seen in the diagram – and the resulting vertical motion of the stylus, known as *pinch-effect*, produces unwanted electrical outputs if the pickup is a stereo type, for, as we shall see shortly, a stereo cartridge has to respond to both lateral and vertical movements of the groove. This may all seem rather dreadful, though the situation is not as black as it may appear, for waveforms with the sort of sharp angles leading to the troubles depicted in **fig. 24** are equivalent to high recorded frequencies, so the harshness of distortion does not occur within the texture of instrumental sound but affects the higher harmonics only.

Related to this is an old bogey of the gramophone called *inner groove distortion*, whereby the quality of sound deteriorates as the pickup moves across the record. Disc records rotate at a constant number of revolutions per minute, which means that the actual linear speed of the groove past the pickup stylus is greater at large diameters than it is towards the middle, the distance round the outer groove of a twelve-inch record being thirty-six inches, falling to fifteen inches at the inner groove. This means that a piece of recorded musical waveform lasting, say, for one second will occupy twenty inches of groove in one position and only just over eight inches in the other. Either way, the number of cycles, undulations and wiggles making up the waveform must still be accommodated, though it is clear that when cramped into eight inches they will all be closer together than when allowed to stretch to twenty inches. In other words, recorded wavelengths are shorter at the inner grooves of a record, and high frequency fluctuations which may be relatively large in relation to a given stylus at the outer diameters can begin to cause appreciable distortion when only two-and-a-half inches from the disc centre.

What are the hi-fi solutions to this tracing distortion problem? Looking at **fig. 24**, it might occur to the reader that if a really minute replay stylus tip were used it would follow the true waveform pattern right down at the bottom of the groove – almost a point, in fact, tracing the path made by the cutter's point. Unfortunately there are two practical snags to this, one on the record and the other in the pickup. Although the front view of the cutter in **fig. 24** shows a sharp right-angled tip, various mechanical factors and the behaviour of the lacquer as it is cut prevent the groove from adopting this precise shape, practical record grooves having a slightly rounded and irregular profile right at the bottom. The dimensions involved mean that the radius of a spherical tip cannot be reduced much below half one thousandth of an inch (0·5 mil or 13 microns) if all makes of record are to be played without risk of 'bottoming', which fault may manifest itself as excessive surface noise and/or a curious fuzzy type of distortion because the stylus cannot settle firmly against the two

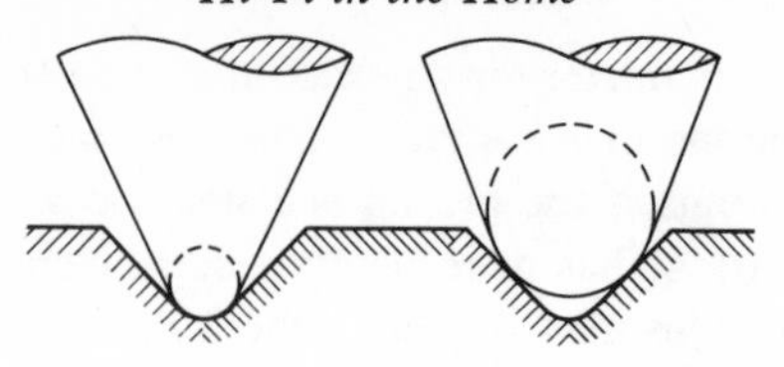

Fig. 25 A replay stylus of correct frontal radius (right) sits comfortably against the two groove walls, while a tip with too small a radius (left) may run on the groove bottom.

groove walls. The beginnings of this unstable situation are shown in **fig. 25(a)**, it being preferable to choose a tip radius which causes the stylus to be securely supported by both groove walls, well away from the bottom, as in **fig. 25(b)**. The other practical objection to very small stylus radii is that even if groove shaping would permit, say, use of a tip with a radius of 0·1 mil, at normal playing weights this would dig into and damage the groove as well as suffering very rapid wear. We shall see shortly that there are other design difficulties when it comes to reducing that 'normal playing weight', so the ultra-small spherical tip radius approach to the problem of tracing distortion must be abandoned.

Note particular use of the phrase 'spherical tip radius', for there is another way of reducing the *effective* radius which involves departure from a simple rounded cone. Ideally, we need a replay stylus which sits safely across the groove as in **fig. 25(b)** when viewed from front or rear, but which nevertheless has relatively narrow corners in contact with the sides of the groove when viewed from above, in order to follow more closely the intricate movements traced out by the cutter's sharp edges. Such a stylus is the elliptical or bi-radial type (**fig. 26**), sitting with its major radius astride the groove for firm support without 'bottoming', and with its minor radii making more intimate contact with groove wall undulations than could be obtained with a spherical

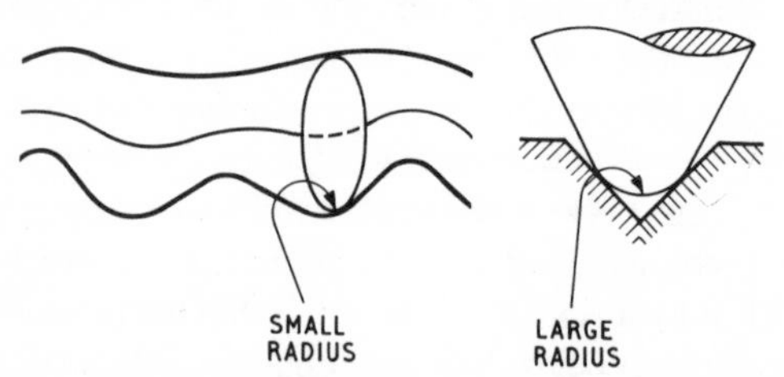

Fig. 26 An elliptical or bi-radial stylus combines small radii at the sides to minimise tracing distortion with a large frontal radius to avoid 'bottoming'. It is shown here sitting in a stereo groove with sharp high frequency modulations on one wall and a less severe waveform on the other.

tip. The combination of large and small curvatures overcomes the 'digging in' problem mentioned as an objection to very small spherical tips, a practical bi-radial stylus having major and minor radii of 0·7 mil and 0·2 mil respectively. To minimise wear and other problems it is wise not to use this type of stylus in a pickup cartridge needing a playing weight of over two grams (2 gms), but the best cartridges all perform satisfactorily at forces below this.

The general situation in relation to tracing distortion and stylus shapes and sizes may be summarised as follows. The spherical tip most commonly used in pickups intended for playing both mono and stereo records has a 0·7 mil radius, this being almost universal in record players and radiograms. A pickup with such a stylus, if used in conjunction with high quality equipment – including loudspeakers which reveal all the subtleties of sound at high frequencies – will not do full justice to the best recordings, particularly at the inner grooves on stereo discs, where tracing distortion will become objectionable during loud climaxes. Pickups generally regarded as being in the hi-fi category therefore employ either spherical tips in the region of 0·5 mil radius or elliptical tips with dimensions near to those mentioned in the previous paragraph.

A quite different approach to the problem of tracing distortion is to standardise the replay stylus at, say, 0·7 mil spherical for all types of equipment, including the most sophisticated, and then distort the recorded waveforms in a predetermined fashion such that the eventual path traced by the pickup stylus resurrects, phoenix-like, the original ideal waveform. Such an approach could in theory finally eliminate all the difficulties created by the inevitable geometrical differences between cutter and replay stylus. Techniques of this sort were pioneered commercially by RCA-Victor on their Dynagroove discs, and early doubts about the resulting recorded quality are now collapsing as methods improve. Other record producers are experimenting with such ideas, and it may well come about that a standard process will be evolved giving minimum distortion for all types of pickup using an agreed stylus tip size. However, this is for the future, and high fidelity at present demands that tracing distortion in general and inner groove deterioration in particular be minimised by paying careful attention to stylus shape and size.

So much for tracing which, though a geometrical problem, is sometimes confused with tracking, a mechanical matter. Even in the absence of tracing errors it does not necessarily follow that a stylus will faithfully adhere to both groove walls during all the complexities, amplitudes and accelerations of actual musical waveforms. The groove has to *move* the stylus in accordance with its own changing shape, and if the stylus is too rigid or too massive it will be disinclined to move when the groove shifts either too far or too fast. When this situation arises the stylus, in trying to resist lateral motion, is forced to climb up one groove wall or the other, and in extreme cases it may leave the groove completely and jump into the next one. This sort of behaviour manifests itself in reproduction as audible distortion, which may be reduced or eliminated by increasing the playing weight of the pickup.

The ability of a pickup cartridge to track heavily modulated records is known in popular parlance as its trackability, which may be measured in terms of the smallness of the playing weight needed to avoid mistracking when reproducing the most severely modulated grooves. The opposition to stylus motion in a pickup is known technically as *mechanical impedance,* the main components of which are stiffness and mass or inertia. Stiffness is more easily expressed by its reciprocal *compliance,* which is simply an indication of the ease with which the stylus may be deflected from its neutral position by application of a known force. This is measured in compliance units (c.u.) and determines the ability of a cartridge to track large low frequency modulations. At the other end of the frequency scale the limits are set by the effective mass (or inertia) 'seen' by the record groove when it attempts to move the stylus very rapidly, and it is common to express this (not necessarily honestly or meaningfully) in milligrams (mg). With good pickup cartridges these parameters permit playing weights in the region of one gram, at which pressure record wear may be completely discounted and wear of diamond styli is extremely slow, permitting the playing of several thousand LP sides before there is significant deterioration.

The sort of hi-fi cartridge to work in this fashion would have an effective tip mass (or 'mass referred to stylus tip') in the region of 1 mg and a compliance around 20 c.u. However, in both cases the manufacturer's specification might qualify such figures in accordance with the direction of stylus movement, for stereo pickups must permit the stylus to follow the independent motion of two groove walls, which necessarily involves freedom of movement both laterally and vertically. How and why stereo groove modulations may be related to movements up-and-down and side-to-side is commonly misunderstood, so it will be worth devoting a few sentences to this.

An ordinary mono record groove simply moves laterally as in **fig. 24**, and a perfect stylus not subject to the minor parasitic vertical pinch-effect movements would, if tracing and tracking correctly, also simply move from side to side. On a stereo disc the two groove walls carry related but nevertheless independent signals, each wall undulating at forty-five degrees to the record surface. The reader may care to refer back to **fig. 16** (page 58) for a close-up view of this situation. What happens to the groove from moment to moment with various combinations of left and right signal is illustrated in **fig. 27**. With modulations on one groove wall only, whether left or right ((**a**) and (**b**)), it will be seen that a stylus perched firmly in the groove as in **fig. 25**(**b**) must move both laterally and vertically. In stereo recording and reproduction those sounds emanating from the centre of the 'picture' are represented by exactly equal signals in the left and right channels, and in accordance with an agreed convention such signals operate the disc cutter so that as one wall goes up the other comes down (**fig. 27**(**c**)), resulting in purely lateral groove movement. This, of course, is the same as on mono discs, which can be regarded from this point of view as stereo recordings of central sound sources only. The last condition depicted in **fig. 27** represents the extreme case of a groove moving purely vertically, where the two signals are momentarily in total opposition or, in technical jargon, out-of-phase. This is a freak situation normally avoided by the recording companies, it being generally safe to assume that the vertical motions of a stereo groove do not reach more than a half or third the maximum lateral amplitude. For this reason it is not strictly necessary for the mechanical impedance at a pickup's stylus tip to be as low in the vertical direction as it needs to be laterally, and with some types employing the moving-iron mechanism this aids design, the associated technical specification making a distinction, for instance, between lateral and vertical compliance.

All this concern to arrange the geometry and mechanics of the stylus for accurate tracing and tracking of the two signals carried by a stereo groove would be to no avail if the resulting stylus motions were not in turn conveyed to two generators or transducers able to respond independently to left- and right-hand modulations. Independence is the key word here, it being important to achieve the maximum possible separation of the two signals. In practical pickups *channel separation* tends to be at its best at middle frequencies, unwanted *crosstalk* creeping in at the frequency extremes. Thus it is common in pickup specifications to give the channel separation, at, say, 1 kHz only, though the better cartridges may claim and achieve a separation of 20 dB or more over most of the audio band, perhaps reducing to 10 dB above about 10 kHz.

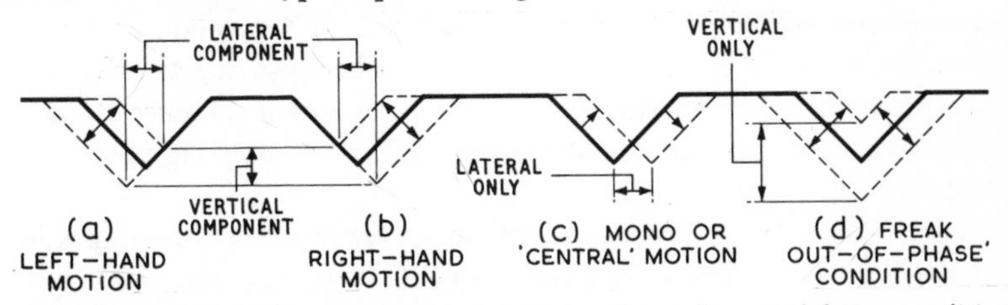

Fig. 27 Cross-section of stereo record groove showing various modulation conditions.

The ease with which the two signals may be separated and yet remain balanced with regard to sensitivity and response depends on the type of transducer employed and the mode of mechanical coupling interposed between stylus and generator. Due to the greater inherent stiffness and mass of crystal or ceramic transducers it is necessary with these to transmit the stylus motions via compliant plastic bars or blocks, tending to introduce an element of inconsistency from sample to sample and creating, for some ears, a type of sound somewhat lacking in inner delicacy. For reasons of this sort the cartridges generally regarded as best from the hi-fi point of view – particularly for transient response – are magnetic types (moving-magnet, moving-iron, variable reluctance, moving-coil, induced-magnet).

Frequency response is also generally somewhat smoother and flatter with magnetic cartridges, though there is a complication here due to an inherent difference between magnetic and crystal/ceramic types in the way in which stylus movements are translated into electrical signals. As the groove – and with it the stylus – moves back and forth it could be said that for any given frequency and amplitude of modulation it oscillates at a particular rate or *velocity*, and it is this characteristic of the recorded waveform to which a magnetic pickup responds. Crystal and ceramic types, on the other hand, give an electrical output proportional to the magnitude of stylus displacement, regardless of frequency. Now, in order to avoid excessive recorded amplitudes at low frequencies and to improve the overall signal-to-noise ratio in disc playing systems, the frequency response of the signals actually cut on to the record groove is tilted according to the internationally agreed curve shown in **fig. 28**. This curve is a measure of the relative modulation velocity, and as magnetic pickups are velocity-sensitive devices they give an output following this pattern, which must be corrected in the associated amplifier control unit. It is therefore assumed when specifying the performance of a magnetic pickup cartridge that the appropriate frequency response correction is applied. The relative depression of low frequency velocity and boosting of higher frequencies involved with the standard recording characteristics means that the actual physical amplitudes of groove displacement are more nearly equal than would be the case with a 'flat' recorded response. This being so, designers of crystal and ceramic cartridges arrange that the effective overall response of disc and pickup combined is reasonably flat, no further correction being needed in the amplifier.

These differences of response are paralleled by differences in output level, magnetic cartridges delivering a few millivolts (mV) only and ceramic types giving something like 50–200 mV. Pickup output levels are often specified as

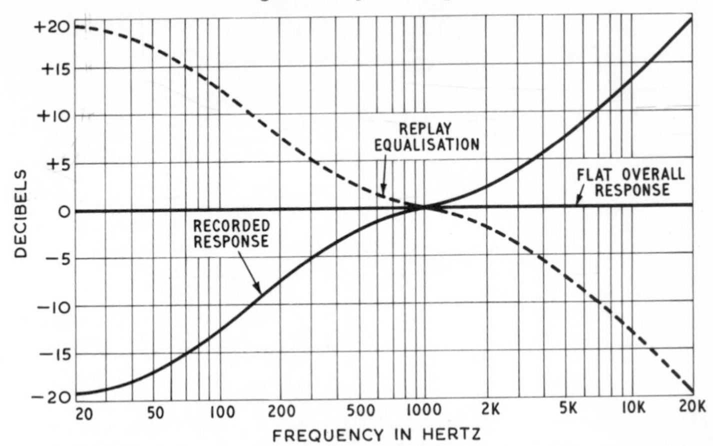

Fig. 28 Standard recording curve for discs (solid line) adopted to avoid excessive amplitudes at low frequencies and preserve a good overall signal-to-noise ratio at high frequencies when corresponding equalisation is applied on replay (dotted curve). This is sometimes referred to as the RIAA curve.

sensitivities in relation to disc modulation, the recorded velocity at 1 kHz being used as a reference. Thus the expression '1·5 mV/cm/sec' means that the electrical output will be one and a half millivolts (1·5 mV) when the recorded waveform has a velocity of one centimetre per second (cm/sec). As a rough guide, it may be assumed that the average recorded velocity on a music disc is about 5 cm/sec, and a pickup with the above sensitivity would therefore deliver an output from this of $1{\cdot}5 \times 5 = 7{\cdot}5$ mV. When choosing equipment, such figures must be considered in relation to the input sensitivity of the associated control unit.

Before leaving pickup cartridges for a look at the arms used to carry them across the record, it will be worth examining the response and crosstalk curves of a typical high quality stereo magnetic device. These are shown in **fig. 29** in the form in which they might appear in a published test report. The frequency response is smooth and flat over most of the range, the output declining a little above 5 kHz but then rising again to a peak at 16 kHz. A response of this sort is by no means universal, though it is generally difficult in design to avoid some irregularity at high frequencies, and the general shape shown in **fig. 29** may be regarded as very good if not perfect. A small difference of overall level between the left and right outputs is evident in this case, but is less than one decibel and would be of no consequence in practice. The crosstalk curves show the extent to which modulation on one groove wall will cause an electri-

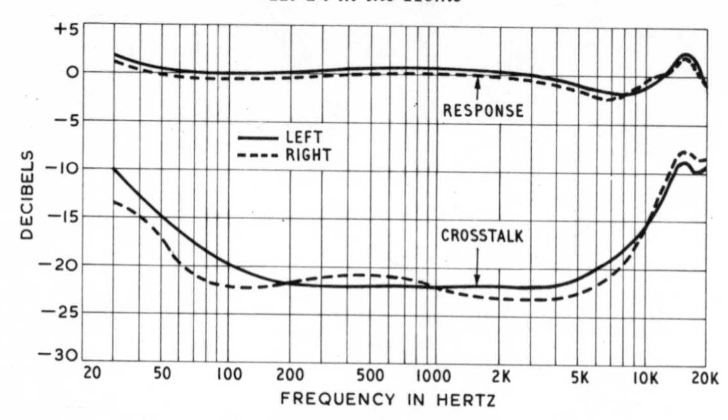

Fig. 29 Frequency response and crosstalk curves of a high quality magnetic stereo pickup cartridge. Note that response is within ± 2 dB over the whole range.

cal output to appear in the wrong channel, such an output indicating imperfect correlation between stylus motion and the resulting generator motion. It will be seen that channel separation is poorest where a peak occurs in the response. This is because high frequency peaks are due to mechanical resonances, and at their resonant frequencies objects tend to get a little out of hand when set into motion. It could be, for instance, that at 16 kHz the stylus cantilever has a natural tendency to vibrate laterally rather than at forty-five degrees to the record surface, and since lateral motion will produce equal outputs in the two channels one could expect to see some decrease of channel separation at this frequency. In inferior cartridges the high frequency resonances are less well controlled or 'damped', resulting in both higher peaks in the response curve and poor channel separation.

At the other end of the frequency band the response is just beginning to rise at 30 Hz, with an associated worsening of crosstalk. Poor channel separation at very low frequencies is not particularly troublesome for the stereo listener as most of the directional information occurs higher up the scale. However, a peak in the response – even if at a frequency too low to be heard – can be troublesome mechanically in connection with pickup arm behaviour when playing warped records. Which brings us to arms.

An ideal pickup arm is neutral, its function being simply to allow the cartridge to move across the record at the correct height and angle with a playing weight appropriate to its tracking ability: it should add very little mass to the cartridge and offer no frictional opposition to its progress. There is a common confusion over the subject of masses and weights, arising perhaps from the usual practice of referring to playing weight rather than downward force, and also because the everyday unit of force – the gram or ounce – is also used for weight and mass. A typical magnetic cartridge would weigh about 7 gms, even though the downward force or playing weight required for good tracking may be only 1 gm. A pickup arm would therefore have to provide an *upward* force of 6 gms to counteract the unwanted part of the cartridge weight, leaving just 1 gm for playing purposes. Although in these circumstances the effective static weight or force on the stylus – as measured by a small pair of scales – would be only 1 gm, the actual mass of the cartridge is still there in a dynamic sense when the pickup head is moved suddenly – as by a warped record.

This may perhaps be illustrated by an analogy with hinged doors. An elderly house has a thick oak front door mounted on massive stable-door hinges, while the entrance to a recently installed toilet off the hall is a light structure made up from hardboard panels. In both cases the full weight is taken by the hinges and the doors are free to swing without scraping the floor – rather like a pickup in which *all* the cartridge weight has been cancelled, leaving the arm free to swing just above the record surface. Imagine that both doors have been left ajar. A child playing in the hall falls against the toilet door, which responds by swinging away fairly swiftly; then the child falls with similar force against the front door, but the latter's much greater mass keeps it in place: child runs to mother with bumped head. Likewise with pickup arms, where the rapid movements sometimes imparted by a warped disc can, as it were, bump their heads if the effective mass or *inertia* of the arm/cartridge combinations is too high, though the 'bump' would in this case be taken by the compliance associated with the stylus and transducer, perhaps causing the cartridge to distort or mistrack if it happens at a moment when heavy groove modulation is already straining the stylus to the limit set by its playing weight.

With any arm/cartridge combination there occurs what is commonly called the 'main arm resonance'. This is at a low frequency determined by the total effective mass at the pickup head and the compliance of the stylus. With a given mass the resonant frequency goes down as the compliance is increased, or with a given compliance it goes down as the mass is increased. It is obviously desirable that this resonance shall not affect the pickup's frequency response, though with high quality cartridges it normally occurs well below the audio band, the slight rise at 30 Hz in **fig. 29** being typical of the minimal effects to be expected. In practice the danger is of the opposite sort, for if the resonance occurs at too low a frequency it can move into the region where certain types of rapid warp or 'ripples' occur on some discs and where floors or furniture might vibrate as one walks across the listening room. Thus there are good reasons for a 'lightweight' approach in the design of pickup arms, it being generally desirable to minimise extra mass added to the cartridge by head-shell, arm and counter-weighting or counter-springing arrangements. This improves general stability, as do facilities found on the more sophisticated designs for balancing the system in various planes.

Bearings must be of high quality so that frictional drag which would oppose the pickup's progress across the disc – or oppose its back-and-forth motion on an eccentric record – is kept to an absolute minimum. With cartridges able to track at a downward force of 1 gm this requires that extraneous side-thrusts, whether from bearing friction or badly arranged connecting leads, be less than about 0·1 gm or 100 mg; otherwise the playing weight must be increased, as any extra lateral force tends to push the stylus away from one side of the groove and thus reduce its effective playing weight on that wall.

There is another type of unwanted side-thrust, this time connected with pickup arm geometry and referred to briefly in the last chapter. When records are made, the cutting stylus moves across the disc in a straight line from the outside edge towards the centre spindle – along a true radius. Most pickup arms, however, rotate about a pivot, causing the stylus to follow a curved path. It can be seen from **fig. 30** that a simple straight arm will therefore position the cartridge correctly at one point only along its traverse, and that in other positions the natural motion path of the stylus will not be at right-angles to the groove, as it must be if it is to follow the implanted waveform accurately. This is called *tracking error* or, more correctly, lateral tracking error to distinguish it from the vertical motion path of the stylus in a stereo cartridge, which is by convention fifteen degrees forward from truly vertical. This aspect of cartridge behaviour is often quoted in specifications as *vertical tracking angle*. The common means of reducing lateral tracking error to something in the region of two degrees in the worst position is also shown, in outline, in **fig. 30**. If the centre-line of the head or cartridge is offset from the arm's main axis – either by bending the arm or suitably angling the head on a straight arm – and the pivot so positioned that when the head is swung to the record centre the stylus overhangs the spindle by a pre-determined amount, it is found that over the recorded portion of the disc the cartridge axis remains very nearly at right-angles to a line drawn to the record centre.

Apart from a few special arms containing pivoted levers or equipped with other ingenious means of achieving parallel tracking, this offset-plus-overhang idea is practically universal on pickup arms – as is the side-thrust which it causes. **Fig. 31** shows a practical arm incorporating the geometry introduced in

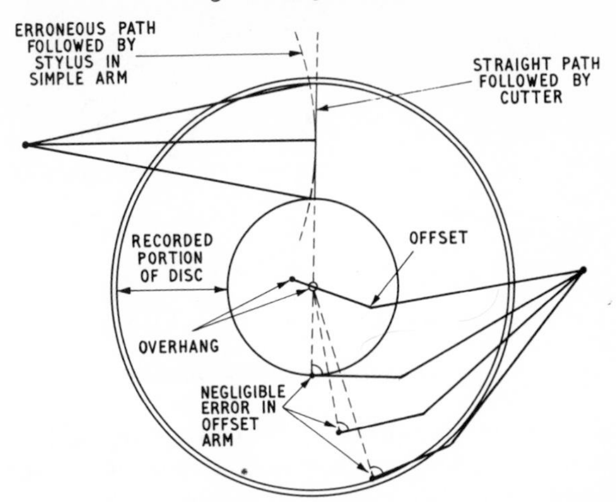

Fig. 30 This shows the unavoidable tracking error with a simple straight pickup arm (top) and how the error can be negligible with an arm employing conventional offset and overhang (bottom).

the previous diagram. With the cartridge oriented correctly in relation to the groove, the latter moving past the stylus strictly in line with the pickup head's axis, it follows that any forward frictional drag exerted on the stylus by the groove does not simply pull directly and neutrally on the pivot, but, when projected back, is seen to apply a clockwise turning force about the pivot due to the offset geometry. The net result is that whenever the stylus is in the groove of a rotating record the pickup is subjected to an additional inward force, and the better arms are fitted with gadgets for overcoming this, known variously as bias compensators, anti-skating devices, side-thrust adjustors, and so on.

Another refinement found on good arms is a lowering device, enabling the user to apply and remove the pickup stylus at any point across a record without the risk of groove damage inevitably attendant upon hand lowering and lifting. Some pickup arms are permanently attached to the associated turntable, though except in the case of auto-changer mechanisms they may be regarded as separate entities when assessing most of the design points so far mentioned. Auto-changers themselves are in many cases designed to satisfy most hi-fi performance criteria, though apart from on one extremely expensive and elaborate device which applies and removes each disc in turn, records are dropped one upon the other whilst moving and inevitably suffer from damaged and/or noisy surfaces after a few playings. Also, the reader is likely to be seriously interested in reproducing good music, most items of which are

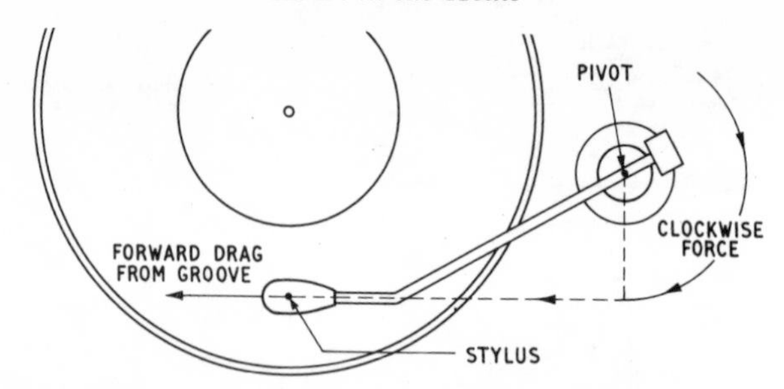

Fig. 31 With an offset arm the frictional drag exerted on the stylus by the groove acts along a line displaced from the pivot axis. This produces a clockwise torque tending to make the pickup skate inwards across the record.

either contained on the two sides of a 12-inch disc, which will therefore need turning over, or are operatic sets, these days rarely offered as auto-couplings. For these reasons the auto-changer facility cannot be treated as a serious part of hi-fi in the home and will from now on be regarded as irrelevant to the purposes of this book. This is very much a British, even a personal point of view, which I hope will be forgiven by North American readers who have come to confuse the irrelevant with the indispensable.

Before moving on to turntables, it will be useful to summarise all the preceding observations on pickups, taking as a framework the relevant items from the technical points listed at the end of Chapter 2. For low non-linear distortion of stereo records the stylus should, if spherical, be no larger than 0·5 mil in radius, or preferably be a well shaped elliptical type. The pickup's mechanical impedance should be low enough to permit tracking of heavily modulated discs at a playing weight below about 1½ gms, and extraneous influences on tracking performance such as arm stability and side-thrust should receive proper attention. To ensure good stereo behaviour there should be adequate channel separation, particularly at middle and high frequencies, with a reasonable balance of sensitivity between left and right. Frequency response should be free from serious peaks and dips, similar in the two channels, and preferably within ± 2 dB over the whole range. On the last point it should be noted that whereas a curve such as that in **fig. 29** would satisfy the ± 2 dB requirement *and* sound musically balanced, it is possible for the response of a crystal or inferior ceramic cartridge to come within the same limits over a shorter range (say 50 Hz – 12 kHz) but sound poor due to a response shaped as in **fig. 32**. Here, natural brilliance due to musical treble in the 1–5 kHz range is depressed, while the peak at 8 kHz might on some types of music add a false and rather 'tinny' brightness giving a superficial impression of proper balance that would eventually tire the ear. Magnetic cartridges are generally more likely to avoid this type of fault, offering also a better transient performance and greater revelation of subtle musical details. However, some combinations of equipment may actually favour what is in theory a deficient response in one element because of reverse effects in another.

Turntables, like pickup arms, should be neutral, providing no more than a horizontal surface rotating at the correct speed about a centre spindle of standard size, the surface being of a material offering sufficient friction to grip records without risk of damage. There should be no mechanical vibration, electrical interference or radiated sound.

Taking first the 'correct speed', this means not only that discs shall be rotated at $33\frac{1}{3}$ or 45 r.p.m. but that from instant to instant there must be no minor variations of the sort named in earlier chapters as wow and flutter. A convenient inclusive word sometimes used for these short-term changes of speed is *wobble*, the total of which should amount to less than about 0·15 per cent of the nominal speed if it is not to be heard as a waver of pitch or roughening of sound in music. This may strike the reader as a surprisingly small change for the ear to notice when one considers that semitones are spaced by 6 per cent and that even people with perfect pitch can seldom differentiate to better than a quarter of a semitone. But recognition of a fixed error in pitch is not the same thing as sensitivity to a change *whilst it is taking place*, and it is the latter which determines the tolerance limits of turntable speed wobble. Some types of music are more easily marred by wow than others, and in the last resort it is the ear that must judge turntable performance.

Achievement of that 0·15 per cent depends on a number of factors, most noticeable of which is the weight of the turntable platter used on transcription units. (The word 'transcription' is often applied to high quality turntables, a practice inherited from a time when the type of unit now regarded as desirable for domestic hi-fi installations was normally only found in professional studios and frequently used when transcribing recorded material from disc to disc).

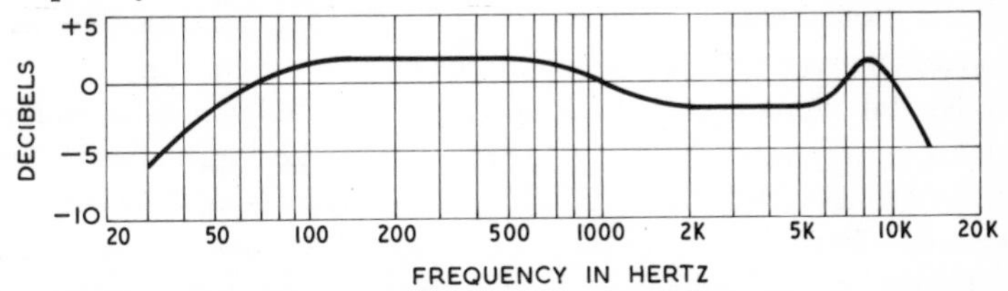

Fig. 32 Possible frequency response from an inferior pickup cartridge. This is within ± 2 dB over the range 50 Hz–12 kHz, but would sound unbalanced because of a 4 dB 'step' between musical bass and treble.

The principle of the flywheel will be known to most readers, the idea being that once a heavy wheel is rotating it tends to carry on due to its angular *momentum,* resisting external attempts to alter either its rotational velocity or its plane of motion. For a given rate of rotation, the larger the moving mass (inertia in the static sense) and the greater its distance from the axis, the more marked this stabilising effect becomes. In transcription turntables the platter is likely to weigh upwards of about 1·8 Kgms (4 lbs) and to have a cross-sectional shape placing most of its mass towards the periphery; thus the platter is itself a sort of flywheel which, once set into motion and driven in the manner described in the previous chapter, maintains a constant speed despite any rapid minor variation that might otherwise occur due to the motor and drive mechanism.

Long-term speed changes pose different problems, being caused by things like variation in frictional drag from pickups and/or disc cleaning devices as they traverse the record, changes in bearing friction as equipment warms up, and fluctuations of mains supply voltage. Solutions come from the use of more powerful motors and from attention to design details of bearings and the drive transmission system. In addition to the main speed selector, a good turntable will have a fine speed control for exact adjustment during use, partly to compensate for any remaining mechanical changes and partly to serve the pitch conscious music lover, who will from time to time find minor differences of key between recordings. Such controls normally provide an adjustment of several per cent and are operated in conjunction with a *stroboscope,* which indicates visually whether rotation is below, at, or above the correct rate. A stroboscope or 'strobe' is essentially a series of alternate black and white bands or dots arranged in a ring on the rotating object, the number of bands being chosen to give an impression that the pattern is stationary, if the speed is correct, when illuminated by light derived from the normal 50 Hz AC mains supply. Because of this need for AC lighting it is not always convenient to use a simple cardboard strobe of the type designed to drop over the turntable spindle, and some units are therefore equipped with a neon light arranged to illuminate a pattern viewable through a small window or mirror.

Mechanical vibration, manifesting itself in reproduction as rumble or thumping noises, is the second evil to be avoided on hi-fi turntables. While the platter's flywheel action may be effective in ironing out speed wobble, it is still possible for vibrations to be transmitted through the metalwork to the record, thus shaking the disc relative to the pickup stylus, a process which the cartridge cannot distinguish from groove modulation. Vibrations can arise from small and imperfectly finished bearings, particularly the main central bearing for the turntable platter itself, from the motor, and from the coupling between motor and platter. Adequate size and attention to detail in design, finish and lubrication characterise the transcription approach to these points, with, in particular, many refinements and variations in motor-to-platter drive arrangements. The simple idler wheel described in the previous chapter (**fig. 15**, page 55) may be supplemented with rubber or plastic belts which act as additional buffers to reduce transmission of rumble or rapid speed fluctuations to the turntable, while in all good designs the idler itself is retracted from pulley and turntable when the device is switched off. If left in contact it is inclined to develop 'flats' which may sound as a repetitive thumping or knocking noise.

With all these precautions a good transcription motor/turntable unit can suppress rumble to a level equivalent to 50–60 dB below normal peak groove modulation. This may not seem very impressive in relation to a possible recorded dynamic range of 60 dB, but in practice most rumble is confined to very low frequencies (below 50Hz), and reference to **fig. 7** (page 30) shows that at 50 dB below maximum music levels the ear is working near its lower threshhold at such frequencies. A noise figure modified to take account of the ear's characteristics in this manner is said to be *weighted*. However, large loudspeaker systems with an extended bass response will reveal rumble from all but the very best turntables, a point to which we shall return when dealing with practical choice of equipment. Most of the precautions aimed at reducing rumble also minimise the actual acoustic noise radiated by the turntable mechanism – an important point, for near-perfection at all other stages in domestic high fidelity in terms of sounds coming from the loudspeakers can be ruined by whirring noise emanating from the equipment cabinet.

Apart from such direct sounds and unwanted vibrations transmitted to the pickup stylus, there is a type of electrical interference which can arise from motor/turntables. This is not of the sort produced by vacuum cleaners or dirty switch contacts, but it can enter the audio system invisibly by means of magnetic coupling to the pickup cartridge or its wiring. The electric motor used to drive the turntable, like all other such motors, incorporates coil windings to generate magnetic fields, and since the motor is energised from the AC mains supply these fields alternate at the mains frequency of 50 Hz. It is difficult to confine such fields entirely within the motor casing – just as the earth's magnetic field extends into space and will actuate a compass in an aircraft. Therefore, in the absence of thorough screening there can be sufficient stray AC field to couple with the windings in a magnetic pickup cartridge, thus producing audible hum in the loudspeakers. This is a further matter for particular attention in the design of transcription units.

Another precaution in relation to the use of magnetic pickups concerns the material used for the turntable platter. Magnetic cartridges incorporate small magnets, and in some cases their external fields are powerful enough to be attracted downwards by a steel turntable, thus altering the effective playing weight. For this reason the pressed-steel platters found on cheap units are avoided for hi-fi use, though if other performance features are adequate employment of a non-magnetic turntable is unnecessary when using ceramic cartridges, which are not influenced by magnetic fields or materials.

Magnetism, of course, is the basis of tape recording, bringing us to the next major audio component for application of hi-fi principles. Six of the seven hi-fi criteria are applicable to tape recorders and we shall start with an interrelated group of three: non-linear distortion, noise and maximum signal handling capacity. We learnt in the last chapter that the tape recording process requires the addition of a continuous high frequency oscillation to the wanted audio signal if severe distortion is to be avoided. This is because particles in the tape coating react to the audio magnetic field applied by the tape-head in a non-linear manner; that is to say, the modulation level achieved is not in strict proportion to the applied signal amplitude. If a curve is plotted relating tape-head input to the resulting recorded signal – a sort of input-to-output *transfer characteristic* – the shape has a kink (**fig. 33**) which distorts the applied waveform in the region immediately near its 'zero' line. By adding a continuous high frequency tone to the input (usually somewhere in the range 30–100 kHz) the audio waveform rides, as it were, on top of these more rapid fluctuations (**fig. 33(b)**), the artificial zero lines thus created moving out to linear parts of the curve. In the diagram the input waveform has the same amplitude in the two cases, though when applied on its own it produces a small and distorted magnetisation pattern on the tape, whereas with bias the recorded waveform is larger and undistorted. The HF bias tone is also implanted on the tape, at reduced level, though from this point on it may be ignored for most practical purposes as on replay the head-gap width imposes an upper limit well below the bias frequency: the bias has performed its linearising task at the instant of tape magnetisation. There are various hi-fi refinements in application of the HF tone, such as *cross-field bias* which uses an extra tape-head simply for biasing purposes, the object being to avoid a partial erasing effect operating at high audio frequencies when bias and signal are fed to the same head.

While the ultimate upper frequency limit of a tape recording system is set, for a given tape velocity, by the replay head-gap width, in practice other factors influence the point at which the response starts to fall. A tendency for the bias in conventional machines to have a slight erasing effect on the high frequency components of audio signals sometimes tempts non-hi-fi manufacturers to reduce the bias level, enabling them to claim rather more impressive frequency responses than would normally be obtained at the speeds in question. The reader may wonder why extending the frequency range of audio equipment should be regarded as a non-hi-fi activity, which of course it is not unless the change degrades some other aspect of performance. Looking again at **fig. 33**, the tips of the audio waveform riding on the HF bias are just accommodated within the straight portions of the transfer characteristic, and if the bias amplitude is reduced the audio extremities will extend into the kink and become flattened. Lowering the bias would therefore in this case increase distortion. The general relationship between bias, recorded signal and distortion is shown in **fig. 34**, from which it is clear that the bias current giving maximum recorded level does not coincide exactly with minimum harmonic distortion; thus some roughness is added to the sound quality when peak music amplitudes fully modulate the tape. In hi-fi practice, therefore, the bias is set to give maximum *undistorted* output from the tape, usually requiring about 20 per cent more bias current than that giving greatest sensitivity, and producing an optimum signal-to-distortion ratio at the expense of some 2 dB in peak modulation capacity. Reduction of bias to improve the audio HF frequency response is a move away from this optimum arrangement, and as small gains in response and signal-to-noise ratio do not add as much to listening pleasure as is taken away by increased distortion, this practice is to be avoided in any tape system claiming to offer high fidelity.

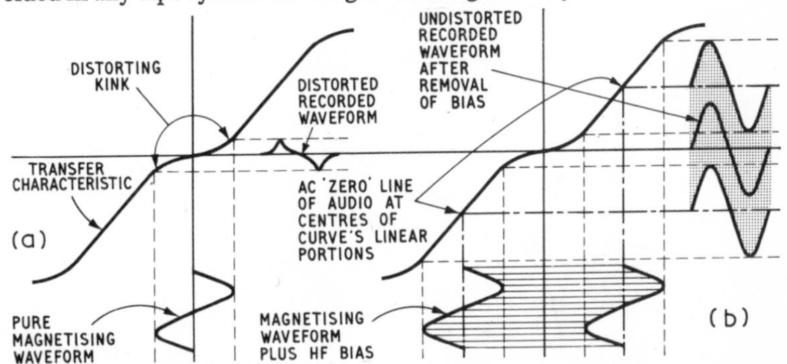

Fig. 33 Magnetic tape recording transfer characteristic (relating electrical signal fed to tape-head and resulting magnetisation pattern on tape) has a kink which causes severe distortion of the recorded waveform (a). By adding a high frequency bias to the signal (b), the audio may be pushed away from the kink to magnetise the tape in regions where the characteristic is linear.

Assuming the bias to be set correctly, it is normal to arrange for maximum music levels not to exceed a modulation producing about 3 per cent harmonic distortion. Some form of level indicating device is calibrated to register visually this nominal overload point, commonly an electronic magic-eye, but a meter on more expensive machines. Sensitivity of the magic-eye is set for closure of the green luminous gap to coincide with peak recording level, while meters frequently have green and red portions on their dials corresponding to 'safe' and 'overload' levels. On stereo tape recorders the left and right signals are sometimes combined for registration on a single indicator, though most machines have separate meters for the two channels. The usefulness of these devices is sometimes limited by an inability to respond adequately to sudden loud transients in music, as the type of meter circuit designed to indicate genuine short-term peaks (peak programme meter or PPM) is rather expensive, and is usually found only on professional equipment. Meters fitted to domestic recorders are often of the VU (volume unit) type, based on certain assumptions about the ear's tolerance of brief overloads. In practice the user soon becomes acquainted with the limitations of a particular modulation indicator on various types of music, the calibrated nominal overload point representing a recorder's maximum signal handling capacity for most of the time.

Having set the upper boundary, the dynamic range that can be accommodated on tape is limited at the lowest level by noise in the form of tape hiss. There has been much improvement in tape over the years from this point of

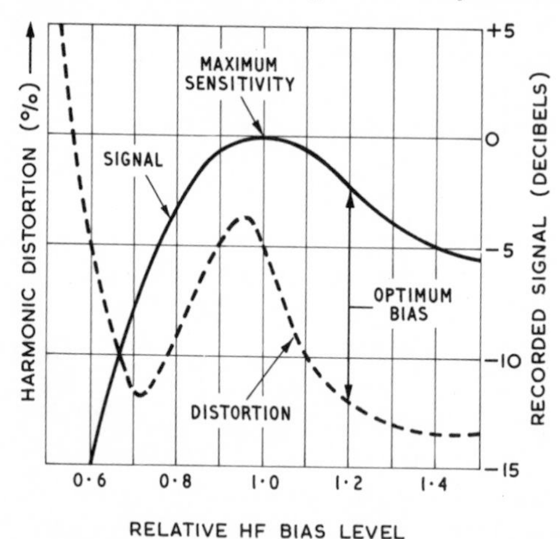

Fig. 34 High frequency bias applied with the audio signal to a tape-head should be set for an optimum signal-to-distortion ratio rather than simply maximum sensitivity.

view, though irreducible random clusters of magnetic particles in the tape coating must ultimately be audible in reproduction as a gentle background hiss. Some tape is inferior in this respect, and tape recorders occasionally worsen matters by providing inadequate erasure of previously recorded signals, the remains of which – though perhaps not audible as music or speech – mix with the hiss to lower the overall signal-to-noise ratio. An imperfect sine-waveform for the HF current fed either to the erase head or with signal as bias to the record head can also degrade the tape's basic random noise. In addition, the tape-head used for record or replay can itself increase tape hiss by becoming permanently magnetised, due either to external magnetic fields or to a small unwanted direct current flow through its windings arising from a faulty circuit component. However, assuming performance in these respects to be at its best, the actual level of hiss in relation to recorded signal depends on track width, amplifier design, and to some extent tape speed.

It will be recalled that quarter-inch tape may carry on its width one, two or four tracks. Each reduction of track width more than halves the signal level induced in the replay tape-head, while at the same time lowering the reproduced hiss level by a smaller factor; the difference arises from the statistics of random particle distribution in the tape coating. The net effect is to worsen the available signal-to-noise ratio by just over 3 dB. In many domestic tape recorders the circuits used to amplify tape-head signals introduce further random noise at a level comparable with that coming from the tape, and as the replayed signal is lowered in level by use of narrower tracks this amplifier noise becomes relatively more troublesome. On all but the very best machines, therefore, changing from two-track to four-track operation worsens the overall potential S/N ratio by up to 6 dB.

Another point relevant to track standards is an irritating tape fault known as *drop-out*, in which the signal disappears or literally 'drops out' for a split second because of some minor discontinuity in tape coating. This fault becomes more noticeable with narrower tracks and/or lower speeds, as offending patches of tape will naturally scan a larger fraction of the tape-head gap if the gap is shorter, and take longer to pass if the speed is lower. Also more evident on four-track recorders are minor departures from intimate tape/head contact, arising from rough-edged, mis-shapen or over-stiff tape and sometimes similar to drop-out in audible effect. The solution, frequently advocated in the literature supplied with quarter-track machines, is use of thinner and therefore more flexible tape; as noted in the previous chapter, there are five thicknesses to choose from. One possible small penalty to pay for the use of extra-thin tape is enhancement of an effect known as *print-through*, whereby intense signals are transferred magnetically at a very low level to adjacent layers of tape on the spool if recordings are stored for lengthy periods. In practice this is not much of a bother to domestic users, though commercial master tapes are sometimes spooled unconventionally to minimise the audible effects of any fortissimo musical climaxes attempting to duplicate themselves in this manner. A minor disc record fault sometimes mistakenly attributed to tape print-through is popularly known as *pre-echo*. This is a mechanical equivalent of the above tape effect, whereby heavy modulation in one groove causes slight deformation in its neighbour, thus producing a small output from the pickup displaced in time by one disc revolution.

Returning to tape matters, there are a few more variables still to be considered in connection with tape hiss and other irreducible background noises. The very small electrical signals induced into replay heads by the tape require that the associated circuits afford considerable amplification or *gain*, making it difficult to avoid adding some extra hiss to the inherent tape noise, as already noted, and also inviting the addition of hum from nearby AC supplies, mains transformers and electric motors. The tape-head itself is very vulnerable in this respect, being a component designed specifically to respond to magnetic fields; on good machines the screening arrangements are carefully thought-out so that when the tape is actually passing the head any stray magnetic hum fields are kept to a very low level in the vicinity of the head-gap. Tape replay speed also affects the overall signal-to-noise ratio, though in an indirect way related to its bearing on frequency response, to be discussed shortly. Suffice to note here that a host of subtle, interrelated losses at high frequencies or short recorded wavelengths are all eased as the recording and replay speed is raised, the net effect being some improvement in S/N ratio at high frequencies.

Finally, the reader must by now be suffering from technical indigestion without yet knowing what dynamic range may be accommodated on and usefully reproduced from magnetic tape. Starting at the top of the tree, with everything adjusted for perfection in laboratory fashion the range between 3 per cent distortion and random noise on the full width of best virgin tape is around 70 dB. Large professional machines of the sort used by recording and broadcasting organisations achieve better than 60 dB at 15 inches per second (i/s), while the very best domestic semi-professional recorders will manage 55 dB or a little more at $7\frac{1}{2}$ i/s. Further down the price/quality scale the noise performance deteriorates steadily, average unsophisticated domestic models offering 45 dB or less, even though purchasers will frequently find very optimistic figures printed in manufacturers' literature. Cassette machines working at a speed of $1\frac{7}{8}$ i/s with two pairs of stereo tracks accommodated on tape only just over one eighth of an inch wide have until recently been regarded as beneath hi-fi consideration. However, improvements in tape and heads are changing this, and S/N ratios of 40 dB or more have been achieved even with compact low-speed cassettes.

As time goes by it is possible that all these figures will be improved by 10–20 dB through the domestic application of electronic noise-reducing techniques previously confined to professional recording studios. Such processes effectively compress the dynamic range of the original signal before it is recorded on tape, thus holding the quietest music passages well above tape and amplifier noise. When replayed for disc cutting the recorded signal is correspondingly expanded, any unwanted additional noise thus being pushed further out of harm's way. This sort of approach is becoming available on a few recorders and some cassette players, and in the future may be applied to complete hi-fi systems to reduce background noise introduced at any point in the chain between studio microphones and domestic amplifiers. A last point on noise: like turntables, tape recorders have moving parts which can radiate direct acoustic disturbances, and unfortunately the more expensive domestic machines are not always better than cheaper models in this respect.

Just as in turntables, the mechanisms required to move the recording medium along at constant speed inevitably fail to do this perfectly, thus introducing wow and flutter. Wear on bearings, belts, clutches, pulleys and pinch-wheel; incorrect lubrication; badly adjusted brakes or pressure-pads; dirty guides, tape-heads or capstan; inadequate flywheel momentum; or just plain bad design of the tape transport system – any or all of these will add to the total speed wobble. Also as with turntables, attention to detail in design and assembly reduces the cumulative effect of these factors to give total wow and flutter figures below about 0·15 per cent on good machines, though normal domestic recorders are rarely able to satisfy a critical ear in this respect on the more searching types of music. Apart from the flywheel, which on some more elaborate machines may be an integral part of the tape-drive motor, many of the rotating parts on a recorder contribute a little extra momentum or inertia to the tape-drive by virtue of their rotation, thus helping to iron out speed changes. However, momentum depends on rate of movement as well as mass, and as most of the rotating parts move faster at higher tape speeds it is found in practice that wow decreases roughly in proportion to increases in tape velocity. Here the similarity to turntables comes to an end, for while disc records have the advantage of being coupled directly to a massive speed stabiliser, the turntable, and thus rarely suffer from audible flutter in reproduction, the small length of magnetic tape passing, at any instant, over tape-heads between feed spool and capstan has negligible mass and therefore depends on constancy of tension and friction to minimise short-term speed changes corresponding to flutter. Such constancy requires care in the positioning of guides, adequate grip on the tape between capstan and pinch-wheel, proper control of any drag imposed by the feed spool, avoidance if possible of pressure-pads, and various other subtle mechanical points.

Moving from mechanical to electrical matters, the next major tape recorder performance feature for discussion is frequency response. In the previous chapter brief mention was made of the need for a particular type of replay response curve because of the way tape-heads react to recorded modulation. If a tape has been modulated by a recording head fed with constant current at all frequencies, output from the tape-head is lowest in the bass, rising steadily in proportion to frequency up to a point determined by the replay head characteristics, and then declining again beyond this 'turnover' frequency. The turnover point is essentially a practical limitation arising from the finite width or thickness of the head-gap, and if this were small enough or the tape speed high enough there would be no more than marginal departures from a response rising in proportion to frequency right up to the top of the audio band. This type of response could be said to have a particular 'slope', with the amplitude halving or doubling for each change of one octave in frequency. On a decibel scale a two-to-one change in voltage or current is equal to 6 dB (how or why this is so is beyond this book's non-mathematical mandate, so the reader must take this on trust), and the basic shape of a tape replay amplifier curve is a line sloping *downwards* with rising frequency at six decibels per octave, representing a mirror-image of the tape-head output curve. This 6 dB/octave slope is shown as a dotted line in **fig. 35**, which diagram also depicts various practical departures from perfection.

With normal heads and tape speeds the frequency above which the head outputs flattens off and then starts to fall is well within the audio band, and if the recorded tape modulation were 'flat' the necessary replay response would be of the type shown by the short-dashed curve in **fig. 35**, where the sensitivity rises above 2 kHz to compensate for replay head losses. However, tape hiss would be so accentuated by such a rise at high frequencies (HF) *after* the recording process, that the replay head's falling HF response normally receives compensation in the form of electrical *pre-emphasis* of the signal before recording, with the result that the replay amplifier may have an approximately flat response above the turnover frequency. In some recorders there is also a little pre-emphasis at very low frequencies to ease signal-to-noise ratio problems when amplifying the extremely small tape-head output. The resulting

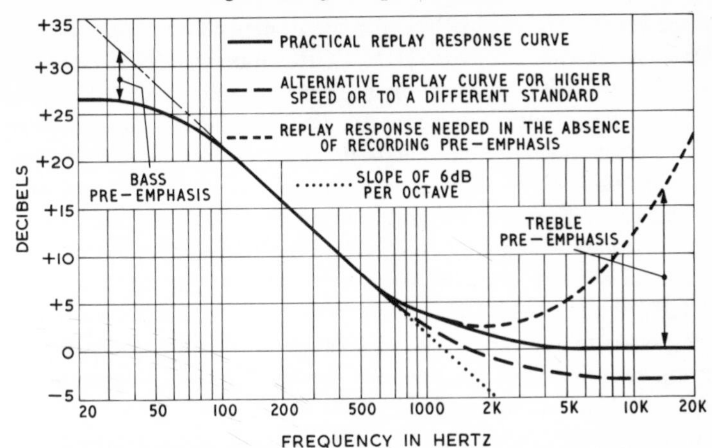

Fig. 35 Solid line and heavy dashed line show practical replay response curves for a tape recorder. Note that these differ in their points of departure from the basic 6 dB/octave slope, corresponding to widely used European and American standard replay characteristics for a tape speed of $3\frac{3}{4}$ i/s. These replay amplifier curves are the inverse of the tape-head output.

replay response curve in a machine using both these corrections during recording is of the type shown as a solid line in **fig. 35.**

As tape speed is increased all recorded waveforms occupy a greater length of oxide and the frequency at which the tape-head output drops to zero is raised accordingly. Also raised is the turnover frequency where the output commences its fall towards zero. This means that the replay curve's necessary point of departure from a true 6 dB/octave slope goes up in proportion to tape speed, with a corresponding reduction in the amount of pre-emphasis needed above this frequency – together with a reduced risk of distortion and overload. For these and related reasons a clean, consistent and extended response at high audio frequencies is always more easily obtained at the higher tape speeds. If the noise, distortion, wow, flutter and drop-out of a recorder are to reach hi-fi standards it should operate at $7\frac{1}{2}$ i/s or more, though as tape, tape-heads and mechanisms improve this is ceasing to be so.

The need for some consistency of performance when reproducing commercial tape records has led to the introduction of certain agreed replay response curves. A recorder adjusted to comply with one of these characteristics should, if used to replay a tape conforming to the same standard, provide a flat overall frequency response. Since, as we have seen, the optimum recording and replay corrections vary with tape speed, there are corresponding differences in the chosen replay curves. In addition to the necessary changes between tape speeds there are various curves for the same speed, depending upon whether a recorder complies with European (DIN, CCIR), American (NAB, RIAA, EIA), or International (IEC) Standards in old or new versions: the reader will gather that the situation is not so simple as with disc reproduction, though the various standards are gradually being brought into line. To ease matters for designers, replay standards are specified in terms of a technical characteristic of circuit behaviour called *time-constant,* given in microseconds (μS). As with decibel ratios, the reader need not be dragged into the mathematics of this, though it may help on occasion to realise that when a recorder is said to comply with, say, CCIR 70 μS or NAB 50 μS characteristics, this simply means that the replay responses follow curves basically similar to the solid line in **fig. 35**, but flattening away from the main slope at various frequencies depending on the standards and tape speeds involved.

It is clearly desirable for tape recorders to follow agreed replay characteristics whenever possible, though the *overall* response of a recorder – measured as a difference between signal in and signal out after recording and replay – should come up to certain minimal standards for hi-fi purposes whether or not particular replay responses are adopted. In this necessarily limited description of tape recording processes a number of technically obscure factors tending to tail-off high frequency response and cause irregularities at low frequencies have not been mentioned. These, together with the problems of balancing recording pre-emphasis and replay correction, proper choice of bias level, etc., make it difficult in practice to obtain an absolutely level frequency response. However, on the best domestic machines – commonly referred to as 'semi-professional' models – the response is likely to remain within 2 dB of the mid-frequency level over a range extending from 30 Hz to about 15 kHz at a tape speed of 7½ i/s, the upper limit being restricted to 8–10 kHz at 3¾ i/s and rising to 20 kHz at 15 i/s. The variation of $\pm$ 2 dB would normally apply to the frequency extremes only, with a steady fall-away at the top end, possibly preceded by a small rise (**fig. 36**), and sometimes a little irregularity below 200 Hz.

The type of variation which can upset musical balance because of a move between two output levels at a mid-frequency, as in the pickup response of **fig. 32**, is rare on high-quality tape recorders; but however flat the overall record/replay response, if a machine is used for replay-only purposes there can be errors of tonal balance when pre-recorded tapes have been made to a standard different from that adopted in the recorder. Returning to **fig. 35**, the solid and heavy dashed curves correspond, respectively, to commonly adopted European and American replay responses for tapes recorded at 3¾ i/s (140 μS and 90 μS time-constants), and if a tape made to one standard is replayed on a machine adjusted for the other there will be either an emphasis or a reduction in musical brilliance, depending on whether the replay time-constant is higher or lower than the recording time-constant. The difference in this case is simply represented by the gap between the two curves, perhaps more easily seen as the deviation from a nominally flat response added to one of the typical overall curves in **fig. 36**. An ultra-refinement on very expensive tape recorders is universal switched compensation for all standards and speeds, though most domestic machines opt for a single standard and arrange for the automatic selection of appropriate electrical responses as the speed is changed mechanically.

Now, a brief look at stereo tape matters before moving on to a general round-up of hi-fi tape points not automatically encompassed by the 'seven strands'. There are no especially tedious or difficult problems related to two-channel operation in tape recording, and provided all the hi-fi tape criteria so far discussed are applied, together with exact electrical duplication for the initially well separated signals from stereo tape-head onwards, there remains simply the mechanical matter of accurate and consistent tape positioning in relation to the head's two gaps. Stereo only differs from mono in this respect because small intermittent changes in response or output, of the sort caused by poor tape/head contact and variable tape coating, are more noticeable in stereo. The magnitude of such changes usually varies across the tape width, thus affecting one track more than the other from instant to instant, thereby shifting the apparent position of the stereo sound image in a rather jarring fashion. It follows from earlier comments on drop-out and track width that

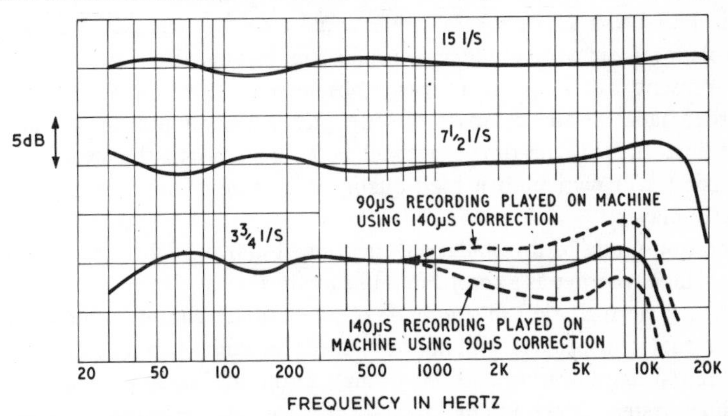

Fig. 36 Typical frequency response curves for a top quality domestic tape recorder or semi-professional machine. Recent improvements in tape and heads enable these performances to be bettered at the HF end, with a potential 1⅞ i/s cassette response better than the 3¾ i/s curve shown here.

the above stereo faults will be accentuated by four-track operation, especially as in this mode (**fig. 23**, page 64) the left-hand signal is carried on a track at the tape edge, where mechanical tape irregularities are generally more prevalent than in the region of the other track near the tape centre. In the best recorders control of tape movement and position is sufficiently accurate and consistent to obviate significant audible defects arising from such causes, especially when using the thinner tapes in good condition. However, roughly handled tape which has been spooled badly several times tends to have creases and turned-over edges, and such faults are distinctly more likely to affect stereo performance on quarter-track machines.

In this context, quarter-track operation refers to the two-way stereo system described on page 64, where two conventional 2-channel recordings are accommodated, one in one direction and one in the other. However, 1970 saw the introduction of some genuine 4-channel tape recorders employing tape-heads with four active gaps in one vertical line, for simultaneous reproduction of the four separate signals needed for quadraphonic sound (see Chapter 10). Returning to two-channel stereo, there is the matter of tape economy to consider, for since each recording of a given length must occupy two tracks instead of one, the user must either purchase twice the quantity of tape for half-track recordings, or tolerate a slightly degraded overall performance from a quarter-track recorder for the same outlay on tape. If a machine is to be used primarily for replaying commercial tape records a four-track model is imperative, as most pre-recorded spool-to-spool stereo material is offered in this form. This being so, only semi-professional recorders should be considered suitable for use in a top grade hi-fi installation. Having said this, we must note that present trends point to early obsolescence for spool-to-spool pre-recorded tapes as the compact cassette increasingly dominates commercial recording. Indeed, the time may well be at hand when the wealthier hi-fi enthusiast will own both a high-quality conventional tape recorder *and* a cassette player.

The 'top-grade hi-fi installation' mentioned above might not, in fact, include a tape recorder as a complete self-contained item, for in cases where the recorder is normally left *in situ* much of its amplifier circuitry, and of course the internal loudspeakers, will not be needed, the audio signal being extracted after initial amplification and fed to an appropriate input part on the hi-fi amplifier system. Machines designed for this sort of application are known as *tape units*, which for a given quality of performance cost less than complete tape recorders. Having no built-in loudspeakers, they cannot be detached from the other equipment and used for playing previously recorded tapes, though recording is certainly possible 'away from home', with facilities often provided for monitoring recorded signals via headphones. In addition to incorporating all the electronic circuits needed for the recording process, including microphone preamplifier and bias and erase oscillator, tape units normally include those early parts of the replay circuitry providing corrections as discussed earlier, with such initial amplification after this as may be necessary to provide a signal suitable for feeding to separate amplifiers. In stereo versions these circuits are duplicated as necessary. Tape units, incidentally, should not be confused with the *tape deck*, which is simply the mechanical part of a recorder detached from electronics and case.

Whether tape recorder or tape unit, many machines will on occasion be used for recording 'live' signals from microphones in addition to storing radio programmes or playing commercial tapes, and for this purpose most models are equipped with appropriate inputs. However, many recorders have microphone input circuits suitable only for use with crystal models, a type of unit employing the piezo-electric principle mentioned in connection with pickups in the previous chapter and not usually capable of high audio quality. There are exceptions, as microphones, like loudspeakers, are to some extent a matter of taste, but it is generally agreed that while crystal types are quite suitable for speech and, perhaps, some solo instrumental music, moving-coil, ribbon or capacitor microphones are necessary for high quality concerted music recording. The hi-fi tape user wishing to capture live musical sounds at the sort of standard obtainable from radio or commercial recordings must therefore look for a machine designed to work with at least a moving-coil microphone.

Also relevant to recording matters is the possible inclusion of a *monitor head*. This is used for replay purposes only, what would normally be the record-replay head then being confined to recording. This eases electrical design to some extent, as the tape-head does not have to be switched between circuits when changing between record and replay. Despite this, accommodation of a third head raises the overall cost of a recorder and this refinement is normally found only on expensive models. Because of uncertainties in the behaviour of recording level indicators and the possibility that a perfect signal fed to the record head may be ruined, unknown to the recordist, by poor tape contact, a monitor head eases matters by enabling the user to listen to recordings a small fraction of a second after they have been implanted on the tape. On some machines there is a switch permitting an immediate comparison between input and recorded signal, a revealing exercise in high fidelity only possible on three-headed recorders.

5. HIGH FIDELITY II: RECEIVING, AMPLIFYING AND REPRODUCING SOUNDS

IMPERCEPTIBLE MAGNETIC patterns imprinted on tape, invisible radio waves passing through space, minute mechanical wiggles in record grooves – all this raw material of high fidelity eventually wends its way to loudspeakers in the form of electrical signals. There it becomes sound in the air to produce, via our ears, music in the mind. In the previous chapter we examined hi-fi techniques for converting groove wiggles and magnetic patterns into electrical audio signals, the three components involved being as much mechanical as electrical and clearly identifiable as separate entities even when coupled together – as with cartridge to pickup arm, and arm to turntable unit.

Once in electrical form, audio signals are subjected to the mysteries of processing by electronic circuits, the detailed workings of which require a knowledge of AC electrical theory for proper comprehension. Education in such matters, beyond the elementary non-technical skirmish attempted in Chapter 3, is outside the scope of this book, and indeed is not really necessary for a practical understanding of hi-fi. However, the influence of circuit behaviour on the 'seven strands' which we have extracted from music, may be usefully explained and examined, with such graphs and response curves as will illuminate musical performance. Reference back to the basic functional hi-fi component scheme presented in **figs. 13** and **14** (pages 52, 54) reveals that we have yet to cover radio tuners, preamplifiers, power amplifiers and loudspeakers. As mentioned there, the obvious physical separateness of pickups, turntables and tape recorders is not necessarily reflected when we come to the electronic components, it being quite normal for a control unit (preamplifier) and a pair of power amplifiers to be manufactured as one integrated item, and increasingly common for the tuner circuits also to be included. In this connection it is worth noting that whereas a separate control unit usually takes its AC and DC power supplies from the associated power amplifier, a separate tuner is very frequently *self-powered*, meaning that it has its own *power pack* energised from the mains supply and may therefore be used with a wider range of amplifiers. However, we shall follow the logic of function rather than construction in this survey, as practical pros and cons in the choice of separate or integrated units are covered in a later section.

First come tuners. The signal fed to these from aerials, though electrical in form, has to be processed considerably before emerging as audio information; in this respect the tuner may be regarded as a transducing component analogous to pickup and tape recorder and therefore joins them, logically, in preceding the control unit. The main stages from RF input to AF output have been outlined, so we shall straight away examine behaviour from the high quality angle, four only of the seven main hi-fi performance features being subject to possible degradation in radio reception: signal-to-noise ratio, frequency range, distortion and stereo.

As already noted, only the FM mode of radio communication can be regarded as relevant to hi-fi. This is because AM reception in Europe on the MW and LW bands is severely limited both by background noise and in frequency range, especially after dusk, when because of the electrical behaviour of the upper atmosphere distant stations are received more easily and raise the general level of interference. For technical reasons connected with the process of amplitude modulation, radio interference manifests itself particularly at the higher audio frequencies. Exclusion or reduction of unwanted background noise therefore involves sacrificing the top end of the audio band, usually quite severely. The sacrifice is normally effected in the IF stages of a receiver or tuner, where it happens that restriction of the IF bandwidth also limits the eventual audio band at high frequencies. Very sophisticated AM tuners have a variable bandwidth enabling the user to make the best of a bad situation, but the combination of reception problems and a deliberate curtailment of the highest audio frequencies as transmitted – to satisfy international radio agreements – means that from the hi-fi point of view AM radio may be discounted. However, long-distance reception of FM transmissions in the VHF band, while sometimes possible, is generally too freakish to be reliable, and an English listener wishing to hear a concert from Paris or Amsterdam must tolerate the technical limitations imposed by the MW/LW system. For the keen music lover anxious not to miss such broadcasts but nevertheless appreciative of good reproduction from other sound sources, tuners are available covering both MW/LW and VHF/FM reception but designed for interconnection as part of hi-fi systems in the normal way. These are commonly known as AM/FM tuners, with AM circuits in some cases adjusted for a somewhat better overall performance than is obtainable from conventional radios, but offering also the superior quality of FM for local VHF reception on Band 2 (87·5–108 MHz).

'Local' is used here to mean within about thirty to forty miles of a transmitter, for radio waves at very high frequencies cannot follow the Earth's curvature or jump over mountain ranges, the FM sound-radio reception situation being rather similar to that obtaining with television, where there are service areas, fringe areas and virtual 'dead spots'. In Britain the situation is eased by the BBC's mandate to project, as best it can, its main services to the whole population, though in many areas really satisfactory FM performance is only obtained by the use of a proper external aerial – again, just as with television. Practical advice on aerial installation will be found in the chapter on installation, though the matter is raised here because of its bearing on S/N ratio, which depends in fringe reception areas on minimum pick-up of car ignition and other interference and the maximum possible radio signal fed into the tuner's aerial socket.

A big advantage of FM radio, quite apart from the use of a frequency band not over-cluttered with interfering stations, is that most of the unwanted spurious signals which would cause noise via an AM radio are not heard via FM because of the way in which the receiver responds to variations of signal frequency (wanted modulation) rather than rapid changes of signal amplitude (interference added to signal). However, an FM tuner can only do this discriminating job properly if the incoming signal is strong enough to make the circuits insensitive to changes of amplitude. All types of radio receiver employ a function known as *automatic gain control* (AGC), the effect of which is to keep the IF signal fed to the demodulating circuits at constant level over a very wide range of RF amplitudes at the aerial, thus compensating for strong and weak stations and for 'fading', without, in the case of AM receivers, removing short-term (audio frequency) amplitude changes corresponding to modulation. When the incoming station is exceptionally weak, or perhaps missing altogether as one tunes across the dial, AGC causes the RF and IF circuits to operate at maximum gain and thus to amplify their own internally produced noise up to levels comparable with the audio obtained from a normal signal. Much of this noise is equivalent, technically, to amplitude modulation of any weak signal that happens to be present, and as an FM receiver's capacity to discriminate against such modulation falls away at very low input levels the advantages of FM reception begin to disappear.

It follows from this that any received signal tends to reduce the level of background noise or 'mush', but that unless it does this to a significant extent the FM tuner circuits are not operating correctly and the signal is barely usable. This noise-reducing effect is called, appropriately, *quieting*, and manufacturers sometimes specify FM sensitivity in terms of so many microvolts (μV) of signal for so many decibels of quieting, the dB figure indicating the amount by which background noise is depressed in relation to its level in the absence of a proper signal. A quieting figure of 30 dB is frequently regarded as defining a tuner's 'usable sensitivity' (in the American IHF standard, for instance) though to make full hi-fi use of the broadcast dynamic range and to enjoy the potentially silent background of VHF/FM radio this figure should obviously be increased, perhaps even doubled. The possible S/N ratio of a good FM tuner is around 70 dB when aerial inputs amount to a millivolt or so, though in very poor reception locations the available signal may be only a few microvolts. In the latter case it is important to use a tuner offering 20–30 dB of quieting for inputs of 2 μV or less, thus ensuring operation above the 'usable' level with a correspondingly reasonable S/N ratio.

Whether a tuner is giving a modest noise performance with only a few microvolts from the aerial, or superb results with a much higher signal (the latter more usual for most BBC listeners in the U.K.), the nature of FM reception is such that the audio output voltage from a tuner's demodulating circuit (known as a *frequency discriminator*) remains constant – musical dynamics apart, of course. This is in addition to the influence of AGC, which has a similar but not quite so consistent effect on AM reception. Thus whereas with an AM radio some stations remain audibly weaker than others despite the automatic adjustment of circuit gain, an initially weak FM station will not demand a higher volume control setting than its powerful neighbour a few MHz away, low signal strength showing up instead as a not so quiet background when tuned in. However, we must get things into perspective, for 'not so quiet' in a VHF/FM context on local stations is still very much quieter than anything achievable after dusk via AM on the MW/LW bands – unless one happens to live near to a powerful transmitter handling the programme of one's choice. Something that is no one's choice is a loud rushing noise when tuning between FM stations, though as the quieting effect of proper signals is removed this is bound to happen unless the precise tuning dial positions of wanted stations are known and the volume control setting can therefore be reduced whilst retuning. Some FM tuners cover this point by incorporating inter-station noise suppression or muting circuits. Misunderstandings sometimes arise as this facility is occasionally called *quietening*, not to be confused with quieting already discussed.

A final point on noise in FM reception, taking us also on to the subject of frequency response: just as disc records are cut to the agreed RIAA frequency characteristic and tape records adopt various degress of pre-emphasis, so FM radio transmissions have a rising response at high audio frequencies. Any unwanted background noise introduced during reception tends to be a smooth hiss rather like that from tape – pure random noise, in fact, sometimes known as *white-noise*. Because of the way the ear reacts to this type of sound, and because other spasmodic interference tends to concentrate much of its energy at high frequencies, noise is most easily attenuated aurally by reduction of high-frequency output, a corresponding boost being applied at the transmitter to give a flat overall response. Unfortunately there are two different standards in use, one for North America and one for Europe, defined, as in tape recording, in terms of circuit time-constants: 75 μS and 50 μS respectively. The equivalent response curves are given in **fig. 37**. A more modest boost is adopted in Europe because it has been found that too much HF pre-emphasis can cause distortion at the transmitter due to *overloading* on some types of music, it being necessary then either to reduce the overall signal level or tolerate reduced quality. All FM tuners include the components necessary for de-emphasis after the discriminator as a matter of course, no further correction being needed in the associated control unit. However, rather more care is taken on the better tuners to get the response exactly right; also, should anyone happen to take an FM tuner across the Atlantic, programmes might sound a trifle dim in London with an American tuner or rather bright in New York with a European model – rather as was noted in the previous chapter with tapes and tape recorders using different standards. Nothing else should upset frequency range or response in FM radio reception, any apparent limitation being due to broadcasting problems discussed in the last chapter. As has already been hinted, overall response with even the best AM tuner used at maximum bandwidth during the day is unlikely to extend usefully beyond 8–9 kHz, and for evening listening in Europe the top end has to come down to about 4 kHz to make interference tolerable.

Turning now to distortion, this should be untroublesome with a good FM tuner properly used in a reasonable reception area. Indeed, total harmonic distortion at peak modulation can be under 1 per cent, making 'live' transmissions via FM radio from this point of view potentially superior to tape or disc, which are subject to higher overall non-linearities arising from factors discussed in the previous chapter. The two qualifications, however, are important: improper use and certain types of reception condition can degrade performance very audibly.

It has been mentioned that the demodulation stage in an FM receiver is called a frequency discriminator, indicating that the circuit discriminates according to change of frequency rather than change of amplitude. This means that as the IF signal wavers either side of its nominal centre frequency according to modulations, the discriminator translates these changes into a continuous audio waveform for feeding via a de-emphasis circuit to the amplifying stages. Now, discriminator circuits are designed to respond linearly – that is, give an output in direct proportion to input – up to the frequency deviation used for peak modulation, the frequency-in to amplitude-out performance being a sort of very straight transfer characteristic. For various technical reasons this excellently linear part of the curve is not extended far beyond the

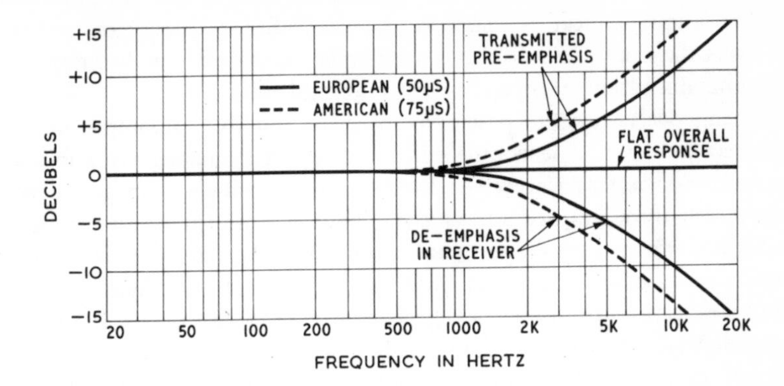

Fig. 37 High frequency response characteristics employed in VHF/FM broadcasting and reception.

peak deviation range, and if the operating centre frequency of the IF signal happens to be off to one side the discriminator cannot do its job properly, leading to distortion. When a receiver is not tuned accurately to an incoming signal the effect is to shift the IF frequency away from optimum, so that simple mistuning can lead to distortion. Tuning 'by ear' can therefore be rather problematic with an FM receiver, and while a weak station may permit this by adjustment of tuning to give minimum background noise – which *should* coincide with optimum linearity – this is seldom definite or accurate enough for normal use and good tuners employ a refinement called *automatic frequency control* (AFC) together with a suitably responsive *tuning indicator*.

AFC 'steers' the circuit accurately into tune when it is brought manually within fairly close range of a signal, so that by finding the dial-points above and below which the desired station suddenly comes into prominence, and then setting the pointer midway between these positions, the user can effect reliable tuning for minimum distortion. Some tuners make AFC optional by means of a switch, when removal of the facility eases location of weak stations situated next to powerful ones and permits use of the tuning indicator for accurate setting without any element of guesswork on any station, the AFC then being switched on to cancel any *drift* which might otherwise occur. At the very high radio frequencies used for FM broadcasting it is difficult, for reasons to do with small changes of electrical *capacitance* in the early stages of the receiver, to avoid some slight shift of tuning (drift) as components change temperature. In an FM tuner without AFC it is therefore usually necessary to keep an eye on the indicator while things warm up, resetting the dial from time to time as needed. Tuning indicators themselves vary in usefulness, though on the better receivers their optimum display point is made to coincide with symmetrical discriminator action for minimum distortion and noise.

Having dealt with the possibility of distortion from improper use (i.e. incorrect tuning), this leaves the odd case of distortion taking place, in effect, between transmitter and receiver. Radio waves, like all electromagnetic radiations, travel through space at the speed of light (300 million metres per second) and it is perfectly possible for a given transmitted signal to arrive at the receiving aerial at different times via different paths – the obvious direct path and by reflections from hills or objects such as gasholders. This is especially so in the VHF band, where the relatively small radio wavelengths make reflections from distinct objects much more likely. When the delayed signal is not altogether negligible in amplitude compared with the direct one – and it can be of comparable level in some circumstances, as when the direct path is over a high hill and the indirect one is unimpeded via flat country over the transmitter-gas-holder-receiver route – problems can arise at the receiver. Path-length delays of this sort can result in quite severe distortion, particularly when differences amount to several miles and when reproducing instruments like the piano which depend on attack qualities for their tone-colour. Distortion arises because modulation on the delayed signal is out of step with that on the direct signal and the audio waveforms combine in such a way as to simulate the effects of bad non-linearity. Comparable strength from signals arriving via routes of more than twenty miles difference – a freak condition – can cause so much distortion that programmes are barely recognisable.

This whole phenomenon is called *multi-path-distortion*. Apart from careful positioning of the aerial, it is minimised in the tuner by a good FM-to-AM sensitivity ratio (high *AM suppression*), a feature adding to circuit sophistication and cost but which also improves the receiver's *capture effect*, a characteristic whereby another FM station of almost equal received signal strength on a similar or even the same radio frequency may be made inaudible. There are other performance features relating to spurious responses at various critical radio frequencies, several of which will be defined in specifications of the better FM tuners.

This leaves stereo. It might be thought that the obvious way to provide two radio sound channels would be to use two transmitters and two receivers, though the reader will appreciate that this is hardly economical either for broadcaster or listener. In Britain there were experiments on these lines some years ago, one signal being transmitted via the VHF/FM service and the other via one of the television sound channels during 'off duty' hours. Enthusiasts would re-position TV receiver and sound radio for this exciting non-visual event, adjusting volume and tone controls for some semblance of stereo balance between reproducers usually markedly different in sound character. But this Heath Robinson age has passed, FM radio now offering left and right simultaneously from one transmitter through a *multiplex* process evolved by the American General Electric and Zenith companies and known, appropriately, as the Zenith–GE system. Officially adopted in the U.S.A. in 1961, this method of broadcasting and receiving stereo signals is now in world-wide use.

It is a *compatible* system, meaning that stereo transmissions may also be received monophonically on ordinary single-channel FM tuners without significant audible degradation, apart from loss of stereo. This is because music modulation applied to the VHF carrier is exactly the same as with mono, stereo 'information' being suitably encoded and applied as a separate package of modulation at frequencies just above the audio band. An explanation of how this is done may usefully begin by reference to the behaviour of stereo record grooves as discussed in the last chapter. We saw there that although the two groove walls carry separate signals, in so far as they represent central sound sources the waveforms are similar and result in purely lateral stylus movements, dissimilarities (spatial separation) producing vertical motions. Pursuing this, it could be imagined that a specially designed pickup generated electrical outputs corresponding, not to movements of the separate left and right groove walls, but to lateral and vertical motions. Such motions, as we have seen, represent respectively those features which the two stereo waveforms have in common and those in which they differ; it is therefore meaningful to designate the corresponding outputs as *sum and difference* signals. Although not immediately available as left and right, all the information about a pair of stereo waveforms is contained within such signals and may be regained by using a simple circuit matrix to perform the necessary sorting out. Indeed, for convenience in mechanical design some moving-iron pickups do in fact work on a sum-and-difference principle, internal cartridge connections converting the signals back to left and right for conventional use.

The point of this digression is to show that normal stereo can be represented as a sum of the two channels (L + R) – which is what we hear when a stereo record is reproduced monophonically – and a difference (L − R), it being quite easy to reconstruct the original L/R regime when required. Record groove and pickup behaviour aid understanding of what might otherwise be too abstract to grasp, though unless working directly from a record groove no mechanical intermediary is needed to produce sum-and-difference equivalents of a stereo signal – just another circuit will do the job. This is what is used at a multiplex transmitter, the *sum* providing normal mono modulation and the *difference* going through an encoding process before joining the sum in a form designed not to upset mono reception. The composite signal, as modulation on a carrier, passes through any FM tuner in quite conventional fashion right up to the discriminator circuit, where demodulation also proceeds as usual to produce a mono audio output plus, in the stereo case, a difference signal encoded at frequencies above the audio band. To reconstitute the left and right information a circuit known as a *multiplex decoder* is interposed between discriminator output and de-emphasis components. Such a circuit will be incorporated as part of any receiver designed for stereo, but mono FM tuners may in many cases be converted for two-channel operation by addition of a separate decoder unit. To save costs for those not able to use the multiplex facility immediately, some mono tuners are manufactured with all the necessary connections for easy addition of decoders at a later date. Such units are often advertised as being 'stereo ready', or something of the sort.

Understanding of the actual decoding process is by no means essential to a grasp of hi-fi at the modest technical level adopted in this book, though a picture of the composite multiplex signal does help to clarify what can otherwise be a complete mystery. **Fig. 38** displays this on a normal frequency scale, the rather alarming shape of the compatible signal simply representing, arbitrarily, an average sound energy distribution that might occur in music. If this diagram were clipped off at 15 kHz it would depict the modulation on a mono FM programme, and when a multiplex signal comes out of the discriminator in a single-channel tuner the de-emphasis circuit attenuates the unwanted material above 15 kHz, leaving the audio to pass on for amplification in the ordinary way.

When handling stereo the transmitter first derives sum-and-difference signals – L + R comprising the mono portion in **fig. 38** – and then encodes the L − R by using it to amplitude modulate a separate sub-carrier at a frequency of 38 kHz. Paradoxically, although AM processes are concerned with changes of amplitude rather than frequency, an AM carrier does in fact undergo an addition and subtraction process whereby *sidebands* are stuck on to it, extending outwards from the nominal carrier frequency by the audio bandwidth at each side. This is why there is overcrowding and interference on the MW/LW bands, as each transmission occupies not just its own single frequency but spreads outwards due to the unavoidable sideband element in amplitude modulation, and by an amount which is a significant fraction of the radio fre-

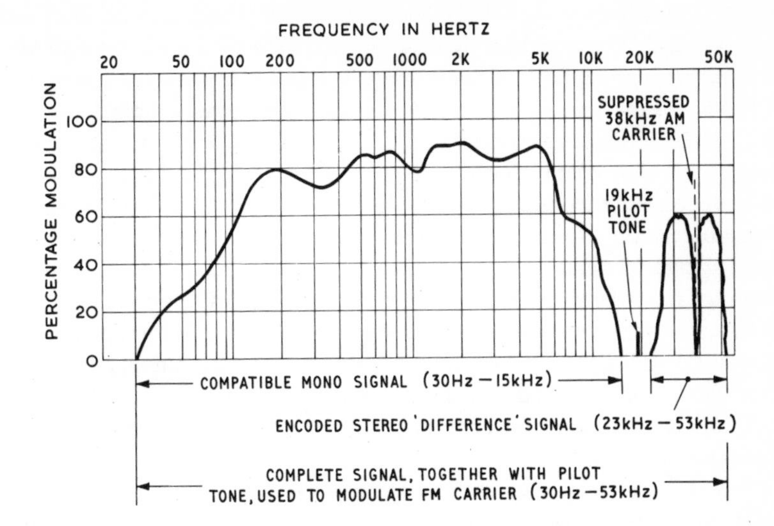

Fig. 38 Derivation of multiplex signal used in stereo broadcasting. This shows all the signal elements making up the eventual modulation, which at peak amplitude (100%) shifts the VHF carrier to ± 75 kHz at a rate corresponding to the modulating frequency, Pilot-tone is fixed at 10% modulation ($\pm 7\frac{1}{2}$ kHz).

quency band. Returning to multiplex, audio difference signals up to 15 kHz add AM sidebands to the sub-carrier, so that in its encoded form the L – R information occupies the band 38 kHz ± 15 kHz, stretching from 23 kHz to 53 kHz. The total encoded stereo signal therefore extends from 30 Hz to 53 kHz, and when applied to the carrier by a process of frequency modulation in which the RF signal is deviated up to ± 75 kHz at a rate depending on the modulating frequency, the overall bandwidth required by a transmission is considerable. For this reason FM broadcasting in general and stereo multiplex in particular cannot be used at MW radio frequencies, though there is no serious problem (so far) on the VHF band.

The theory of multiplex reception requires simply that the encoded signal be demodulated by an ordinary AM detector, thus releasing the stereo difference information for processing with the main L + R signal to regain the original left/right stereo. That is in theory, though in practice there are technical difficulties if the 38 kHz sub-carrier is left intact and added with its sidebands to the main audio modulation, so this carrier is removed or 'suppressed'. However, with no carrier there can be no detection, so means must be found to recreate it in the receiver. To this end a 19 kHz *pilot tone* is transmitted as a fixed part of multiplex modulation (see **fig. 38**), this frequency being chosen for easy conversion to 38 kHz by doubling and because it is exactly half-way between the highest L + R audio frequency (15 kHz) and the lowest frequency of the lower L − R sideband (23 kHz). This eases unambiguous isolation in the receiver's decoder circuit. The decoder's first job is to recreate the missing sub-carrier with correct phase and amplitude, using the pilot tone for reference, and then to carry out the idealised task outlined at the beginning of this paragraph. 'Correct phase' is a key point here, as decoders must be designed carefully and receiver IF circuits aligned accurately to achieve this within a close tolerance; otherwise the stereo difference information comes out falsely and combines with the sum signal to give, in effect, reduced channel separation. With a good tuner and decoder in correct mutual adjustment stereo channel separation can be around 40 dB, and better than 20 dB should be obtained in practice, which is comparable with that available from the best stereo pickups.

There are two remaining points on stereo radio reception. Although the Zenith–GE system is compatible in the sense that multiplex transmissions may be received as mono on mono tuners, a tuner switched for stereo operation will not necessarily receive non-stereo programmes, as circuit functions and interconnections will in some cases block the mono signal in the absence of a 19 kHz pilot tone. However, there are various circuit dodges used to overcome this minor limitation, and many tuners will be found with automatic adjustment to suit the transmission, while others must be switched manually between the two functions. As an additional aid, in some tuners the pilot tone is used also to actuate an indicator light so that the user knows when to switch to stereo operation. The final stereo point brings us full circle in this survey of hi-fi FM tuner criteria, for a possible snag with stereo reception is a worsened signal-to-noise ratio in comparison with mono in otherwise identical circumstances. For reasons connected with modulating signal bandwidth, the presence of a continuous pilot tone, and other related factors, the effective S/N ratio is lowered such that areas only just within noise-free reception distance of an FM transmitter may not be suitable for stereo. It is the receiving process as much as the transmitted signal itself which causes this, so that earlier remarks on noise problems in FM reception are largely unaffected except when the receiver is switched for stereo operation. The matter therefore involves tuner sensitivity, local reception conditions, and type or position of aerial – matters to be approached again later at a more practical level. Here we leave tuners and the mystifications of audio signals attached in the form of modulation to radio waves in space or IF signals in receivers. The reader suffering from mental indigestion may well care to feel the *terra firma* of simple(!) musical waveforms as we move on through amplifiers and loudspeakers, perhaps forgetting for a while the intractabilities of amplitude modulation that insists on varying frequency, and frequency modulation that changes according to amplitude.

We now have respectable hi-fi signals in electrical form at audio frequencies from disc, tape and radio. These must all be brought to a similar voltage level and made easily selectable; replay corrections must be applied where necessary; some manual alterations to tonal balance may be useful; any small remaining deficiencies such as low frequency rumble or 'edgy' distortion at high frequencies could be filtered away; sound volume must be adjusted to taste; the stereo image may need shifting slightly to right or left; and the signal thus processed and adjusted must be suitable for feeding to two power amplifiers in order to drive a pair of loudspeakers. All these functions are performed by a hi-fi preamplifier, making this the nerve centre of any sophisticated home-music system; hence the very apt alternative name of 'control unit' applied when the relevant circuits are housed as a separate entity. These circuits could in theory degrade an audio signal in five respects – distortion, noise, frequency response, transient response and stereo – so we shall examine hi-fi requirements accordingly.

The preamplifier is a sort of buffer between signal sources and power amplifiers. The latter are not normally adjustable, simply requiring input signals of a certain voltage amplitude for a particular power output. If a preamplifier can deliver such a signal without running beyond the limits of linearity there is no distortion problem at this end of the buffer, and unless a unit designed for use with a sensitive power amplifier is coupled instead to a very insensitive one, thus raising the necessary signal voltage, there is normally no difficulty. At the input end things are not so simple, as a practical preamplifier must be able to cope with a wide range of signal levels from various pickups, tape recorders and tuners. This is dealt with partly by providing separate inputs of appropriate sensitivity for these various sources, and partly by using circuits with a large *overload margin* so that signals must rise well above the nominal input voltage before causing distortion. At some point in any preamplifier there is a volume control, and as the user will obviously always set this to avoid driving the power amplifier into audible distortion there is no risk of overloading preamplifier stages *after* this point. However, before this control the circuits have to be more tolerant, a fact best illustrated by the following example.

It was explained in the previous chapter how pickup cartridges come to have their sensitivity expressed in millivolts per centimetre-per-second (mV/cm/

sec) of recorded modulation velocity, 'average' velocity being 5cm/sec. A typical high-quality magnetic cartridge might have an output of 1·5 mV/cm/sec and would probably be used with a preamplifier rated at around 4 mV input sensitivity. Such a rating indicates that the preamplifier will deliver its full nominal output signal to the power section if an input of 4 mV is applied with the volume control turned fully up. Now, while a recorded velocity of 5 cm/sec may be a convenient working figure, the loudest musical passages on commercial discs frequently result in modulation levels of 20 cm/sec or more, resulting in a pickup output of $1{\cdot}5 \times 20 = 30$ mV. Since the preamplifier needs only 4 mV in order to drive the power amplifier to its full output, the user automatically operates with the volume control turned down from maximum by an appropriate amount. But the first circuit still has to handle that 30 mV, which is more than seven times its nominal input figure, this ratio representing the minimum safe overload margin needed in such a case. Great care therefore has to be exercised in the design of preamplifier input stages – particularly those dealing with pickups. To minimise the risk of overload some units incorporate such refinements as high and low sensitivity input sockets, pre-set gain controls, plug-in adaptors to suit various pickups, and so on. Such matters being dealt with adequately, no respectable control unit or preamplifier will introduce significant non-linear distortion, which may be expected to remain below 0·1 per cent in terms of harmonics or intermodulation at all signal levels and over the whole audio frequency range. Practical input sensitivities and the very important related matter of *impedance* will be covered when we come to consider matching and interconnections in later sections.

Also confined in the main to input stages is the troublesome matter of unwanted background noise. Such noise, whether of the random sort caused by minute electrical agitations in components or simply hum induced from valve heaters or nearby mains transformers, must be considered in relation to audio signal levels. If the noise-level is, say, more than 60 dB below maximum music peaks the dynamic range available on disc records can be accommodated without very quiet passages or subtleties of ambience becoming masked. Clearly, with a given noise-level it becomes progressively more difficult to maintain this desirable S/N ratio as the signal amplitude is reduced, and since signals are always weakest at the input to an amplifier system it is inevitably the first stage that requires particular attention in design. While carefully conceived valve circuits were good in this respect, the transistor has eased matters considerably, with internally induced hum now almost a thing of the past and random noise (hiss) low enough to grant S/N ratios of over 60 dB even when using very insensitive magnetic pickups. Despite this, many users of hi-fi equipment have difficulties, particularly with hum, which often seems to occur in the preamplifier but in fact usually arises from the way in which pick-ups and other feeding devices are connected to the input.

Frequency response comes next, to be considered in two lights: fixed corrections and tonal control. Corrections of response must be applied to any signal which by convention arrives complete with an unequalised recording characteristic. This encompasses disc records and tape recordings, though in most cases the latter will be dealt with by preamplifiers in the recorder designed to suit the particular tape-head employed. Some amplifiers have input points intended to be fed directly from tape-heads, with a frequency response of the sort shown in **fig. 35** (page 93), but except in the hands of a technical experimenter this approach tends to be rather inexact. Gramophone records are more straightforward, as the recording characteristic (RIAA) has been standardised for some years and any pickup with an output proportional to recorded velocity will offer to the amplifier a signal following this standard. If plotted on a graph, correct replay equalisation is a 'mirror-image' of the recorded curve (**fig. 28**, page 77). It is comparatively easy to arrange for such a response in a preamplifier: an accuracy of ± 1 dB may be expected in good equipment.

This all applies to magnetic pickups, which, because they respond to modulation velocity are called *velocity pickups*. Crystal and ceramic cartridges, on the other hand, are sensitive to stylus displacement or modulation amplitude and are thus sometimes known as *amplitude pickups*. They are largely self-compensating and have a higher output voltage than magnetic types; preamplifiers therefore usually cater for them with appropriate separate inputs. Experience indicates widespread confusion over the related matter of sensitivities and impedances, another matter for the practicabilities of later chapters.

Apart from several input sockets for connection of various signal sources, all preamplifiers have some means of switching between such sources – sometimes a simple rotary switch, sometimes a row of press-buttons. After selection and, where necessary, equalisation, it may be thought that audio signals could then be passed on to the power amplifier without more ado. But this is not always the case, as several factors make it desirable that the user shall have some additional control over tonal balance if maximum musical pleasure is to be had from all programme material. Some recordings and radio broadcast may sound over-bright or dim, may lack fullness in the deep bass or seem thick-textured due to excessive low-frequency reverberation. Studio engineers obviously attempt to minimise such shortcomings, though whether or not they are regarded as faults is often a matter of personal taste arising from various microphone techniques or the chosen replay loudness. An instrumental and tonal balance acceptable to one listener may seem unnatural to another. Also, acoustics of the listening environment and loudspeaker performance will affect the sound in this respect. These are matters to be covered in more detail in later chapters, the immediate point being that preamplifiers should offer a range of tonal control permitting independent adjustment of bass and treble.

For various technical reasons it is likely that once a musical performance has been captured in the studio, with whatever overall balance the producer deems appropriate, any further changes due to deficiencies in circuits or transducers are likely to involve attenuation at the two frequency extremes. From this point of view tone controls should facilitate boosting of bass and treble extremities only, points at which the response starts to rise being brought progressively nearer to middle frequencies as the controls are rotated. **Fig. 39** shows how this might look in terms of frequency response, with corresponding mirror-image attenuations.

This seems all very neat and logical but it is by no means the whole story; technical losses of this pure and simple kind, while certainly occurring from time to time, are usually masked by differences of musical balance due to microphone techniques, and these demand a rather different approach. If a record sounds generally over-brilliant the response is likely to need lowering in the 2–5 kHz region, but if this is done with a treble control of the **fig. 39** type set, say, somewhere between −2 and −3 the output will be rather severely cut at 10 kHz. The result could be a correct musical balance in terms of instrumental brilliance resulting from overtones below 5 kHz, but a lack of subtlety and 'bite' of the sort due to sound energy at very high frequencies; reference back to **fig. 2** (page 17) shows that nearly all instruments have some output up to 10 kHz. In the converse case of a dim sound quality, arising perhaps from use of fairly distant microphones in a hall or studio with a rather heavy acoustic, a boost may be needed in the upper-middle region. However, the very highest frequencies will now be lifted out of proportion, and as most of the more objectionable distortion in domestic sound reproduction – particularly from discs, due to pickup tracing limitations – occurs above about 7 kHz, we see that the attractions of very simple tone controls begin to fade. At the bass end there are analogous problems, with undue loss of deep fundamentals when attenuating, and troublesome emphasis of very low frequency noise when boosting.

This being so, the reader may wonder why we have bothered with **fig. 39.** The reason is partly that some amplifiers do in fact adopt tone control re-

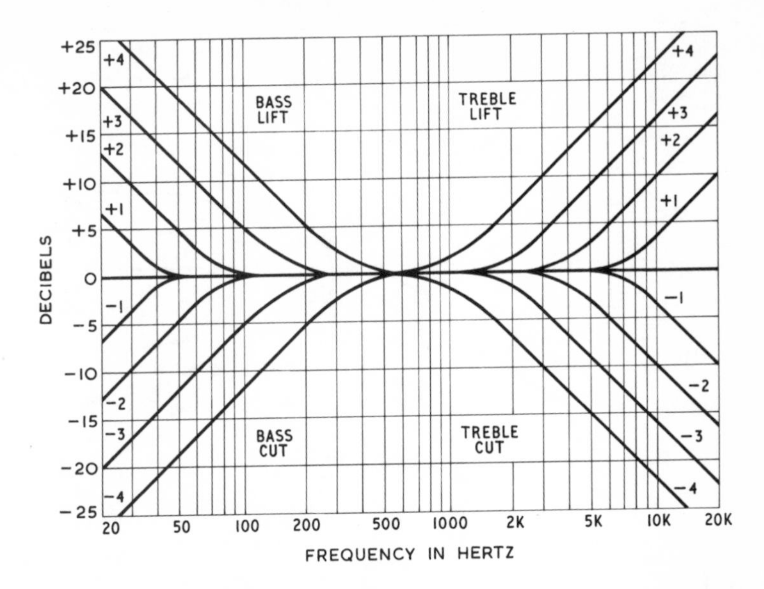

Fig. 39 Response of a possible tone control circuit offering boost or cut with constant slope and variable turnover frequency. Not very satisfactory musically, as explained in the text.

sponses of this sort – and we shall see in a later chapter that this is not always so unreasonable – and partly to ease explanation of the more suitable curves shown in **fig. 40**. Here it is possible to brighten a dim sound without over-emphasis of the upper harmonics and any associated distortion, with equal facility to reduce over-brilliance while retaining some high frequency subtlety. At the bottom end a small boost lifts the deep bass only, this commonly being the initial requirement in home reproduction. Further upward adjustment continues to boost the lowest frequencies whilst also extending the rise into the tenor region above 200 Hz, where an exceptionally 'thin' sounding signal might benefit from some accentuation. Conversely, bass cut can be used to lighten a thick musical texture without undue sacrifice of fundamental tones.

Practical tone controls are usually continuously variable between extremes of lift and cut, with a nominally 'flat' response position somewhere in between. The curves in **figs. 39** and **40** are drawn simply to show the sort of variations that might be expected at various settings, the plus and minus figures being quite arbitrary and not representing actual calibrations met on commercial units, a matter in which there are no standards. Also, the maximum limits

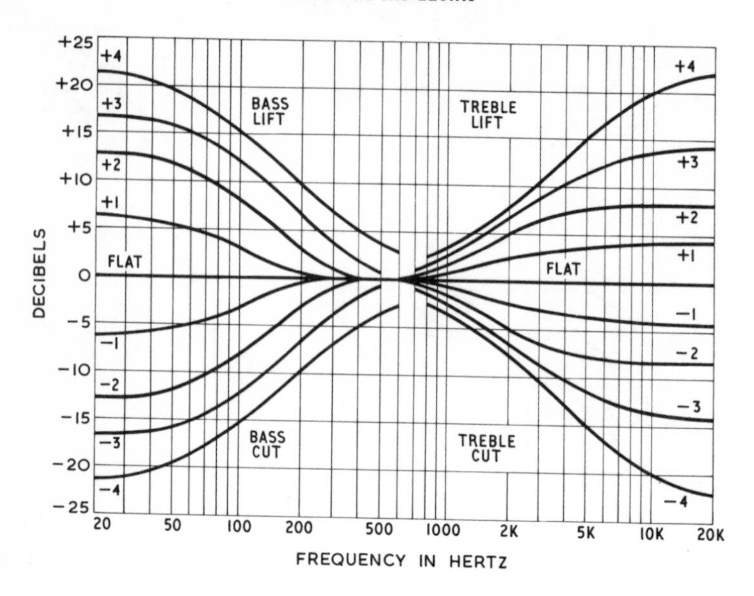

Fig. 40 Desirable type of tone control response, giving shifts of musical balance without excessive effects at the frequency extremes.

shown are rather more drastic than will usually be needed in practice, the curves labelled +3 and −3 representing the sort of extremes found on domestic equipment. Before leaving tone controls mention should be made of another type of treble lift provided on some amplifiers and called *presence*. This is a switchable boost that raises the region above about 2 kHz by a few decibels, rather in the manner of a +1 treble lift in **fig. 40**, but starting perhaps a little lower down the frequency scale. The effect is to brighten dim or 'fuggy', sounding music, to bring it forward or make it more nearly 'present' for the listener.

Now we come to *filters*. These are designed to remove undesirable noise and distortion with minimal degradation of musical quality, rather as the mesh in a sink runaway should hold back potato peelings without slowing down the flow of water. In both cases there are limitations, for too many peelings will themselves clog the flow, and excessive noise or distortion will extend too far into the musical frequency band to be attenuated without some loss of balance or tone-colour. Filters used to cope with these problems are, like tone controls, circuits within the preamplifier arranged to have particular types of frequency response. However, they differ from such controls by attenuating very

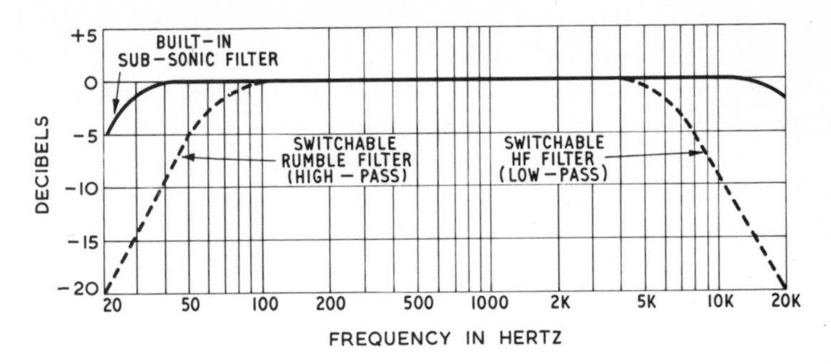

Fig. 41 Some typical preamplifier filter responses.

rapidly above or below particular frequencies, and because of the associated steep slope in the response curve are sometimes known as steep-cut-filters.

At low frequencies the offending noises usually come as rumble from inferior turntables, and the corresponding *rumble filter* must offer a reasonable compromise between adequate bass response and elimination of such noises. **Fig. 41** shows the sort of LF fall-away (dotted curve) that might be provided as an optional response at the touch of a switch, and although such a filter would be said to have a *cut-off frequency* (by convention, frequency at which response is down by 3 dB) of 60 Hz, it is surprising – until one looks again at **fig. 2** – how little this affects most music. On amplifiers of very high calibre, the sort not normally used with any but the most expensive ancillary equipment, a switched filter of this sort will often be omitted or arranged to work at a lower frequency. The full line in **fig. 41** shows a possible resulting response, designed not so much to eliminate audible rumble as to remove sub-sonic noise-signals that might otherwise pass through the power amplifier to damage loudspeakers or create distortion by causing excessive cone movement. Various labels will be found attached to the type of filter so far discussed: rumble, sub-sonic, low, low frequency, LF, Lo and *high-pass*, this last being the technically correct engineer's name even though slightly confusing at first glance.

At the other end of the frequency scale we find *low-pass* devices, known variously as scratch, hiss, high, high frequency, HF and Hi filters. The 'scratch' label is an inheritance from the days of 78 r.p.m. shellac records, though hiss is still with us from a minority of LP discs and from most tapes played at domestic speeds. However, unless removing interference attached to programmes coming from AM tuners, an HF filter is these days used in the main for reducing the effects of distortion rather than coping with noise. It has been explained in earlier chapters that amplitude distortion produces intermodulation components, and that these cannot help sounding harsh and unpleasant. A signal subjected to such distortion over the whole frequency range is irretrievably degraded, and it might therefore be thought that nothing can be done about this in a preamplifier. Now this is true, but it happens that the type of non-linearity operative in pickup tracing distortion is confined by groove/stylus geometry to the higher audio frequencies, the majority of objectionable extra harmonic and inharmonic waveform components being generated, as noted earlier, above 7 kHz. Likewise, various factors in tape recording and broadcasting tend to introduce a little HF distortion while leaving the rest of the range unblemished. It follows that a filter designed to attenuate high frequencies steeply without cutting very far into the musical spectrum is often useful, and is a fairly common provision on all but the least ambitious amplifiers. A single switched HF filter of this sort is likely to have a response similar to that shown in **fig. 41** (dotted line). More elaborate preamplifiers enable the user to alter the steepness of fall-away by rotating a knob, thus making it possible always to obtain maximum musical information at minimum distortion. The greatest sophistication combines variable slope with a selection of cut-off frequencies.

Excessive use of the HF filter can upset transient performance in a preamplifier. In theory, any deviation from a level response degrades musical transients, which depend on an extended frequency range in addition to freedom from resonance. But the ear seems to prefer a balance of modest deficiencies to a combination of one markedly unnatural characteristic and perfection in all other respects. Thus modest filtering to remove some distorted edginess from string tone is acceptable despite a slight loss of impact in cymbal clashes, whereas too much filter introduces a 'strangled' sound quality, indicating that the transients need a little more freedom.

Returning to frequency response, the overall performance of a good preamplifier with all controls set at their nominally 'level' positions can and should be within ±1 dB out to 30 Hz and 15 kHz. Response should not be affected by volume settings unless a *loudness control* is incorporated; a conventional volume control is not frequency-sensitive. On most amplifier systems sporting a loudness control the facility is switchable, the same potentiometer becoming a plain volume control when the relevant switch is off. Reference to **fig. 7** (page 30) and the associated comments in Chapter 1 will help understanding of the theory behind loudness controls. If music is reproduced to create at the ears a lower sound level than would occur if the listener sat in the recording studio, tonal balance is distorted. This is because as the overall sound intensity is lowered there is disproportionately greater reduction of subjective loudness at low frequencies – at least as revealed by tests conducted with isolated pure tones. Thus when we listen to music at low volume levels it is necessary to apply some bass lift to regain the correct balance. Since, it is argued, most domestic music is indeed heard at lower than concert hall levels, it is appropriate to incorporate some automatic compensation in the amplifier, with maximum departure from a flat response at low control settings. The response adopted varies between amplifiers, but resembles the lower loudness-lines in **fig. 7**, rising at low frequencies.

On the face of it this all seems very fair, though some of the underlying assumptions are questionable. Firstly, it does not necessarily follow that patterns of hearing sensitivity revealed with sine-wave tones in the laboratory apply simply and fully to complex music waveforms. Personal observations from various positions in concert halls suggests that while there is indeed some relative loss of bass at greater distances from the orchestra, it is not quite of the magnitude assumed in the design of loudness controls. Secondly, while it is obviously true that most home listeners using cheap record players or radiograms do not reproduce music at the full natural level, this is not always the case with people using good hi-fi equipment. Thirdly, if the listener adopts a replay level that places him, aurally, near the back of a concert hall but uses loudness compensation when reproducing a recording offering the balance heard from a seat in the stalls, he will presumably have sensations corresponding to a loudness as heard at the back and tonal balance as heard at the front. No doubt an enlightening experiment in psycho-acoustics, but quite unnatural and certainly nothing to do with true high fidelity. In good stereo recordings the apparent distance of the performers is more or less set by the recorded ambience, and this determines a natural replay loudness. If reproduction at this level happens to coincide with a loudness control setting still giving some bass lift (and it might easily do so, for no such control can anticipate or cover a wide range of recording levels or signal source amplitudes) then the extra 'compensation' is clearly quite irrelevant.

In short, the loudness control is no more than a bright pseudo-scientific idea unrelated to natural music reproduction: its application is either irrelevant on its own reckoning, or, if relevant, bound by definition to produce an unnatural sound quality. The reader may object that by the same sort of logic the subject should not have been mentioned in this chapter on high fidelity; but in some quarters loudness controls are accepted as part of the natural hi-fi order, and to refrain from comment might be regarded as a serious omission –

rather as with auto-changers in the previous chapter. Finally on this loudness question, for those who like some bass lift for listening to background music late at night, there is, after all, the ordinary bass control.

Now to the last preamplifier performance feature: stereo. This is mainly a matter of duplicating all the circuits while avoiding interchannel crosstalk which could arise between adjacent wiring or via common power supply components. Tone and volume control potentiometers are coupled in pairs (ganged), and to preserve proper channel balance at all settings these components have to be accurate and reliable. In really good units left/right errors will not exceed 2 dB at any frequency with practical combinations of control settings. Channel separation should easily be maintained at better than 40 dB. Apart from a few unusual amplifiers in which the controls are concentric but not rigidly ganged together, thus permitting differential channel adjustments, all stereo units have a *balance control*. This is arranged to give equal gain to both channels in its central position, tipping the balance of gain towards the left channel in one direction and towards the right in the other. As we shall see later, this has the effect of shifting the sound-stage to compensate for any unbalance in stereo signals.

Various switching arrangements are adopted to cope with mono and stereo inputs, it being customary to arrange internal connections so that a mono signal may be reproduced *double-mono* via both channels, and also, in some cases, via one channel only when preferred. For those who have muddled their loudspeaker wires or who like members of the orchestra to change sides between symphonic movements, some amplifiers even provide a channel reversal switch. Another refinement is a signal outlet point providing a *derived centre channel*, the derivation being a simple addition of left and right signals for feeding a third power amplifier. This in turn operates a loudspeaker to fill the proverbial stereo hole-in-the-middle, which really only arises, as will be explained in the next chapter, with ping-pong type recordings or improperly placed or unsuitable loudspeakers.

The power amplifier, next item in this hi-fi survey, may be treated as a functional entity even though preamplifier circuits will frequently be integrated with it rather than housed separately as a control unit. The PA is fed with an audio signal voltage at its input and delivers audio power to a loudspeaker at the output; it has no adjustments or controls apart possibly from a few pre-set devices not to be touched by the unqualified; it is normally the largest of the three electronic units; and it is likely to generate most heat. Two aspects of performance may be dismissed as unlikely to degrade musical signals: background noise and stereo. The latter simply involves circuit duplication on a single chassis, and as only the power supply is common to both amplifiers, with no ganging of controls or need for a channel balancing circuit, there is normally no bother with crosstalk. Unlike preamplifiers, power circuits do not have to handle very low voltage input signals and are therefore not usually subject to difficulties with internally generated noise. Taking the full output power for reference, S/N ratios of over 80 dB are commonly achieved, though some models may produce an audible hiss if used with certain types of highly efficient loudspeaker. This leaves frequency response, transient response, non-linear distortion and power capacity.

Frequency response may normally also be dismissed as this can easily be flat within much less than one decibel over the whole audio range; but there are some minor qualifications to this arising from distortion factors, to which we shall return shortly. Transient performance, as was noted in connection with preamplifier filters, depends on frequency response, and it might be thought that as this is so flat in power amplifiers there cannot be any transient difficulties. Well, this is not the whole story, for even though the sharp edge of a transient musical sound – perhaps a drum beat, cymbal clash or plucked harp string – will not be blunted if the response is wide and flat, that very edge may excite circuit resonances beyond the lower and upper audible limits. If, at the same time, the amplifier is being driven by the reproduced music signal towards its power limit it is possible for any 'ringing' thus introduced to interact with the signal in various subtle ways. Despite the transient impact there can be a certain fluffy imprecision. Valve amplifiers were generally more inclined to this sort of trouble than are their transistor successors, mainly because a valve power output stage uses a transformer, and this component must be designed and used with the sort of care devoted only to the very best equipment.

The reader may come across references to *square-wave testing* in assessments of power amplifier performance. The square-wave is a rather unnatural man-made type of signal which happens to be the last word in transients. **Fig. 42** shows what it looks like if displayed as a vibration-curve, together with the sort of thing done to such a signal by a poor amplifier. A first-class power amplifier will leave the waveform almost unscathed. An important point here is that transient performance is dependent in many amplifiers on the impedance of the *load* offered by the loudspeaker. We shall see soon that this never has the quality of pure electrical resistance, and amplifiers therefore have to be very tolerant. Related to this is *stability* or refusal to develop self-oscillation. Some amplifiers will, in certain conditions of use and with certain types of loudspeaker load, actually generate continuous oscillations at supersonic frequen-

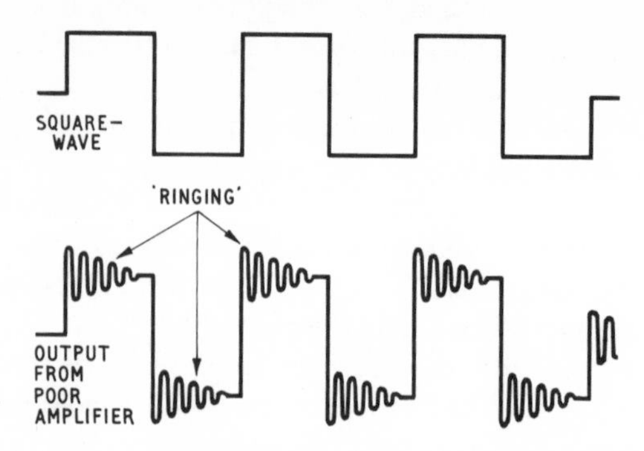

Fig. 42 Square-wave and the sort of output it might produce from a power amplifier with poor transient response.

cies. Unconditional load stability is therefore a desirable feature in amplifier specifications. Also connected with load values and output conditions is a feature called *damping factor,* to be explained shortly when we investigate loudspeakers.

Very low non-linear or amplitude distortion is desirable in power amplifiers as in other links in the hi-fi chain. A distinction here is that this must be achieved at high levels of output power (watts) rather than with the small voltage signals handled so far. As the power rises so does the risk of distortion, and to confine this to a negligible percentage up to the desired power level a circuit technique called *negative feedback* is employed. To apply such feedback, some of the output signal is injected into an earlier part of the circuit at a point before any significant distortion would have occurred. It is applied so that the fed-back waveforms are always in phase opposition to the input signal waveforms. Extra input is needed to overcome the negative influence of the feedback, though any distortion added by the circuit after this point is *not* overcome by the increased input signal because it naturally contains no corresponding anti-phase waveform components. Thus while input and feedback find a point of equilibrium corresponding to a certain reduction of overall voltage gain, the full measure of any distortion is fed back into the circuit and, having an inverted phase, tends to cancel itself at whatever point it was originally introduced. Such introduction occurs mainly in the output stages, and with feedback applied can be held down to a low level until just before the unavoidable overload point, as depicted in **fig. 43** (full line). In Britain a total harmonic distortion (THD) of 0·1 per cent at full rated output power is widely

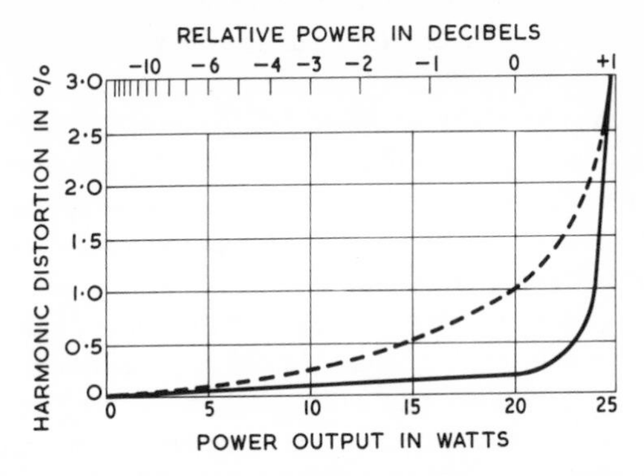

Fig. 43 Two ways in which distortion might change with output from power amplifiers. Both amplifiers would have similar power ratings, though the solid line performance is preferable. The overload point is clearly above 20W in this case, but is rather indeterminate on the dotted line.

regarded as the ideal at which to aim, and while this is probably unnecessarily stringent, the American 1 per cent is certainly too permissive, particularly for types of distortion which can arise in some transistor amplifiers. If the full line curve in **fig. 43** applied at all audio frequencies it would be reasonable to rate such an amplifier at, say, 23 watts (0·5 per cent THD).

Distortion figures should be treated with care, as it would be easy, for instance, to say that the two amplifiers in **fig. 43** will produce 24 watts and 20 watts respectively for 1·0 per cent distortion. True as well as easy – but misleading. In one case distortion remains negligible right up to a point only fractionally below the rated output, whereas in the other it might impart a very slight roughness to the sound on musical passages reaching to within a few decibels of the nominal maximum power. More important than this, the solid line curve might only apply at some mid-frequency convenient to the manufacturer, power and distortion performance at high and low frequencies not being mentioned. Power can be limited at both audio extremes in cheaper valve amplifiers due to shortcomings in output transformers, and was also restricted at the HF end in some early solid-state circuits due to use of cheap output transistors. Also, there is a tendency for circuits to shift the phase of waveforms at high frequencies, which means that negative feedback is not quite so negative and thus becomes less effective at reducing distortion. In some circumstances it might actually become positive, leading to the instability troubles mentioned earlier. It is in these respects that specifications of power amplifier frequency response sometimes have to be qualified, for while a circuit could have a flat wide-range response when measured at a lower power output, it may not be capable of delivering the *full* power over such a wide band. In technical parlance its *power bandwidth* may be somewhat restricted, this bandwidth normally being defined by the lower and upper frequencies at which an amplifier will deliver half the full rated power. A plot of power output capability against frequency in known as the *power response.*

The type of distortion so far considered is of the simple overload variety, which falls progressively with output level and only becomes serious when an amplifier is operated at the limits of its power handling capacity. All output stages in hi-fi amplifiers employ balanced pairs of valves or transistors in *push-pull* circuit configurations wherein components work in phase opposition for lower distortion at greater efficiency. There are variations on this push-pull theme, the associated circuits being designated in 'classes' – Class–A, Class–AB, Class–B, etc.

To examine these in detail would take us too deeply into technical circuit problems, but in general terms it may be noted that with Class–A operation (normal in high quality valve amplifiers) the signal is so disposed in relation to the overall transfer characteristic that distortion must be highest when signal amplitude is high and rapidly becomes immeasurably small as the signal level decreases. With Class–B operation (normal in transistor amplifiers) achievement of low distortion at low levels depends on fairly critical circuit adjustments. This is because positive and negative portions of the signal waveform are divided between the pair of output transistors, and it is important to ensure that signals cross the polarity line smoothly. If the crossing-over process suffers any discontinuity it produces *crossover distortion*, and as this takes place at the AC 'zero' line it is bound to be effective at low rather than high amplitudes. Good transistor power amplifiers employ sophisticated techniques to give distortion performances indistinguishable from those obtained with Class–A circuits, though many early solid-state amplifiers suffered from a characteristic called 'transistor sound'. This was almost certainly due to crossover distortion, though human nature being what it is, it was for a time heralded as a great new hi-fi revelation, many users managing to convince themselves that the change was due to improved transient performance. Crossover distortion would be represented on curves like those in **fig. 43** by a tendency for the distortion to stop falling at some fairly low power level, perhaps even climbing again as the power output approaches zero.

Power handling capacity is the one remaining amplifier feature to be discussed before we break away from electronic circuits to look at loudspeakers. It was noted in Chapter 2 that a power somewhere in the region of 5–30 watts might be required domestically to drive loudspeakers of average efficiency up to natural peak music levels. The word efficiency is used here in the engineer's sense to denote that percentage of the electrical power fed in which comes out as sound. We shall see later that this can vary even more widely than the six-to-one ratio implied here, though for present purposes it will be convenient to take a figure of 20 watts for some explanation of wattage ratings. In any case, differences of a few watts in maximum output capacity are not very significant in terms of corresponding loudness changes, especially when it is recalled that halving subjective loudness requires a fall of 10 dB, which reduces our 20 watts to 2 watts as noted on the top scale in **fig. 43.**

There are various ways of measuring power output, some meaningful, some not so meaningful. The basic, traditional way normally adopted in Britain is to measure the continuous sine-wave power that the amplifier will deliver into a stated load at various frequencies for the claimed total distortion. This is simple and foolproof, and is really the only completely honest way of doing it. A simple sine-wave (**fig. 3**, page 21) varies its height above the 'zero' line as it passes through its cycle of change, and the power dissipated when a signal with such a voltage waveform is applied across a load depends on the mean or effective voltage. Due to the mathematical procedure for determining this, it has the rather alarming name of *root mean square*, normally abbreviated to RMS. Inevitably, the RMS figure is below the peak figure and in ordinary AC voltage measurements it is RMS which is quoted. However, it has been known for amplifier manufacturers to use sine-waves for power output measurement, and then to quote that power produced at the instant of the peak, which happens to give just twice the power, or 40 watts in the **fig. 43** case. This is rather like rating a 1 kW electric fire at 2 kW with the claim that while the heat fed into the user's room is equivalent to only 1 kW, it will actually be handling a peak of 2 kW for a few microseconds twice in every cycle of mains frequency.

Another variation on a related theme is called *music power*, which is the maximum sine-wave power obtainable before presence of the signal has time to degrade performance. It is in the nature of Class–B amplifier circuits that current drawn from the power supply varies according to signal level. Things are normally arranged so that short bursts of current consumption do not upset the supply, though sustained current drain will lower its voltage and thereby reduce the power capacity of the amplifier. For this reason some Class–B circuits will deliver much more music power (length of time unspecified) than genuine continuous power, which of course is why this particular type of rating was thought up. The best amplifiers incorporate stabilised power supplies which do not affect power output performance, though it can normally be taken that a model whose specification includes a music power rating is not so equipped. If such a specification also includes a continuous rating (sometimes called RMS power) and the power quoted therein is sufficient for the user's purpose, then there are of course no objections. Amplifiers quoting music power only are probably not worthy of a hi-fi label, and any quoting nothing more than *peak* music power should be treated with grave suspicion: our honest 20 watt device could receive a label of 80 watts or more in such a case.

Back to music, it is sometimes objected that neither sine-waves nor sustained high powers are ever met in practice. Well, some organ-pedal notes are certainly pure enough to satisfy the first point for practical purposes, and the mighty apocalyptic climax in Tchaikovsky's 'Manfred Symphony' – with a large orchestra playing furiously yet completely drowned by a sustained organ chord – might well be rather risky for an amplifier rated in music power. And one could find many other examples of sustained high powers in organ or choral/orchestral music. However, this only matters when amplifiers are run near their limits during very loud musical climaxes, quite possible and therefore justifying insistence on rather severe standards, but dependent domestically on listening-room size, loudspeaker efficiency and other factors.

Now we move from electronic circuits to loudspeakers, the former a mystery to the layman but nevertheless amenable to accurate measurement, the latter somehow more understandable yet very subject to personal taste when it comes to assessing performance. A diaphragm buzzing back and forth to radiate sound into a room seems a relatively simple idea, though the reader will recall that some problems come up even at an elementary non-hi-fi level. These same design difficulties occur with varying intensity in all grades of loudspeaker and may be summarised under two headings: (a) extension of frequency range at the two audio extremes, and (b) subjugation of resonances and colorations to achieve a smooth overall response and acceptable transient performance.

A major point mentioned in Chapter 3 was the need for some sort of structure or enclosure to isolate the two sides of a speaker cone, thus avoiding cancellation of sound at low frequencies. The simplest device for this is a plain baffle, though for this to be effective down to the lowest bass notes it must be very large, requiring speaker units to be mounted in walls, floors or cupboards. Such practices not being normal for the majority of music lovers, complete commercial speaker systems are housed in cabinets, these employing a variety of principles to 'load' the speaker cone at low frequencies.

Before describing these it will be useful to look back for a moment at the drive unit itself in **fig. 11** (page 46). The diaphragm on a moving-coil speaker unit is free to move in and out, being constrained only by the combined stiffnesses of rear suspension and cone surround. These latter may be likened to springs, and the cone – having some mass – may be regarded as a small weight. If a weight (mass) is hung upon a spring (compliance) and set into motion it will oscillate up and down for a while at a particular frequency, this being the fundamental or natural resonance of the system. Another example of this is found in pickups, where the total effective mass of the arm springs against the stylus compliance to give an LF resonance. We saw in the previous chapter that in this case the resonant frequency depends on both mass and compliance, going down as either increases. The same applies to speakers, any moving-coil unit having a fundamental resonance which must be taken into account when designing an enclosing cabinet. Unlike pickups, in which this critical frequency is usually below the bottom end of the audio band, hi-fi speaker units designed for acceptable efficiency and reliability usually have their resonance somewhere between 25 Hz and 80 Hz. Cabinets therefore serve a double purpose: they isolate front and rear of cone in the manner of a simple baffle, and they restrain

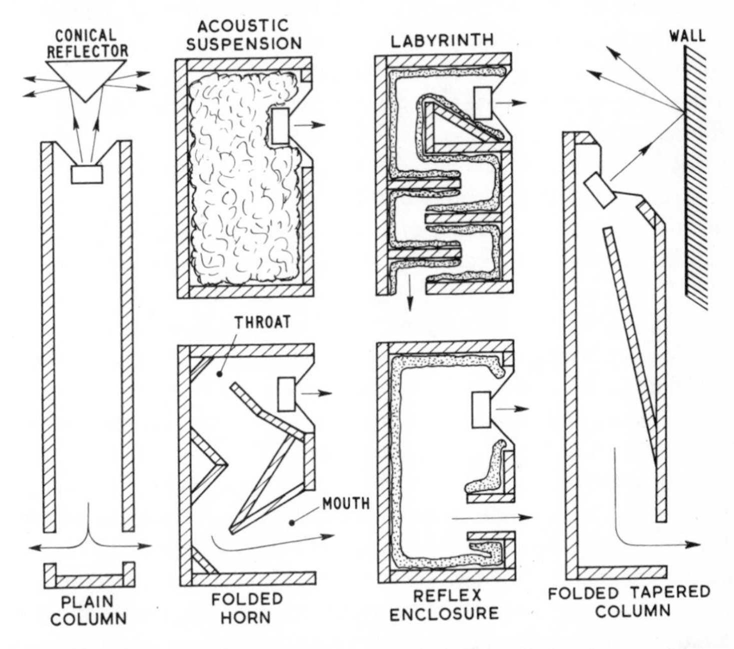

Fig. 44 Six types of low frequency loudspeaker loading. Enclosed space in acoustic suspension ('infinite baffle') cabinets is usually filled with absorbent material, and absorbent lining is applied to inner walls of labyrinth and reflex types.

diaphragm motion in the frequency region where it might otherwise get out of control due to the resonance.

For many years the most popular type of speaker cabinet was the *reflex* enclosure (**fig. 44**); this makes use of an acoustical resonance to combat the essentially mechanical one of the drive unit. There is an outlet from the cabinet – port or vent – which makes contact with the internal volume of air either directly or via a short tunnel as shown in the diagram. The air in the tunnel or around the port behaves rather like the mass in a mechanical system, and that within the enclosure is analogous to a compliance, the two working together in a resonant fashion. Whenever the speaker cone vibrates to produce sound it must alternately compress and rarefy the air inside the cabinet as well as that outside, and since the internal pressure-changes are coupled directly to an acoustical resonant system there is a complex interaction of forces. The net result is that neither the cone resonance nor the enclosure resonance is particularly troublesome – thus avoiding severe peaks in the frequency response – yet efficiency is maintained despite the comparatively short path via the port between front and rear of cone. This is because one effect of the masses and compliances is to reverse the phase of sound coming from the vent at low frequencies, thus augmenting acoustic output from the cone front instead of opposing it. The term *phase invertor* is therefore sometimes applied to reflex enclosures.

A measure of the extent to which a simple cone resonance might be modified by reflex loading is illustrated in **fig. 45**. This shows the way in which electrical impedance, measured in a unit called the *ohm*, changes with frequency in a moving-coil loudspeaker. In this case the cone's natural resonance is at 60 Hz, and in the absence of any restraining influence from a cabinet it vibrates with notably greater vigour when fed with a signal at this frequency. For technical reasons to do with electromagnetism the correspondingly increased amplitude of voice-coil movement produces a sharp rise in impedance, as shown by the dotted curve. At some frequency or other this is the basic natural effect with any moving-coil drive unit. When the unit is mounted in a properly matched reflex enclosure the resonances interact by dividing into a sort of 'double hump' (solid curve), still well above the nominal speaker impedance of 8 ohms but not representing particularly drastic undulations in the associated frequency response. There are variations on this reflex theme, most commonly involving addition of acoustical resistance by means of absorbent material within the enclosure (resistive reflex) or fitting a special acoustic resistance unit (ARU) across the port. Modifications of this sort reduce still further the resonant element in overall bass performance. Another variation, with

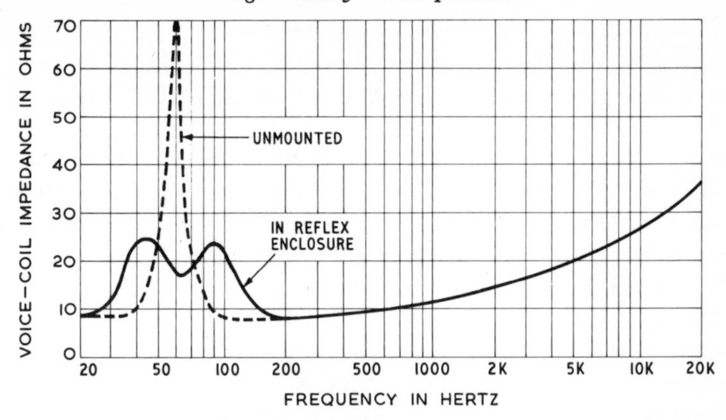

Fig. 45 Typical electrical impedance curves of moving-coil speaker unit, showing effect of reflex cabinet at low frequencies, and rise due to inductance of voice-coil at high frequencies.

possibilities for miniaturisation not otherwise feasible with reflex enclosures, employs an auxiliary cone unit (without magnet) in place of the port. This extra cone, referred to as an auxiliary bass radiator (ABR) or passive radiator, raises bass efficiency for a given cabinet size.

To be effective in the way intended, a reflex cabinet should have relatively dumpy proportions, otherwise it starts behaving like a pipe or column, which has rather different acoustics. The *labyrinth* enclosure (**fig. 44**) can be both dumpy and pipe-like, the 'pipe' being formed by a succession of internal folds. The idea here is to have a long duct which, by virtue of an absorbent lining, only lets through the lowest frequencies, the total column length being chosen for a modest resonance to maintain response in the extreme bass. The principle has a variant known as the *transmission line*, a folded pipe filled with absorbent material which has proved to be an excellent form of bass loading. The folded *horn* uses a quite different acoustical principle, for here the space behind the cone expands according to a mathematical law whereby the diaphragm is matched to the air more efficiently. The cone therefore needs to move less vigorously than in other types of enclosure to produce a given sound level, acoustic output at low frequencies coming mainly from the horn mouth. A disadvantage is that a considerable length of horn and a large mouth have to be accommodated for the idea to be really effective at low bass frequencies; this makes for complex and expensive folding and lowers the useful upper frequency limit of the range covered by the horn. However, good design produces outstanding results, and horn-loading can introduce a certain solidity to loudspeaker bass performance which makes the idea attractive to hi-fi perfectionists, including the author: I confess to a mania in this respect, having built massive concrete horn-speaker systems in two houses in succession.*

Column speakers are sometimes used domestically, and the simple type shown on the left in **fig. 44** was popular in the early days of domestic stereo due to its provision of a reasonably extended bass response without occupying too much floor space – an important point when using two speakers in a small room. A plain column with the drive unit at one end tends to be rather resonant, leading to the sophistication of a tapered column with the speaker situated some distance from the closed, narrow end, with a fold included to reduce overall external length. This is shown on the right in the illustration and may be regarded as acoustically somewhere between a horn and a plain column in performance. Both column types are shown with drive units facing upwards and depending on reflecting surfaces for conveying the higher frequencies to listeners, a point to which we shall return later after considering stereo in the next chapter.

This leaves the *acoustic suspension* system, probably the type of speaker enclosure most widely used in modest hi-fi installations. There has been a constant quest during the history of domestic audio for satisfactory bass performance without gargantuan loudspeaker structures, and there was a breakthrough in the early 1960s with the emergence of miniature speakers using this principle. If a moving-coil unit is mounted in a genuine infinite baffle of the sort mentioned earlier it will radiate sound with fairly steady efficiency down to the main cone resonance. In this region there will be a rise of output due to the ease with which the diaphragm vibrates at that particular frequency, the response falling away fairly rapidly below this point. An objection to this arrangement – apart from the need to knock holes in walls – is that in delivering an approximately steady acoustic output with falling frequency, diaphragm excursions become very large compared with those obtained in the types of enclosure so far considered. There is no acoustic 'load' to constrain cone movement. Large cone excursions of this sort can cause amplitude non-linearity in two ways: (1) the voice-coil moves beyond front and rear extremities of the magnetic air gap, thus suddenly limiting the driving forces generated by peak audio currents; and (2) mechanical suspension devices tend to have compliances which vary with magnitude of deflection, again interfering with cone motion. The coil/magnet problem is solved by using coils considerably longer

* 'Five speakers – How to Make Them', by Baldock, West and Crabbe; and 'A Concrete Horn Loudspeaker System Mk. II', *Hi–Fi News*, October/November 1967.

than the gap so that the same number of turns remain active all the time, but provision of a really linear suspension compliance by mechanical means is extremely difficult. This is the function of acoustic suspension.

When a closely-fitting cupboard door is shut rapidly one senses a springy cushioning effect from the enclosed air; if a speaker is mounted in a totally sealed cabinet the air inside behaves similarly, like a permanently sprung cushion. When air is subjected to the sort of alternating pressures found behind a loudspeaker cone in a small cabinet its compliance or stiffness remains far more constant than that achieved with mechanical cone surrounds or rear suspension systems. An enclosed volume of air, then, is the thing to use for suspension of a speaker cone which has to undergo large deflections, and as the latter are still necessary – indeed even more so – when the 'infinite' part of a baffle takes the form of a closed box rather than a house wall, the small so-called infinite baffle cabinet (IB) automatically solves its own problems. The cone still has to be mounted on a frame and sealed around the edge, otherwise air would simply be pumped back and forth rather than alternatively compressed and rarefield; but this sealing is made so compliant that it is very much less stiff than the enclosed air, the air itself providing most of the spring – hence *acoustic* suspension. (In fact there has to be just a tiny amount of air leakage to accommodate long-term changes in atmospheric pressure, otherwise the speaker behaves like an aneroid barometer!)

It was explained earlier that mechanical and acoustical resonances arise from the presence together of mass and compliance, and as the enclosed air stiffness in an IB cabinet is greater than that contributed by the mechanical cone suspension, the total effective compliance 'seen' by the cone mass is much lower than when the drive unit is unmounted. This pushes the fundamental cone resonance up the frequency scale, and it becomes necessary to avoid *too much* air stiffness if bass performance is to be maintained when using very small cabinets. It is easier to depress the surface of a toy rubber balloon at one point with a finger than over a larger area with a hand; likewise, with a given enclosure volume a small diameter cone will find the air less stiff than a larger cone. The popular miniature IB speaker systems therefore use very small bass drive units to achieve a correct admixture of mechanical and acoustic compliance, with appropriate 'long throw' moving-coils to provide the necessary large amplitudes of diaphragm movement. A long coil means that only a fraction of the winding is within the magnetic gap, current through the remainder not contributing any driving force. The small IB is therefore rather inefficient as a transducer, requiring more amplifier power than other types for creation of a given sound level. By including a generous quantity of absorbent fibrous material within the cabinet air space (see **fig. 44**) the main resonance is damped down to an acceptable level and the useful response extended somewhat below the resonant frequency.

How extended can or should loudspeaker bass response be? We shall see in later chapters that speakers cannot really be judged satisfactorily away from their proposed environment, which has a particularly strong influence at low frequencies. In general, it may be said that overall size determines how far into the deep bass a speaker's output will go. Whatever may be claimed in manufacturer's literature, very few have a useful output below 40–50 Hz, and most smaller speakers start to fall away in the 70–100 Hz region. The listener's ear does not necessarily register this as a lack of bass in music, partly due to subjective bass arising from harmonics, and also because room and speaker resonances often seem to make up for what may be missing in the bottom octave. It is usually possible to find commercial speakers that will do justice, from the viewpoint of bass response, to everything except deep organ pedals and the bass drum, and even with these a response extending down to 40 Hz can create a pleasing enough impression for most people – again due to the ear's knack of adding the missing fundamentals.

So far we have concentrated on low frequencies, though high frequencies do not look after themselves but raise different and equally recalcitrant problems. A speaker cone does not move like a simple rigid piston all the time, but as the frequency is raised starts to break up into separate vibrating sections, an effect known as *cone breakup*; this leads to a ragged response. Also, the inertia contributed by cone mass inhibits rapid vibrations, and HF output therefore tends to fall away. Various techniques are used to overcome these limitations: diaphragm stiffening through use of metal or sandwiched plastic foam; diaphragm lightening by enabling the moving-coil to drive a small inner cone at high frequencies; controlled and properly graduated compliance across the cone from coil to edge, confining HF to the inner parts and thus reducing the mass to be moved as the frequency rises; or use of entirely separate HF *tweeters* to take over from the bass unit (*woofer*) at a suitable frequency. The last technique is most common, and although involving extra drive units it is not necessarily more expensive as the separate units have only to handle a limited range of frequencies and therefore do not need to be so elaborate. In very sophisticated speakers there may be a third drive-unit designed to handle the mid-range, though simple multiplication of units for the sake of sheer quantity should be treated with suspicion, especially if a great point is made of the quantity in advertisements.

In two or three-way speaker systems it is not possible simply to connect the amplifier output to the moving-coil terminals of each unit, as the signal first has to be divided into suitable frequency bands. This division takes place in a special multiple filter included within the complete speaker and called a *crossover* (not to be confused with crossover distortion in amplifiers). The frequency at which the signal is split between units is called the crossover frequency, usually somewhere between 500 Hz and 3 kHz.

Most woofer units are essentially variations on the simple moving-coil arrangement so far taken as our reference, with differing diameters, diaphragm thicknesses and shapes. Tweeters, however, usually look rather different, even though most use the same basic moving-coil drive. Cones are abandoned in favour of small domed diaphragms (see **fig. 46**), and sometimes a short horn is used to provide acoustic loading. The *ribbon* tweeter is an interesting variation on the moving-coil theme, with a single strip of aluminium alloy ribbon carrying the audio currents instead of a coil, the ribbon running between two parallel magnetic poles and facing into a horn. A type of sound generator without any moving parts is the *ionophone* or ionic tweeter. In this a continuous electric (spark) discharge is modulated by the audio signal and sound is generated literally out of thin air. Absence of mechanical elements permits perfect transient performance, as there is nothing to resonate or 'ring' apart possibly from an incorrectly designed horn coupled to the ionic chamber. Also with very little material to move independently of driving forces is the *electrostatic* speaker (ELS). In this, a coated light plastic diaphragm is stretched on a frame and positioned, sandwich-like, between perforated sheets of metal, movements being induced by electrical charges. The audio output from the amplifier is stepped up to a suitably high voltage by a transformer and applied, together with a DC polarising voltage analogous to the permanent magnet in a moving-coil speaker, to the diaphragm assembly. The diaphragm is driven evenly over its whole area and cannot therefore suffer from the breakup problems which plague conventional cones, and the mass is so low that HF response is fully maintained. The electrostatic principle is applicable, not only to tweeters but to the whole audio range, and one or two full-range ELS systems have been produced. These are large flat structures, particularly noted for purity of sound quality and freedom from boxy colorations.

The overall sound quality of a complete speaker system depends on the manner in which the individual units are matched together, small differences in efficiency influencing tonal balance. Because of variations in taste and room acoustics some models are fitted with controls permitting individual adjustment of level from mid-range and treble units. Apparent tonal balance can also be upset by minor resonances, sudden changes of directional characteristic,

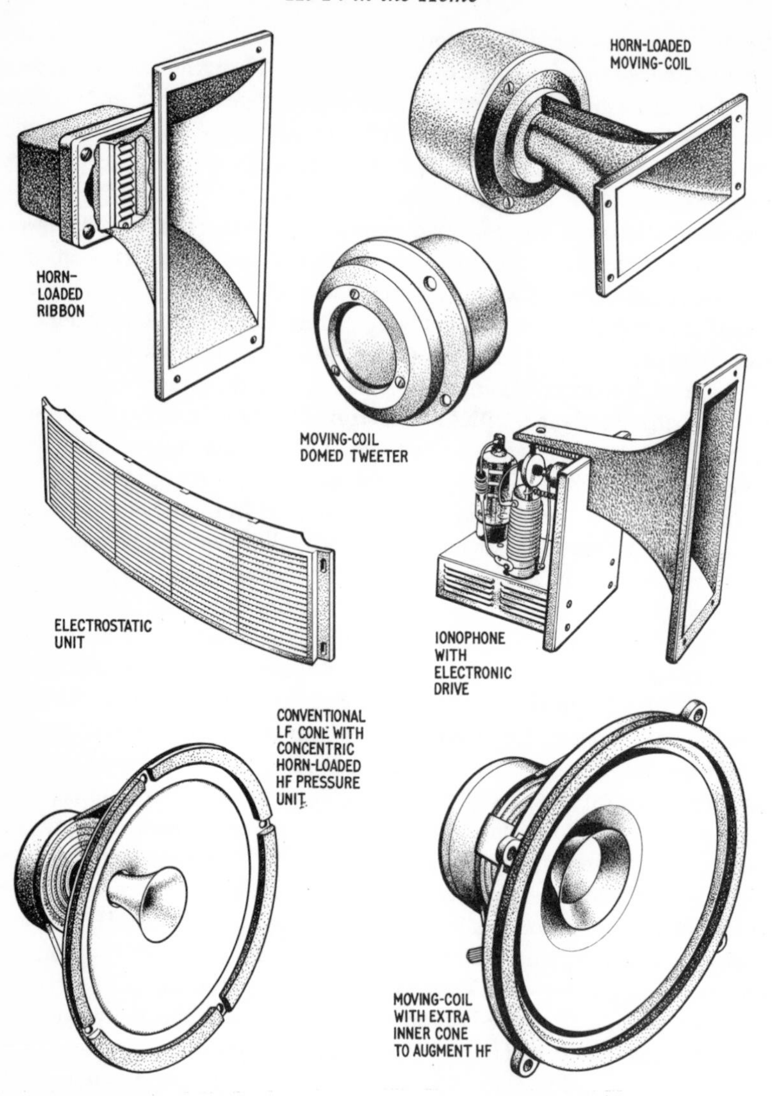

Fig. 46 Various high frequency speaker systems. Horn is shown cut away on ribbon unit to reveal corrugated ribbon stretched between magnet poles. Diaphragm in electrostatic unit is metalised plastic suspended in a wire grid structure. Ionic unit includes electronic circuit to create spark discharge which generates sound in throat of a horn (illustration based on commercial type called 'Ionofane' made by Fane Acoustics). Other units all employ basic moving-coil drive, with various diaphragms, some horn-loaded.

poor transient response, bad placing in relation to listener and/or room, and a host of minor points in the performance of drive units. Also important is an effect whereby sound diffracts around the cabinet more easily at low frequencies, leading to changes in effective sound output level at points in the response related to cabinet size.

The main LF resonance of the system must obviously be properly damped for good transients and freedom from coloration in the bass, and in addition to the various types of cabinet mounting an important factor here is electrical damping. It was stated earlier in connection with **fig. 45** that the rise of electrical impedance in a moving-coil speaker at its main resonance is due to excessive cone motion. The electromagnetic factors determining this also ordain that if the impedance 'seen' by the speaker when looking back at the amplifier output is low in relation to its own impedance, then the LF resonance is brought under better control. In these conditions what the moving-coil (and hence the cone) does mechanically is always in closer accord with what it is told to do electrically by the amplifier. This is why an amplifier output impedance or source impedance is often quoted in terms of damping factor, the higher the ratio of load to source impedance the better the damping.

Closely related to this, and one of the keys to high quality in all moving-coil speakers, is magnetic *flux density*. The stronger the permanent magnetic field against which the coil 'pushes', the stronger the push and the greater the control exerted by amplifier over speaker. Flux refers to imaginary lines of magnetic force passing through the moving-coil via the air gap, and density indicates how many lines there are per unit area, i.e. strength of magnet. This quantity is measured in *gauss* (G), and on all high quality speaker units the figure will be over 10,000 and in many around 14,000; flux densities approaching 20,000 are unusual and only a few outstandingly sensitive units go above this figure. Overall transient performance of a speaker is related to this parameter, but unfortunately it is expensive because strong magnets must be big magnets, and big magnets cost money. They also happen to be heavy, as do well made speaker cabinets, the walls of which must be tough and thick – especially IBs – if further colorations and resonances are to be avoided.

Apart from noting the need for a pair of practically identical speakers, stereo requirements at this tail-end of the hi-fi system must be left for piecemeal consideration in later sections. The angular pattern of sound radiation into the listening room governs all else here, and this will be covered in the next three chapters in terms of: (a) theory; (b) choice of speakers; and (c) practical setting up.

Loudspeakers are far from perfect: proverbially the weakest link in the hi-fi chain, they are at best a cocktail of minor imperfections mixed with a skill which can nevertheless fool the ear into hearing 'through' them to seemingly very realistic musical sounds. Colorations abound, a fact easily observed when a musical signal is switched to a succession of speakers – they will all sound different no matter how good each may be judged by experienced listeners. At the end of a whole book on the subject by an acknowledged expert* we find the following passage: 'Loudspeaker design is amenable to scientific methods, but nowhere nearly so accurately as amplifier design. Compromise and experiment figure very largely and, as the final judge is the human ear, the whole exercise is as much an art as a science . . . and it looks like being so for many years yet.' Despite this, good loudspeakers sound very fine indeed, with a transparency of sound, freedom from boom, absence of shrillness or boxiness and a vividness without artificiality that please the ear and seem to make nonsense of the ragged acoustic response curves and poor transient performance revealed by objective measurements.

* *Loudspeakers in Your Home,* by Ralph West, page 108.

6. MUSIC TO LISTENER OR LISTENER TO MUSIC?

THE ABOVE title poses a rather academic seeming question, perhaps striking the reader as something of a diversion coming as it does between chapters on hi-fi specifications and choice of equipment. The excuse for this interpolation is that one important aspect of high-quality music reproduction has so far received but scant and passing mention. This is stereophony, which should be explained and justified before proceeding with recommendations based on an assumption that monophonic sound is inadequate.

When listening to reproduced music in the home it is obvious that in a basic physical sense music has been brought to the listener, only a visit to the concert hall or opera house taking listener to music. There are also psychological and aesthetic barriers separating these two conditions. However, for the present purpose this chapter's title must be taken metaphorically, arising, as we shall see, from acoustic considerations.

The typical domestic sitting room used for listening to music might accommodate, 'live', the solo sounds of such instruments as guitar, recorder, spinet, flute, mandolin, oboe or a small piano. But even for some of these the acoustic would be too confined for more than quiet practice: the solo voice could give pleasure with lieder, but would not be able comfortably to expand to the levels normal in a recital room; a violin would produce an unflattering sound at any level above mezzo-piano; and a string quartet would be out of the question for all but the most palatial homes. From this it is evident that if by bringing the music to the listener we mean creating the sort of sounds that would be heard if the artists were playing in his room, then the exercise is acoustically untenable for all but the smallest scale solo instrumental or vocal music.

Thus high fidelity is not concerned with bringing orchestras – or even string quartets – into our homes, but with presenting the players or singers performing in acoustic settings appropriate to the music, allowing us to 'listen in' to both the music and the surrounding ambience almost as if we had been transported to the studio. In practice this transportation includes the listening room in which we are seated, and unless we wear headphones the sense of space around and with the music can lie only between and beyond the loud speakers, almost as if the end of the room had been opened up on to the concert hall.

This window-into-the-studio effect can only be achieved really convincingly with stereophonic reproduction, employing two or more signal channels. The reasons for this are various, but may be summarised under two main headings:

(1) the ear needs some spatial distinction between instrumental and vocal sounds on the one hand and the encompassing reverberation and ambience on the other; and (2) a broad 'window' will not offer a natural aural view unless it is obtained without also broadening individual sound sources within the picture. In other words, unless the apparent sound-stage presented to the listener has approximately the same spatial make-up as that offered to the microphones in the recording or transmitting studio, it will be unconvincing as an illusion of the real thing – even though it might still be quite enjoyable as a limited reproduction.

No matter how many microphones may be used and however carefully they may be positioned, the resulting signal in a single-channel audio system is our old friend the lone unilinear waveform carrying all the information about music and ambience alike, without any code enabling us to reconstruct the spatial disposition of one in relation to the other. If the microphones used for such a recording are placed at a sufficient distance from the performers for the multiple reflected sounds which make up the ambience or 'sense of the hall' to be captured at a reasonable level in relation to the direct sound from the instruments, it might well be found that in reproduction the sound is much more clouded or fogged with reverberation than that heard by a listener present at the original performance. Similarly, it is a common experience in tape recording that when a voice is recorded with the microphone at a moderate distance in a room which would not normally be regarded as particularly reverberant, on play-back it almost seems that the recording was made in a bathroom. Single channel (monophonic) reproduction makes the voice and reverberation come out together, whereas in real life our hearing separates them. This is because the real-life listener has aural clues from his two ears – of which he may not be conscious but which nevertheless contribute to his subjective impression – which lay out hall or room before him, allowing them to add their normal quota of ambience without muddling the music or speech by becoming directionally confused therewith.

In single-channel reproduction the instrumental sounds come *with* rather than *within* the studio ambience, and while a satisfying sense of depth or distance may be conveyed, no arrangement of multiple loudspeaker systems and reflected sound (at one time popular with hi-fi enthusiasts in an attempt to recapture some sense of lateral space by artificial means) will regain the original reverberation pattern, and any extra spaciousness that might be achieved by such expedients is gained at the expense of broadened or ill-defined sound images for instruments or voices which should clearly remain small in relation, say, to a full orchestra. Therefore a means must be found to recreate, at least over a reasonable lateral angle, an impression of the spatial arrangement of instruments and voices employed within the studio or concert hall, it following automatically from this that if the microphones are not too close to the performers they will also capture – in natural distribution – the pattern of reflections comprising the acoustic ambience. Such a means is known as stereophony, which, in the happily apt words of B. J. Webb,* seems to present the ambience 'separately from the music, much as a frame is separate from a picture, yet forms with it a single entity'.

The subject of stereo is riddled with confusions and half-truths, some historical in origin and some arising from the use of poor recordings or misuse of equipment. First, then, a little disillusionment. An early misunderstanding which persists concerned the supposed mutual exclusiveness of stereo and hi-fi – the question is sometimes asked: 'Is it hi-fi or stereo?' When stereo records first became available there was a rush to produce suitable playing equipment, much of it distinctly inferior from the hi-fi point of view. Claims were made that stereo so revolutionised sound reproduction as to make the traditional laborious pursuit of lower distortion, wider frequency range, etc, relatively unimportant, thus creating a popular image of stereo as a discovery somehow superseding rather than enhancing hi-fi. There was in fact a grain of truth in this, for the added sense of space detectable even on a cheap record player (using a detachable second speaker for stereo) can, for a while, distract the unaccustomed ear from other limitations – especially if the recordings used employ the two-channel medium in a gimmicky fashion for left/right dramatics with popular light music. But this is all history, for while records of this sort may persist, it is now generally accepted that stereo is just another technical advance on the road to better music in the home – an important advance, certainly, but also a necessary ingredient of hi-fi and not an excuse for neglecting any of the other stringent requirements listed in the two previous chapters.

Another common misconception is that the signals fed to the two loudspeakers in a stereo system are derived, not according to the distribution of sound sources in an imaginary concert hall, but according to frequency. We have seen that in many loudspeaker systems the musical signals are divided into two or more frequency bands in accordance with the efficacy of various drive units, and stereo is sometimes confused with this practice. Those who demonstrate hi-fi equipment are sometimes asked by people previously unacquainted with stereo reproduction which of the two spaced speakers carried the treble and which the bass, the answer that both carry the whole range

* *Stereo for Beginners*, page 7.

throwing the questioner into confusion. We shall see shortly how stereo signals are made up and why, therefore, it is necessary to have the well-balanced performance between the two channels noted in connection with pickups and speakers in previous chapters. There have indeed been commercial attempts to produce a pseudo-stereo effect by dividing a mono signal into two parts with carefully contoured overlapping frequency spectra, the idea being to achieve at least some apparent separation of instrumental groupings. Some commercial mono recordings are 'enhanced' by similar means and issued as stereo versions. However, there is often much ambiguity in the sound, with instruments moving across the space between the loudspeakers according to the note being played, and without any separation of orchestra and ambience.

Also exhibiting this latter limitation is another type of reproduction sometimes mistaken for stereo: use of two loudspeakers with an otherwise single-channel system. Simple addition of an extra speaker to a mono set-up, with the same signal reproduced via both speakers (double-mono), may well give a more pleasing and open sound on orchestral music; but this gain should not be attributed to any stereo effect, which can only arise with two isolated signals all the way from microphones to loudspeakers.

Next on this list of false suppositions, and arising – paradoxically – partly because of that very isolation, is the notion that stereo does no more than split the music into two parts, with instruments or voices simply grouped to come from left or right, with an aural hole-in-the-middle. Previous paragraphs have shown that this is certainly not the object of stereophony, and light music transcribed – sometimes even composed – for this type of 'ping-pong' recording is in most cases probably offered to make some sort of impact on the listener when reproduced via one-piece stereograms. Loudspeaker units in these devices are seldom spaced sufficiently to achieve good lateral analysis of the sound picture for more than one or two people sitting uncomfortably close to the instrument.

The last piece of common stereo misinformation is summed up by advertisements like one declaring: 'suddenly, you're sitting in the middle of the orchestra'. It is difficult with conventional two-channel stereo to convey the impression that one is in the middle of the concert hall, let alone in the middle of the players; a very large open window looking into the studio, yes, but not a perch on the conductor's rostrum. Apart from being a rather over-strenuous situation for most music lovers, the latter would require the listener to sit *between* the two loudspeakers, thus producing certain psycho-acoustic anomalies such as an impression that central parts of the orchestra are in or just above one's head instead of one's head being in the centre of the orchestra.

This statement may raise some eyebrows, and serves as a reminder that the actual process of stereophonic perception has not so far been elucidated. The reader has withstood a lengthy declaration of what stereo is claimed to achieve, followed by five explanations of what stereo is not, and must by now feel entitled to some information on what it is and how it works.

A sense of depth may be conveyed by reverberation recorded with a musical signal, loudness of instruments and voices relative to the ambience providing the ear with clues for the dimension of distance. Whether heard 'live' or from loudspeakers, this characteristic of musical sounds may be observed and appreciated by one-eared listeners, whereas awareness of the lateral disposition of sound sources – other than in a rather vague manner unless accompanied by experimental head moving – depends on the use of two ears, or *binaural* hearing. Any sound source not lying exactly in a plane determined by straight-ahead, overhead and straight-behind must be nearer to one ear than to the other, and it is primarily the resulting small differences between the sonic patterns at our two ears which give us an ability to localise sounds in space. Referring to **fig. 47**, it will be seen that when a sound comes from straight ahead (A) it follows an exactly similar path to each ear, while one from a point modestly to the right (B) has to traverse a longer path to the left ear than to the right, at the same time having to 'bend' or diffract slightly before

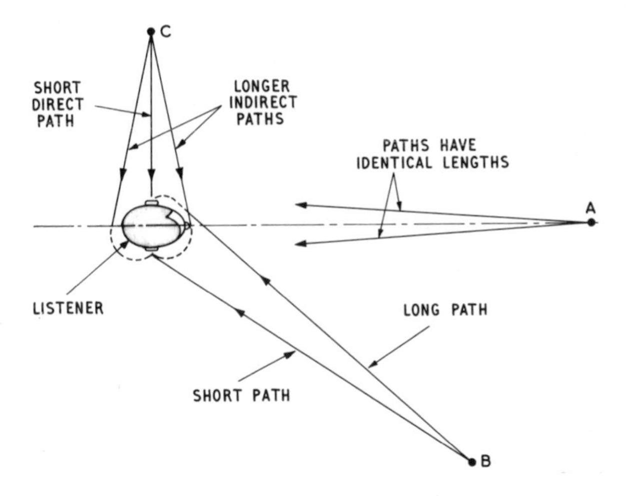

Fig. 47 Distances from the two ears to a sound source are identical only for sounds coming from straight ahead (A) or from somewhere in a central plane passing through the head and 'A'. Left/right differences around the head vary according to direction of sound.

entering the more distant ear. In the extreme case of a sound coming purely from the left (C) there is the maximum possible left/right differential coupled with considerable diffraction to reach the further ear. Musical sounds are richly laden with small changes of physical structure, not only from note to note but sometimes even from cycle to cycle; every one of these changes may be likened to a small transient, each of which will arrive at one ear before the other if emanating from a point away from the listener's centre-line. Apart from these arrival-time differences, known as interaural time delays, the head tends to cast an acoustic shadow whose severity increases with frequency, and as sound is diffracted around the head to reach the far ear it becomes attenuated roughly in proportion to the angle of 'bend' and – above about 1 kHz – to the frequency. Thus the ears are fed with a continuous stream of differential data relating to time, amplitude and frequency spectrum, from which, automatically and without thought, the aural mechanism throws up into consciousness an awareness of direction.

The object of stereophony is to create a similar awareness of direction, though it must do this by a process of subterfuge in order to escape, subjectively, the objective fact that all the sounds heard come from two loudspeakers only. How is the object achieved and what is the subterfuge? **Fig. 48** shows a listener situated before two loudspeakers, and it is instructive to consider what

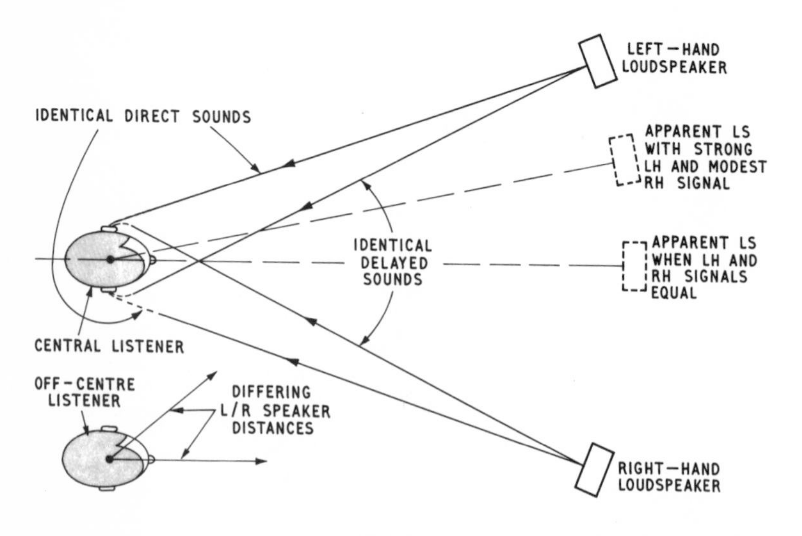

Fig. 48 Derivation of apparent sound sources between a pair of loudspeakers.

happens at his two ears as various sounds are emitted. If a particular sound comes from the left-hand speaker but not from the right the hearing mechanism responds in the normal way and the source is 'localised' in the speaker; likewise in reverse when the signal is fed to the right-hand speaker. Now, if the same signal is fed to both speakers simultaneously each ear will receive first a direct sound from its nearer speaker and then a slightly delayed sound with some high frequency attenuation from the further one; but *both ears are treated identically,* which is what happens when a single sound comes from straight ahead, and in this situation the listener judges the sound source to be out in front and quite detached from the two speakers. Thus by switching a given signal from left-only via both to right-only we can cause the apparent sound source to move between three distinct directions, and it might occur to the reader that by apportioning a signal such that it is neither confined to one speaker nor equally divided between the two, one could create an impression of sounds coming from other points across the space between the extremes. This is indeed possible, and is the basis of stereo reproduction: referring again to **fig. 48**, if a signal is fed at full strength into the left-hand speaker and at a modest level into its opposite number, our hypothetical centrally placed listener will 'locate' the sound source at some intermediate position between left and centre.

For purposes of explanation we have fed the two speakers with suitable fractions of a monophonic signal, though with practical stereophony – sometimes involving a multitude of instrumental or vocal sound sources – the left and right channels handle separate complex musical waveforms, always intricately interrelated but changing one against the other from instant to instant. Just as every nuance of sound from a full orchestra can be carried as minute undulations on a single waveform, so the tiny differences of amplitude, time and shape between two versions of those same undulations can register in the listener's mind as awareness of voices and musical instruments spread out in space. Pursuing this analogy, high-quality single-channel reproduction preserves sufficient detail in a unilinear waveform for the musical listener to pick out the characteristic sound of a particular instrument from a tutti orchestral passage; stereo takes this further by preserving enough small differences between two such waveforms for the listener then to give the instrument an approximate spatial location within the orchestra. Likewise, just as it is possible with good mono not only to isolate, aurally, one instrumental sound, but also to hear many separate instruments or groups of instruments playing simultaneously, so good stereo enables the listener to witness without confusion all the major sound sources within the orchestral array variously positioned before the conductor at the same moment in time. Thus it follows that in stereo listening the two 'apparent' loudspeakers shown in **fig. 48** could be joined by very many others, and that at times a whole multitude of instrumental sound sources would appear to be displayed across the space between and beyond the two real loudspeakers, the direction of each source being dictated by some small difference between the waveforms in the left and right channels of a stereo recording or transmission.

Actually, this last statement is not quite correct, for the direction of those sounds coming from exactly mid-way between the two speakers – or apparently in or above one's head if one chooses that curious conductor's rostrum position by sitting half-way between the speakers rather than forming a triangle with them – is determined not by a difference between the left and right signals, but by their exact similarity. Our discussion of the stereo illusion started with reference to a special double-mono cases of this sort, and indeed it follows from this that whenever a central sound image occurs in a stereo 'picture' there must be a waveform pattern common to both channels. This accords with the explanation of stereo record groove and pickup behaviour given in Chapter 4, where it was noted that whenever both groove walls carry the same modulations (lateral only) the resulting sound image is central and is literally created by a sort of double-mono technique. For this reason a mono disc played on a system otherwise adjusted for reproducing stereo records should, ideally, produce a single central sound image; as we shall see later, this fact can be of use when setting up equipment in the home.

The importance of two identical channels with good electrical isolation has been mentioned before, and we saw particularly that the best pickups are designed to maintain a close balance between left and right while at the same time minimising crosstalk between the two channels. Arising from the identity of the two signals whenever an image is to be created at the centre of the sound-stage, it follows that the differences between left and right waveforms become progressively more subtle and minute as one moves from signals representing sound sources at the extremes to those associated with modestly-spaced sounds near the centre. The more subtle the differences, the more easily are they confused or obscured by leakage of signal from one channel to the other or by fortuitous changes of channel balance as the frequency varies. Also, any appreciable overall bias of amplitude or sensitivity in favour of left or right will move the whole sound-stage to one side, but in a differential manner affecting the near-central sounds more radically than those from the sides. Similarly, a drastic diminution of channel separation (large amount of crosstalk) at a particular frequency would mean that any instrumental sound with a large energy content at that frequency would seem to come from near the centre regardless of where the conductor had placed the player in relation to his colleagues or – in the extreme case – regardless of where the instrument might appear to be when producing notes not near the offending frequency.

But these are rather remote contingencies unlikely to arise with good equipment properly used, a far more relevant point being the manner in which actual sounds are radiated into the listening room from the loudspeakers, and the extent to which the listener's position influences the subjective stereo sound picture apparently located between the two objective sources. In order to explain the process of stereo perception we have so far assumed, as in **fig. 48**, that the listener sits exactly on a centre-line bisecting the space in front of the two loudspeakers, thus equalising sound intensities and interaural time delays so that identical left/right signals give the precise central image which is the starting point of stereophony. But of course if listening to music in the home required solitary confinement with one's head in a clamp this book would never have been written, so it will be expedient to spend a few paragraphs explaining how and why stereo listening may in practice offer reasonable geometrical freedom. Referring again to **fig. 48**, if the listener is positioned well off to one side he is automatically nearer to one loudspeaker than the other, and sounds emitted from both units at the same instant will arrive at different times. This tends to upset stereo perception, the more so as only the slightest movement away from the centre-line introduces a larger interaural time difference than can ever occur between the two ears in 'live' listening (approximately 0·6 of a millisecond, the sound-time distance around the head from one ear to the other). Thus the aural mechanism is robbed of some potent information and succumbs to what is known as the *precedence effect* whereby differences in arrival-time of otherwise similar sounds register as differences in loudness, the whole sound as heard tending to come from whichever source has temporal precedence.

We make unconscious use of this effect when conducting a conversation with someone at the far end of an acoustically-lively room. In that situation the total loudness of the other person's speech depends very largely on multiple reflections within the room, yet the voice still seems to come from the one direction. This is because as the speech pattern changes from instant to instant each fresh wavefront arrives at the listener's ears from that direction *first,* however much the total loudness may depend on all the other sound energy arriving later by reflection. In this natural situation the aural localising faculty is *aided* by the precedence effect, but in the unnatural off-centre stereo situation it is stereo ambiguity that is aided, the ears being confused by the arrival-time information trying to register, say, as 'right' while the other data says 'centre' or 'slightly left'. The impression that eventually enters consciousness is of a sound image shifted towards the nearer speaker, somewhat broadened or blurred, and leaving a slight hole-in-the-middle, however precise it would be if the listener were in the 'stereo seat'.

How, then, do we get away from that special seat? One approach, widely known in the U.K. as the 'Hugh Brittain method',* makes use of loudspeaker directional effects. Just as the false arrival-time data tend to shift sound images towards the nearer speaker by converting time into loudness, so we may fool the ears by presenting them with a sort of reverse loudness tending to hold the stereo image in its original position. This is achieved by so positioning or angling the speakers that the more intense sound comes from the further unit, employing such 'beaming' as may be inherent in the design. Most speakers have some tendency to concentrate sound in a forward direction at

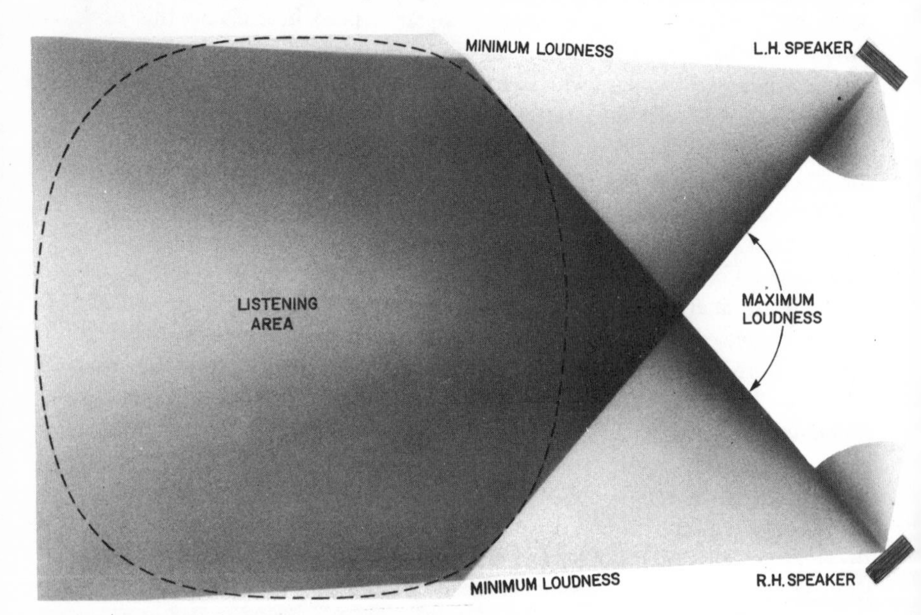

Fig. 49 Using polar response of loudspeakers to improve stereo perception for listeners away from the centre line. A listener moving to the left enters a region of greater sound intensity from the right-hand speaker, and vice versa.

* 'Two-channel Stereophonic Sound Systems', by F. H. Brittain and D. M. Leakey, *Wireless World*, May and July 1956.

right-angles to the cabinet front, the resulting beam usually narrowing at high frequencies. By angling the speakers to give their maximum sound level to the furthest listeners as in **fig. 49**, there is some automatic compensation for the unnatural anomaly of widely staggered arrival-times. A listener sitting on the extreme left would first receive, in the classic double-mono test case, a signal of modest amplitude (light shading) indicating a left-hand sound source, followed by a further signal from the right whose much higher intensity (dark shading) compensates for the time precedence that would normally accord greater loudness to sounds from the nearer speaker. The net effect is to shift the subjective sound image back to the centre. At the centre there is no problem, as the ideal equal-loudness, identical arrival-time situation still holds, while to the right the above argument applies in mirror-image.

This is all rather idealised, as no speaker yet made has an exactly correct radiation pattern, and individual reactions to loudness/time compensation seem to vary. Also, at frequencies above 3–4 kHz the ear is sensitive enough to differences in left/right frequency spectra for such information in this region to actuate the localising faculty despite the time anomaly, and speaker systems combining a loudness pattern of the **fig. 49** type at middle frequencies with very even and symmetrical coverage of the audience above 3–4 kHz will in practice give a sufficiently stable and unambiguous stereo image over a wide enough listening areas to satisfy a large majority of music lovers.* Partly because of the highly subjective nature of stereo perception, but mainly because of an aesthetic preference for facing speakers straight down the listening room rather than angling them for optimum stereo performance, the reader may find occasional reference to the derived centre channel mentioned previously. This is an attempt to compensate for vagueness of central images by feeding the more nearly common waveform components from the left and right channels via a third amplifier to a centrally placed loudspeaker. Also, the matter of loudspeaker types and placings can be further confused by debates between proponents of directional and omni-directional speakers, the latter depending very much for their stereo performance on listening-room acoustics. However, all these further points relating to speaker behaviour are essentially practical matters which will fall into place in the next two chapters, it remaining for us here to look first at the studio techniques employed in producing the raw material for our two channels, and then to examine a few purely musical justifications for stereophony.

* 'Loudspeakers for Stereo' and 'Stabilising Stereo Images', by Joseph Enock, *Hi-Fi News*, January 1964 and July 1967.

Although commercial stereo recordings were not introduced to the U.K. until 1955 (tape) and 1958 (disc), with not even a limited regular two-channel radio service until 1966, the world's very first public stereo experiment took place in 1881, a mere five years after the invention of the telephone. This was at the Paris Opera, where an engineer, Clément Ader, positioned two groups of telephone microphones or 'transmitters' on the stage to left and right, these being connected by telephone line to subscribers who listened binaurally by means of headphones to whatever operatic or dramatic production was being enacted. There were a few other investigations along these lines before World War I, followed in the 1920s by several attempts to use pairs of radio transmitters. However, the first really thorough experiments took place in the early 1930s, and nearly all modern stereo recording techniques can be related to the theory and practice of microphone placings, signal mixing and disc cutting evolved at that time.

In Britain, A. D. Blumlein of the Columbia Graphophone Company demonstrated in 1930 a comprehensive stereo recording system covering everything from microphone arrangement to the cutting of a single record groove carrying the left/right information on its two walls, and had it not been for severe practical limitations imposed by the then standard 78 r.p.m. shellac disc we might have had stereo records twenty years earlier. Blumlein had a fully worked-out theory of stereo perception whereby the interaural time differences which we experience in real life are simulated – for loudspeaker listening – by means of inter-channel amplitude differences in the recorded signals. To avoid ambiguity or confusion his system demands that microphone placings give, simply, inequalities of amplitude between left and right signals and *no* differences in arrival-time. The only way in which this can be achieved is by placing the two microphones at the same point in space, using their directional sensitivity to change the signal amplitudes according to sound source disposition. In practice of course objects cannot be both separate entities and occupy the same position, though by placing two small microphones one directly above the other they may be regarded as coincident in relation to lateral sounds. Simple stereo recording employing the Blumlein method therefore uses a single *crossed-pair* or *in-line* coincident microphone placed centrally in front of the performers as in **fig. 50**. Each of the two microphone capsules is arranged to have maximum sensitivity towards one boundary of the proposed sound stage, so that with the orchestral layout shown in the diagram the left-hand microphone is most sensitive to sounds coming from the violins and least sensitive towards the double basses, and vice versa for its partner. Being centrally placed, the violas – or the conductor if he shouts at the players

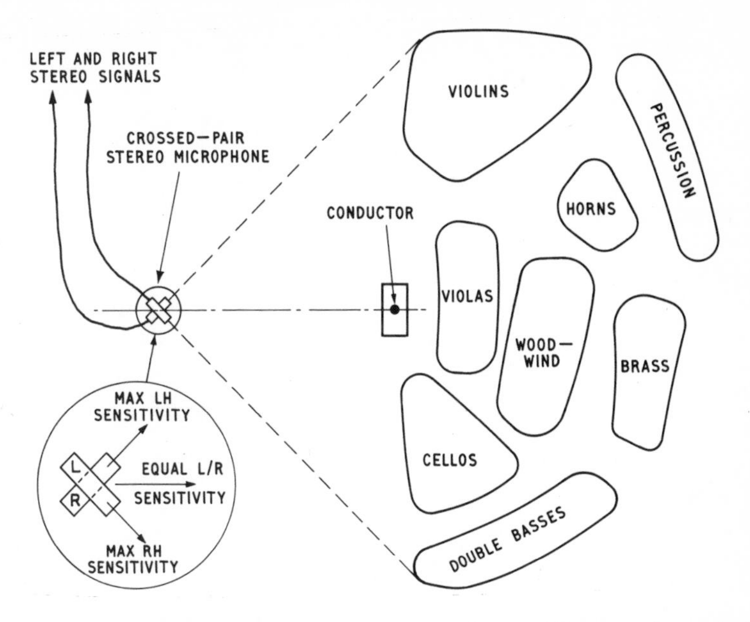

Fig. 50 Using a coincident crossed-pair microphone for recording an orchestra in the manner of Blumlein.

– will actuate both capsules equally, giving our familar double-mono signal.

A quite different approach to stereo was evolved in the U.S.A. a year or so after Blumlein's work in Britain. Starting with microphones positioned as if they were ears on a dummy head (a technique found to be excellent when the two signals thus obtained were heard via headphones – but unsuccessful for reproduction by loudspeakers), engineers at the Bell Telephone Laboratories eventually adopted what is known as the 'curtain of sound' concept. This argues, quite simply, that if many microphones are placed in front of the performers and a corresponding set of loudspeakers with the same layout is presented to an audience, with each microphone feeding its equivalent speaker, the original sound pattern passing the 'curtain' of microphones will be re-created to emerge from the curtain of speakers (**fig. 51**). Ideally there should be an infinite number of channels, though one object of the Bell experiments was to ascertain how few circuits are needed for satisfactory results. Using three channels, a number of anomalies were encountered but overcome sufficiently to justify, in 1933, first using the Philadelphia Orchestra under Leopold Stokowski for relaying a concert from one room to another within the

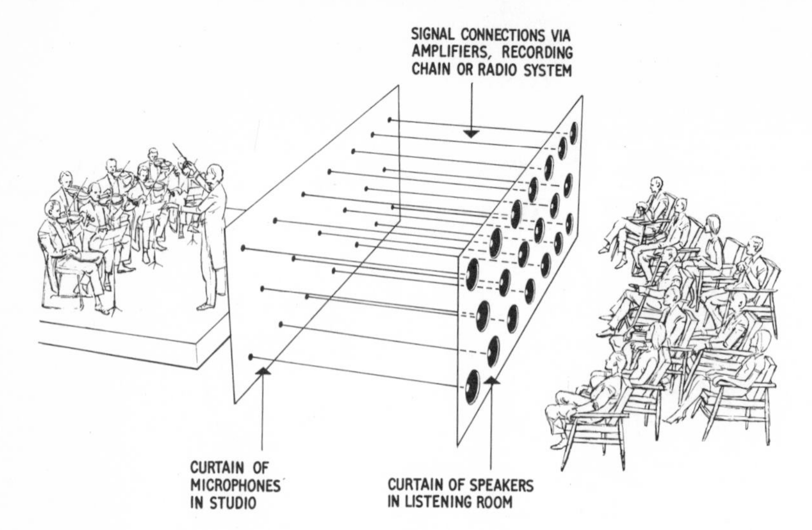

Fig. 51 The 'curtain of sound' recording concept, in which original wavefront from the orchestra is recreated at home.

Philadelphia Academy of Music, and then repeating the event with the orchestra's deputy conductor in charge at Philadelphia and Stokowski balancing the three channels in Constitution Hall, Washington, after the signal had passed through specially corrected telephone lines.

It is interesting, in the light of this, that while it has not so far become practical to offer more than two channels on commercial stereo recordings,* very many American master recordings are made on three-track tape machines, the nominal centre channel subsequently being divided equally between left and right, double-mono fashion, after the overall balance has been set by the producer. Achieving this 'overall balance' is often a very complicated business, especially in the recording of large musical forces where it is common to employ a dozen or more microphones. Each of these may have to be carefully positioned for a convincing ratio of direct to reverberant sound, then matched-in electrically to give both the correct left/right amplitude allocation for spatial position of the instruments or voices which it covers, and the right total amplitude to avoid over-prominence or weakness within the overall musical texture.

The microphone techniques used in recording and broadcasting may follow almost any pattern between the extreme simplicity of a single crossed-pair hung in front of the players as in **fig. 50** to the seventeen instruments used in the Walthamstow Assembly Hall (**fig. 52**) for recording Mahler's 'Symphony

* But see page 256.

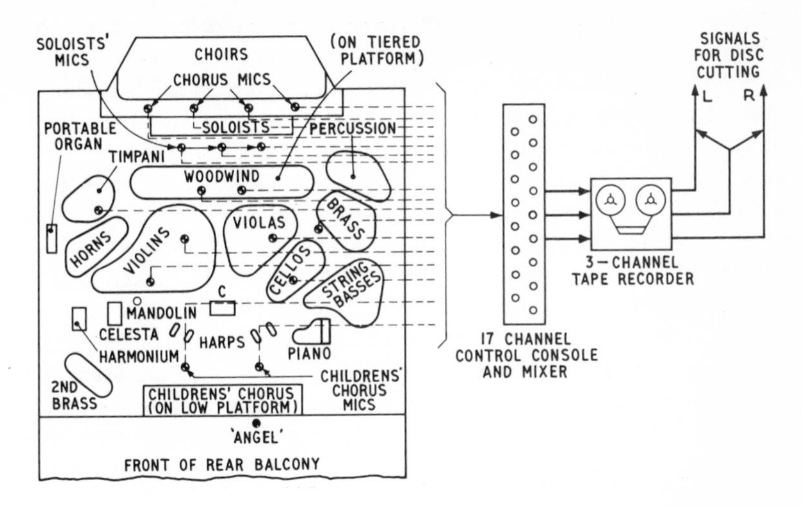

Fig. 52 Layout of performers and microphones used by CBS for recording Mahler's 'Symphony of a Thousand' at the Walthamstow Assembly Hall in 1966. The 17 signals were mixed and allocated to give basic left, right and centre channels for tape recording. Later the centre channel was split equally to left and right (double-mono) for two-channel disc recording. Only a 'portable' organ was used at Walthamstow, the main organ part being recorded separately in Zurich Cathedral and mixed in later.

of a Thousand' in 1966.* The single microphone will, like the human hearing system, pick up the sound pattern present at one point in space and will not therefore produce signals containing aural contradictions which the ears might designate as 'unnatural'. But for all except small-scale chamber music the simple unaided Blumlein microphone usually offers a curiously unstimulating sound, perhaps partly due to other limitations inherent in two-channel stereophony (to be discussed in Chapter 10), and partly because finding the precise position for perfect tonal, instrumental, spatial and reverberant balance for various assemblages of musicians in various halls would use up more valuable time – the musicians all being paid for every practice run – than is lost when setting up a multi-microphone system.

Single microphones, then, are rare, though some of the earliest E.M.I. commercial 'Stereosonic' recordings used a simple crossed-pair, and the BBC is inclined to favour this technique for small groups if the acoustics permit it. Various arrangements of three microphones (left, right and centre) placed out in front of the performers were at one time commonly used by British and

* 'Project-Mahler 8', *Hi-Fi News*, July 1966.

European recording companies for uncomplicated classical works, though refinements of balance these days demand up to eight instruments for recording a normal orchestra, the number mounting further as soloists, choruses, off-stage brass groups, etc., are added. American companies are inclined to use more microphones right at the outset, usually closer to the players and often involving – as a result – a less convincing acoustic ambience or 'bloom' to the eventual sound. Sometimes the master recordings sound too stripped of reverberation to give pleasure and are 'enlivened' with a touch of electronic ambience. The question is largely one of aural house style, though the legitimacy or otherwise of various 'unnatural' techniques in the balancing of commercial recordings is a subject for perennial debate and heated feelings. Let it suffice here to note that however many or few microphones are used, all the signals are eventually mixed, cross-fed and amalgamated to give two only, the differences between them – great or small, in time or amplitude – representing the element of space in whatever music is to be preserved or transmitted for our pleasure.

That pleasure can indeed be enormous, for good studio engineers have an excellent feeling for the subtleties of balance and the right ambient framework. Whatever the special problems in recording or broadcasting a fresh work, one can sense within moments of switching on the natural touch of an accomplished team familiar with a particular setting. The mighty set of Wagner 'Ring' recordings made by Decca in the Sofiensaal at Vienna, and acclaimed throughout the musical world, show few acoustic seams to give away the innumerable microphone channels employed on the giant control desk shown in **fig. 53**. Likewise, the supremely natural and convincing series of E.M.I. orchestral recordings made over the years in London's Kingsway Hall do not suggest, aurally, the widely-spaced microphones – side-to-side and fore-and-aft – used in a typical session (**fig. 54**). In its short history of stereo broadcasting the BBC has also shown that some engineers seem to know the Royal Albert Hall as intimately as the keenest Prom-goer, offering at times some of the most satisfying concert hall sounds ever heard via loudspeakers.

Now, to round off this discourse on stereophony there are a few musical arguments in favour of the technique, which must be listed in a book attempting to present high fidelity as the servant of art.

In discussing the nature of musical sounds in Chapter 1 it was mentioned that acoustic ambience must be regarded as a necessary part of those sounds, there being an appropriate acoustic for each type of music. It is part of the essence of a building's acoustic that it can only be sensed fully by those enveloped within it, providing a subtle three-dimensional frame from all direc-

Fig. 53 Decca's famous recording of Wagner's 'Ring' cycle took place in the Sofiensaal, Vienna. Here we see Georg Solti conducting the Vienna Philharmonic Orchestra during a Götterdämmerung session (top), with engineer Gordon Parry at the mixing console in an adjoining room (bottom). Note the television screens used for intercommunication.

Fig. 54 Many fine E.M.I. recordings have been made in London's Kingsway Hall. Here, Sir John Barbirolli is conducting the New Philharmonia Orchestra in a performance of Mahler's sixth symphony.

tions as a setting for the music, which may itself come from a few specific directions only. We have seen earlier in this chapter that while two-channel stereo cannot recreate a sense of space or ambience in *all* directions, it can offer a very large open window apparently looking into the space containing the musicians, thus permitting at least some differentiation between the 'picture' and its 'frame'. In this sense stereo brings the listener nearer to actual musical sounds as they would be heard if one were really *in* that appropriate hall, church or studio. As many composers have written works with particular types of acoustic – even particular buildings – in mind, we can safely claim that there is some musical point in all this.

Just as an orchestra might be presented within an ambient frame, so individual instruments are presented within or against the orchestra. Size as well as loudness is involved in that part of the musical drama within violin concertos, for instance, in which the massiveness of orchestra is set against the smallness of solo instrument; and a large chorus occupying, aurally, the same small space as a lone singer is obviously not quite what composers have in mind. Likewise, operatic characters are more human and therefore more convincing if heard in natural proportion to each other and to choruses. Another advantage of the ear's enhanced ability, in stereo listening, to locate and concentrate upon a solo instrument is that less inflation of the soloist by means of separate close microphones is necessary in concerto recording. This makes for a more relaxed, concert-hall type of balance.

From the strictly musical angle the most important advantage of stereo is that it can help in the appreciation of complex scores. When listening to music with many contrapuntal strands it is easier to concentrate on a particular instrumental, melodic or harmonic line if those strands are themselves opened up somewhat by spatial displacement. This is not meant to imply that stereo presents music with a coldly analytical inner clarity, just that the texture is more easily probed by those who do consciously analyse, and more readily enjoyed by the average music lover who listens to and absorbs a complex web of sound without really knowing how or why. In the extreme case of multi-part music for string orchestra where the ear is not aided by contrasts in instrumental tone-colour, it is sometimes very difficult to hear one's way through the musical 'argument' without a score unless stereo (or a real orchestra!) lets that little extra aural daylight through the parts. Also, much late Baroque and early Classical music was composed on the assumption that first and second violins would be divided to the left and right, a practice to which many conductors are now returning in order to reveal subtleties in the music which are necessarily lost via single-channel reproduction. Then there are antiphonal effects used in much church music and assuming, for instance, separation of decani and cantoris halves of the choir. Also, many composers make use of spatial contrasts in major and minor ways, quite apart from very specific and bold instructions about the disposition of forces issued, for instance, by Berlioz for his 'Requiem' and Bartok for his 'Music for Strings, Percussion and Celeste'. It is true that for most of the time most composers could have no thought of patterns in space when conceiving their music, yet even the string quartet, that most abstract and refined of musical forms, seems to gain from being presented in stereo – somehow the intimacy of a small group of players is captured more cogently.

But finally, whether or not arguments and speculations convince, the experience of good stereo reproduction usually does. There is a certain openness, freedom and ease which, after a while, makes even the best monophonic sound seem strained and unnatural. One is just that little nearer to the sound of real music-making, which is why stereo is now accepted as an essential element in domestic high fidelity. Likewise, the future may bring us yet nearer to reality by enabling us to simulate the sense of all-round reverberation experienced *in* the concert hall. Current developments in 4-channel stereo promise advances on these lines and are discussed in Chapter 10.

7. HOW TO CHOOSE EQUIPMENT

HAVING ABSORBED the gist of the last three chapters the reader might now feel confident enough to enter a shop, select a few units satisfying hi-fi criteria, take them home, connect up, switch on and relax to musical perfection. That reader would be very lucky, for 'hi-fi criteria' may be satisfied at various levels on the road to perfection – particularly with pickups and loudspeakers – and it is therefore necessary to strike a sensible balance between component parts on a cost/quality basis. Also, interconnections involve technical things like impedance, sensitivity and output levels; and the purchaser simply must hear loudspeakers for himself, however impressive the printed specification or the dealer's enthusiasm. It would be reasonable to claim that music lovers setting out on the road to 'hi-fi in the home' are more likely to suffer disappointment and frustration from an unsuitable combination of units than from individual components failing to live up to their price. A large proportion of the perennial stream of letters received in the editorial offices of *Hi-Fi News* arise from attempts to use together items which are essentially incompatible or of widely differing quality; and most of the remainder ask for advice on the choice of equipment.

A rather extreme example of the way in which things can go wrong through choice of one very high quality item without regard to other links in the hi-fi chain appeared as a note in a daily newspaper in 1967: 'The other week I spent a vast sum of money on a tape recorder made by . . . The tape recorder is the "Rolls Royce" of hi-fi, but the sound wasn't as good as it might be. The manufacturers advised me to replace my faithful old radio with a new stereo tuner. "Your recorder is only as good as your radio" they said. That didn't help much, but as the hi-fi expert who installed the tuner said: "Your tuner is only as good as your aerial." So at great expense, two men came yesterday to fit an aerial on the roof of the block of flats where I live. Still my fi wasn't as hi as it might be, but then a friend explained the problem. "Your equipment is only as good as your loudspeakers" he said'. An informed dealer would have dropped a few hints on these lines at the outset, for it is clearly pointless to spend money on stereo recording facilities, a flat frequency response extending beyond 10 kHz, and a signal-to-noise ratio of 50 dB, in order to record mono AM radio programmes received with a restricted frequency range and severe background noise. Likewise, however good the recorded signals when tuner and aerial have been installed, loudspeakers with a constricted bass output and boxy coloration will still sound bass-less and boxy. One could anticipate a further possible disappointment for this particular customer, as he may have assumed that because his tape recorder incorporated a pair of stereo power amplifiers it could be used *as* an amplifier with a pickup/turntable unit for playing disc records. If he chose a high quality magnetic pickup cartridge this would not work, for very few tape recorders have preamplifier type input points with the sensitivity and built-in equalisation appropriate to such a pick-up; and while a few simple circuit changes might facilitate use of a cheaper ceramic cartridge, this could well turn out to be the weak link when he did eventually up-grade the loudspeakers.

Such are the dilemmas likely to arise from hasty purchase of an isolated link in the reproducing chain, suggesting a need for calm and caution. Three main factors should influence choice of equipment: (1) sound sources likely to be needed – disc, radio and tape, the last possibly including 'live' recording if there are instrumentalists in the family; (2) domestic convenience and aesthetics related to room sizes, family habits, furnishings, etc; and (3) how much money is to be spent, bearing in mind the possibility of building up a hi-fi system step by step. The intending purchaser with a large room set aside just for listening to music and proposing to have every refinement regardless of cost will not be troubled by such mundane points, though if the reader be such, pray forgive a presumption that this is unusual. At the other extreme there are many who, having budgeted for hi-fi, still need to progress in easy stages, whether total expenditure is to be £100 or £400.

A few, like the man in the newspaper report, start with a versatile stereo tape recorder, perhaps because of an additional interest in tape recording as a hobby and a belief (or hope) that the two modest detachable speakers often provided with such machines will suffice for the forseeable hi-fi future. Others begin with a tuner-amplifier and a pair of speakers, using radio as the sole sound source until savings permit purchase of turntable and pickup. This is the cheapest scheme initially, but only makes hi-fi sense for those living within good reception distance of an FM transmitter offering regular stereo music programmes. A third group, probably the majority in Britain, spend rather more money at the first stage on pickup, turntable, amplifier and speakers, hoping to add an FM tuner later but depending in the meantime on the household transistor set for radio reception. This has the great advantage of making the user independent of broadcasting schedules and other people's tastes in music, though it does sometimes so strain family finances that only two or three first-class records are bought in as many months, and these tend to be played to death to all and sundry in the cause of high fidelity. With any of these buying schemes the performance and facilities of the first item purchased should be considered with care to minimise problems at Stage 2: signals that might be fed in, or other units that may be connected, must be compatible, and we shall in due course take a look at possible snags on these lines.

Whatever the price, and whether components are bought *en bloc* or in stages, most users will have matters of style, finish, size, position and domestic convenience to contend with. There is also the risk that a key member of the family will have a fit of audio atavism and demand that if money must be spent on reproduced music it should be invested in a good old-fashioned radiogram. One must be careful here, as the best present-day 'stereograms' incorporate many refinements pioneered in hi-fi equipment, and the objection that such instruments are bound to be inferior in every respect cannot always be sustained, though it may still apply to run-of-the-mill models from the big general radio manufacturers. At one time the chief distinction between hi-fi and non-hi-fi was employment of a separate loudspeaker system, though over the years a growing band of audio enthusiasts stimulated manufacturers to produce high-grade components for each link in the chain, making overall quality the special distinction of hi-fi, while producers of conventional radiograms and record players continued to plod along for twenty years without significant change. However, high fidelity has now become a respectable ideal for ordinary music lovers rather than an esoteric technical hobby, and with the appearance of very sophisticated stereograms (only a tiny minority, though) we have come full circle, as component systems once again differ from consoles primarily in the matter of separate speakers.

The original and basic snag with one-piece instruments was the proximity of speaker and pickup and the resulting near-impossibility of avoiding undesirable interaction in the form of *acoustic feedback* when playing at a reasonable sound level. Modern materials and methods for acoustic and mechanical isolation have eased this problem, while the miniature IB speaker principle facilitates hi-fi performance at the output end without the earlier need for large – and therefore external – speaker enclosures. Transistor circuits have removed a difficulty called *microphony,* in which vibrating valve electrodes cause ringing noises in the loudspeaker; and while auto-changers are still fitted, the insidious record grinding process need not be used on one's best discs, which can be played singly. Why, then, opt for a component system? There are three reasons. First, the small number of really sophisticated stereograms or consoles worth hi-fi consideration can easily be outclassed for the same expenditure, or equalled for much less, by careful choice of separate units. Second, with any one-piece device there is the operational absurdity of having to stand practically on top of the loudspeakers in order to adjust volume and tonal balance. Third, speaker spacing on these devices is quite inadequate for

Fig. 55 The social consequences of choosing a stereogram.

most rooms and it is not possible to position or angle speakers one against the other to achieve reasonable stereo over a sensible listening area. This means that the listener must either tolerate a very small sound-stage at the far end of the room – hardly stereo at all, in fact – for some degree of lateral seating freedom, or be confined to a barely movable 'stereo seat' for a broader angular sound spread. Some possible social implications of the latter choice are depicted by Anscomb in **fig. 55**.

Related to the one-piece console idea, and possibly raised in response to the above points as a plea for *mono* radiograms, is the notion that stereo reproduction is not worth attempting in small rooms because it is somehow impossible to break away from a cramped acoustic to the 'big' open sound associated with good stereo. From various points made in the previous chapter the reader may have gathered that this is simply not true. Indeed, the confined situation of a tiny listening room can benefit most dramatically from the 'window-into-the studio' illusion possible with stereo, and if it is objected that the listener cannot reasonably be expected to accommodate *two* loudspeakers in such a room, one can but point to the ultra-compact bookshelf units which will tuck into odd corners quite unobtrusively. Also, when discussing choice of speakers a few pages hence we shall see that in addition to the convenience of smallness, miniatures can sometimes make good *acoustic* sense in miniature surroundings.

In such situations the other components may also be selected with an eye to using odd shelves, window sills, cupboard tops, etc, and most audio manufacturers now offer free-standing versions of amplifiers, tuners and player units for those wishing to house hi-fi in this way; **fig. 56** shows a set-up arranged on these lines. With such 'open plan' systems there could be an em-

Fig. 56 A 'free-standing' hi-fi scheme, with amplifier, tuner, turntable plinth and one of the two stereo speakers neatly positioned on shelf surfaces. With care, interconnecting cables may be tucked back unobtrusively into corners or behind woodwork. Equipment by Rogers.

barrassing number of interconnecting wires and cables to tuck out of sight if all the functional parts covered in earlier chapters were physically separate, though it is now normal practice to amalgamate preamplifier and power amplifiers into a single component (integrated amplifier), in many cases including the tuner also (tuner-amplifier). It is probably wise in expensive installations to keep the tuner separate, as really elaborate equipment is inclined to be rather large, making for somewhat cumbersome units requiring considerable shelf depth for convenient accommodation. Also, possible future changes in radio reception or design improvements in either amplifiers or tuners might be met with greater economy if the initial purchase is of separate units. Users with more space will not need to weigh up such fine points unless adopting a shelf-standing scheme purely for reasons of appearance. Where floor area is available hi-fi equipment is most easily housed in a single cabinet, only the loudspeakers remaining separate. Extra space also releases one from the need to use tiny speakers; though larger models are generally more expensive, reminding us that choice depends on cost as well as accommodation.

Newcomers to hi-fi are often bewildered by the enormous span of prices in domestic audio equipment, ranging between about £80 and £2,000 for a complete installation. The former would be an extremely modest 'budget stereo' set-up providing disc reproduction only, while the latter would have every possible disc, tape and radio facility with reproduced sound quality at the very peak of present-day standards, probably via professional monitoring loudspeakers. A similar situation is found in cars, with a Mini Moke at £400 set against the grandest Mercedes at £10,000. Just as a well-chosen £80 stereo scheme can qualify for a hi-fi label which a stereogram (perhaps costing more) does not justify, so the Moke can be accepted as an automobile while a motorcycle with sidecar is not. At the other end of the scale, the vast majority of motorists could not even consider the Mercedes even though they might well enjoy driving it; the price is above that regarded by most people as reasonable for a house, let alone a car. However, there is an important difference here between transport and high fidelity, for while anyone will grant the significance and desirability of a completely silent engine, perfect springing and colossal acceleration, it takes time and listening experience on a wide range of musical material to appreciate the increasingly subtle improvements in the higher hi-fi price regions. A rather arbitrary assessment of the situation is offered in **fig. 57**, where it will be seen that a quarter of the possible maximum expenditure secures 90 per cent of the maximum performance. Not everyone would agree with the precise shape of this curve, which does not claim to be more than a personal assessment on the basis of overall sound quality achievable at various prices, but it should underline the important fact that the law of diminishing returns operates with a vengeance in the realm of high fidelity.

The automobile analogy is also included in the illustration, and the reader will see that in some respects it is not far-fetched. In any field there is a point above which a practical man of middle income does not normally go, and in terms of hi-fi equipment available in the U.K. in 1970 a figure of £500 seems reasonable for this. An equivalent point on the car scale cuts across the £2,500 boundary into a realm of luxury, on the way to Cadillacs and the Rolls-Royce, where it would be argued by many that the very good is being replaced by the unnecessarily good. But one cannot be dogmatic here, for subtleties of perception and size of bank account seem curiously interrelated. However, just as most people end up with a good but unostentatious family car in the £600–

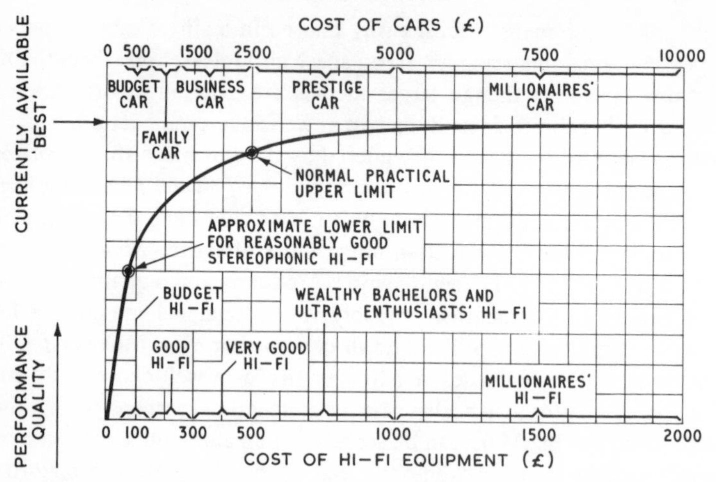

Fig. 57 The law of diminishing returns for hi-fi and cars, based on 1970 prices. Record players and one-piece stereograms are omitted from this scheme for reasons given in the text, though their approximate spans would be, respectively, £10 to £90 and £30 to £400.

£1,100 range, so a similar region of dependability with good performance is found in hi-fi for £140 to £300. It is interesting to note that although the car is now a quite normal family possession, it costs something like five times as much as hi-fi equipment for a similar general degree of perfection. Music loving families might bear this in mind when budgeting for travel and for pleasure.

Having apportioned a fund for hi-fi in the home, there comes the vital matter of choosing the component units, both for individual quality and in relation to each other. The better type of retailer and the technical advice services of audio magazines can be very helpful here, and many of the points made in this chapter should be taken as a guide for possible amendment in the light of the dealer's suggestions, comments of established technical reviewers in the audio press, the reputations of manufacturers, and the reader's particular requirements. Most of the chapter will survey performance and choice of the separate units, commenting as necessary on interconnection and compatibility, and concluding with a brief look at the make-up of some balanced complete systems at various price levels, taking 1970 U.K. prices as a reference. When in doubt on any of the points under discussion the reader may refer back to Chapters 4 and 5 for technical foundations.

Apart from the entirely personal matter of preferences in styling and finish, all links in the hi-fi chain except the tape recorder can be chosen on a basis of performance alone. A radio tuner can be used only for receiving, an amplifier only for controlling and amplifying, a turntable only for turning discs, and so on; but as a tape recorder may be used for recording in addition to being a plain replay device its *function* can be changeable. If recording of live music with good microphones is to be included in the hi-fi repertoire it is important to choose a model with appropriate input facilities, as the microphone socket on most domestic recorders is intended to take the output of a relatively inferior model and will not be sensitive enough for the better moving-coil or ribbon types. Such recording activities might take place away from home, for which purpose the recorder must be portable and, if normally housed with the other units in a cabinet, easily removable therefrom. If the user is quite confident that any 'live' recording can be accomplished with the recorder *in situ* as part of the hi-fi installation, and that otherwise all material to be recorded will be programme signals from radio or disc passing through the system, then the logical choice will be a tape unit rather than a complete recorder.

Detailed description of techniques for recording 'live' via microphones is not within the scope of this book, which is concerned with how to make the best of commercial recordings or radio broadcasts; however, one common misunderstanding should be mentioned in passing. A surprising number of people are unaware that programmes to be recorded may be fed into a tape recorder *electrically,* without need to place a microphone in front of the loudspeaker. It is much more sensible – and from the hi-fi point of view essential – to extract wanted signals as voltage waveforms from an appropriate point in an amplifier or radio circuit, as this avoids recording all the deficiencies and colorations of a loudspeaker and the ambience of the room. Although a speaker system may seem very good, when the sounds it produces are recorded and reproduced a second time all the limitations of this weakest link in the hi-fi chain are doubly emphasised and become very apparent. Most hi-fi amplifiers therefore have outlet sockets providing a feed for tape recorders to facilitate direct recording without microphones, the signal currently passing through the system being intercepted at some suitable juncture before the power amplifier section. This incidental signal tapping point is sometimes known as the *line output*. Similarly, most tape recorders have a line input socket or sockets to take such signals in addition to those from microphones. For replaying purposes signals are fed from a tape unit or recorder to the tape or auxiliary (AUX) input points on an amplifier system. The outputs from tape units are designed specifically for this, but if a complete recorder is to be used the model chosen must have a signal outlet point for this purpose and not simply an extension speaker socket. Also, there must be arrangements for muting the recorder's internal speaker when playing back via the hi-fi system.

These various electrical interconnections should take place at appropriate impedances, signal levels and sensitivities, and generally speaking tape input and output points on amplifiers and the corresponding line output and input points on recorders match satisfactorily in this respect. However, there are some general rules which should be observed, and when choosing a tape recorder – or any other hi-fi unit for that matter – it is wise to check that they are satisfied by the specifications of any units to be coupled together. When feeding a signal from A to B, the amplitude at the output of A (in millivolts or volts) should at least equal the quoted input sensitivity of B, though it should not generally exceed it by a factor of more than about five. Also, the input impedance of B (in ohms, kilohms or megohms) should generally be greater than the output impedance of A by a factor of five or more. The second rule is by no means rigid or universal, and whenever the manufacturer of a transducer, amplifier, recorder or tuner specifies particular matching conditions it should be ascertained that they will be met by the equipment in question.

When in doubt on points of this sort, and especially if the dealer also seems unsure, the items in question should be connected up in the shop and be seen and heard to be working together correctly. If the recording level control on a recorder has to be turned right down to prevent overload of the tape as indicated on the magic-eye or meter, make sure that recorded loud passages do not distort despite a modest reading on the indicator during recording. Conversely, if the control has to be turned right up to achieve a reasonable indicated recording level, check that the resultant tape has no more background noise than a good pre-recorded tape running at the same speed. The earlier condition of too much signal possibly overloading the recorder's input amplifying stage can be cured by a good dealer, who would incorporate a simple signal attenuator made up from a couple of resistors; but the second condition has no simple solution and indicates choice of an alternative combination. The plugs, sockets and cables used for interconnecting equipment follow various patterns, again generally calling for guidance by the retailer. For the reader prepared to do-it-himself at least to the point of connecting up, some practical advice will be offered in the next chapter.

If some of the following notes on tape recorder performance seem rather severe, or entail a more vigorous and fussy approach to the retailer than is the reader's wont, it should be remembered that this item is in many respects the most troublesome and difficult component from the hi-fi point of view. The reader wishing to make a tape recorder the centrepiece of his domestic hi-fi installation – despite the man whose plight opened this chapter – must therefore exercise extreme care when choosing a model. The internal replay amplifiers (assuming stereo, hence the plural) must be available for use independently of the tape transport mechanism, this facility being referred to as *straight through amplifier*. Also, it should be possible to switch off the recorder's driving motors to avoid unnecessary heat, noise and wear when the amplifier is being used in this way with a pickup or radio tuner. Other non-tape performance points to watch on such a machine naturally come under the amplifier heading, to be covered later.

As hinted before, the buyer must be wary of one feature on all tape recorders and tape units, regardless of price: mechanical noise. There are few things more irritating in domestic audio than a background of purring motors and scraping belts when listening to otherwise excellently reproduced music in a quiet room. The 'quiet room' is vital for judgement of such matters, but unfortunately is seldom found in retail shops. This is a problem when choosing any link in the chain, for hi-fi performance is ultimately judged when listening to music, and this should be done in quiet, relaxed surroundings. Never be hurried or bulldozed into purchase of expensive equipment, and if you find a dealer either with a quiet, relaxed listening room, or willing to demonstrate items in your own home, playing music of your choice, at your volume control setting, patronise him. A retailer providing such services may not be able to offer heavily cut prices, or alternatively he may charge for the facilities; but either way, if hi-fi equipment can be judged in domestic or near-domestic surroundings – ideally your own – this is a service worth having.

Returning to tape recorders, it is important when choosing a machine to hear some music of the type most likely to reveal wow and flutter. Piano music with sustained notes or chords provides a searching test for wow; the dying tones of the piano, not normally subject to any interference following the initial transients, can sound distinctly unnatural if wavered by tape speed irregularities. The higher rates of waver called flutter are less easily distinguished as they are not necessarily heard as changes of pitch, but rather add some roughness or uncertainty of tonal character to the sound. Woodwind instruments are generally most revealing here, and the customer should insist on hearing some music with prominent parts for clarinet or flute; the 'bubbly' quality added by flutter may sometimes resemble natural instrumental pitch fluctuations, but the ear soon picks out the difference – especially if recordings are made from discs for direct comparison. The technically wily purchaser will arrive equipped with a test-tape carrying pure, single-frequency tones for assessing performance in these respects, though neither this nor pre-recorded music tapes will cover the matter of wow and flutter completely. This is because a tape machine is used in two stages, to record as well as replay, and any pitch changes introduced during recording will at times be doubled when Stage 2 speed irregularities aid rather than oppose those arising from Stage 1. To cover this, a recorder should be heard doing the complete record/replay job, preferably with signals at the beginning, middle and end of a tape on the largest spools that that machine will accommodate, and at the lowest speed likely to be used for serious music. This takes account of mechanical difficulties that sometimes arise due to inadequate grip on the tape at the pinch-wheel, excessive drag at the spools, generally degraded mechanical performance at low speeds, and other factors.

Signal-to-noise ratio should receive close attention, with an ear open for hum in addition to excessive hiss from tape or circuits. Only the best machines working at $3\frac{3}{4}$ i/s or more will be satisfactory in this respect if orchestral music is to be heard at a reasonable loudness in quiet surroundings. As noted in Chapter 4, some of these 'best machines' have separate tape-heads for the recording and replay processes, the additional monitor head making it possible to hear recorded sound quality an instant after modulation is on the tape. With these one can make what is known as an *A–B comparison* at the touch of a switch, 'A' being the input signal and 'B' the resultant recording. With music of wide dynamic and frequency range this is a searching test from the hi-fi point of view, as the degree of fidelity or infidelity to the original may be judged immediately without reliance on aural memory.

Having satisfied any doubts about overall performance in terms of sound quality – including observation of the recording level indicator to ascertain that the loudest music passages do not distort when set to give an acceptable deflection – it is wise to assess a recorder's general mechanical finish and robustness. Also, convenience in use of the controls should be noted. Following all this, and after taking into account the reputation of various models and the tone of published technical reports, the prices of stereo recorders and tape units suitable for hi-fi installations will be found to lie somewhere in the £120–£200 region (see **fig. 58**). If the machine with most appeal from every point of view happens to be an imported model it would be sensible to check on the service facilities available and the general repute of the manufacturers' agent. Do not always take the retailer's assurances on such matters at face value, for while dealers are favourably disposed towards manufacturers or importers who co-operate readily over guarantees and repairs, it does sometimes happen that the trade is given specially favourable terms for products to be cleared rapidly when a firm is going out of business. A letter to one of the established audio magazines could be useful in obtaining an independent opinion. This of course applies also to other audio components, but with complex electro-

Fig. 58 Some high quality stereo tape units and recorders. Top-left, Chilton 100s; right, Revox A77; middle-left, Ferrograph Series 7; centre, Tandburg 12; middle-right, Akai M10. Bottom-left, Telefunken M 250; right, Bang and Olufsen Beocord 2000. Prices range from £130 to £245 from a basic tape unit to model with full hi-fi amplifier facilities.

mechanical devices like tape recorders it does not follow that goodwill and skill in the dealer's service department can solve all problems – there may be a vital mechanical bit which cannot be duplicated.

Turntables are rather less likely to give mechanical trouble in the sense that bits and pieces may need replacing, though all the possible deficiencies except hum induced into the pickup can be traced to mechanical causes. As with tape recorders, wow and flutter should be judged using recordings of appropriate music – at $33\frac{1}{3}$ r.p.m. particularly, for here again things get worse as the speed is reduced. Audible flutter is far less likely than on a tape recorder, but wow needs close attention. Low frequency rumble is also a perennial problem, though one must be wary here on three accounts: many disc records seem to carry a little LF background noise, the less expensive turntables are inclined to vary from sample to sample, and some loudspeakers in some rooms will be more revealing of rumble than others. A possible approach to this when assessing performance in a shop is to express an interest in deep bass reproduction, when the dealer is likely to oblige by playing organ records via his largest loudspeakers. Having picked speakers giving the deepest and fullest organ pedal tone, and after finding a record free from LF noise when rotated on a very expensive turntable and reproduced through such speakers, ask for a demonstration of the turntable in which you are interested using the same disc, pickup, amplifier and speakers. Also, make sure that the equipment is switched for stereo reproduction, as the pickup will then be sensitive to the vertical vibrations which usually cause most mechanical rumble. If the purchaser anticipates using only modest loudspeakers this whole approach may be unnecessarily ruthless and waste the dealer's time, as small speakers do not reproduce the very low frequency components forming a major part of rumble. A modest turntable may reasonably be used with modest speakers, but when there is a chance that the speakers will eventually be upgraded it may be a false economy to buy a turntable with the slightest sign of audible rumble.

Apart from rumble, there are sometimes knocking noises high enough up the frequency scale to be audible whatever the loudspeaker response, though these usually arise from minor faults on individual units rather than inherently poor design. Acoustic noise radiated by the motor/turntable mechanism must also be assessed with care, again demanding a quiet listening room. There should be no more than a very gentle hissing or 'breathing' noise a couple of feet away from the turntable, and this should be effectively eliminated by a lowered cabinet lid. An irritating feature of some turntables is inadequate torque or turning force to cope with the changing frictional drag imposed by pickup and record cleaning devices without alteration of speed. While a fine speed control may help in this, it is unreasonable that adjustments should be necessary during the playing of a record if the mechanism has had ten minutes or so to warm up before use. This is an important point to watch – and vital for those with perfect pitch – as some units whose performance is otherwise exemplary are troublesome in this respect.

It was mentioned in an earlier chapter that any hum associated with turntables normally arises when coil windings in magnetic pickup cartridges find themselves in an AC magnetic field emanating from the electric motor – the larger the motor, the more powerful the field. To achieve high torque and good speed stability the motor must have adequate turning power, while to achieve a very low magnetic hum field it must be carefully screened and positioned. This raises costs and is one of the many reasons why the very finest turntable units are so expensive (£30–£80). Practical judgement of the not-so-expensive (£15–£20) is easy in this particular respect, for the customer need only ask the dealer to adopt the following procedure. With the proposed pickup cartridge and turntable in use, set the volume control for fairly loud reproduction of a record via loudspeakers with a good bass response. Without alteration of the amplifier controls, raise the pickup just above the record surface and check that switching the turntable off and on makes no difference to background hum with the pickup arm at any normal position across the disc. If it is intended as a first step to use a ceramic cartridge, remember that a magnetic type may reveal hum to which the ceramic will not respond. Particular care should be taken if the proposed cartridge is of the variable reluctance (moving-iron) type, as this is likely to be most sensitive to magnetic hum fields. Note also that some magnetic pickups require the use of non-magnetic turntables. If the dealer seems evasive on this, carry a small magnet with you to see for yourself; it should not attract the platter. If it does, make sure that the proposed cartridge does not incorporate magnets powerful enough to be sensed by a hand-held miniature screwdriver passed near the base.

A few good quality turntables are shown in **fig. 59**, and inclusion of a pickup arm on two of these as an integral part of the assembly brings us to this next component. Because of the need for low frictional drag from the bearings, minimal inertia, ease of accurate alignment, compensation for the side-thrust introduced by arm geometry, and relative insensitivity to jarring forces applied to the player unit, the better pickup arms are characterised by a 'precision engineering' approach to design and finish. Look for accuracy of construction; for evidence that the designer has attempted to minimise the masses of metal parts near the cartridge (a heavy, cast head shell is a bad sign); for an efficient lowering device to obviate the delicate and dangerous business of

Fig. 59 Turntables. Top, Thorens TD 125 with electronic speed control. Left-centre, Goldring GL 75 complete with arm on lidded plinth. Below this, Garrard 401 with stroboscope engraved on rim. Right-centre, Bang & Olufsen Beogram 1000, with arm on plinth. Bottom-right, inexpensive Connoisseur BDI, also supplied as a kit. Prices from £15 to £76.

hand lowering at small playing weights; and for exact and comprehensive setting-up instructions which indicate awareness of the various dynamic and geometrical problems discussed in Chapter 4. Also, ensure that the arm can be positioned correctly in relation to the turntable that you intend to use, and that when so positioned the counterweighting assembly at the rear may be accommodated within the proposed cabinet. The purchaser unable to afford the most expensive arms (about £35) but wishing nevertheless to use a high-grade magnetic cartridge capable of working at a playing weight of less than $1\frac{1}{2}$ grams, would be well advised to discuss the matter carefully with his dealer and possibly write to one of the hi-fi journals for comment on the advisability of other combinations (see **fig. 60**). Arms with tolerable performance but much reduced flexibility and sophistication are available down to around £10.

Choice of the pickup cartridge itself is one of those things dependent to some extent on personal taste, for while quality generally improves with price, the differences between cartridges within a given price grouping are often too subtle for equally careful listeners to agree in their performance ratings. At the budget end of the hi-fi scale the position is further complicated by a tendency for ceramic cartridges (about £5) to vary somewhat in frequency response, crosstalk and tracking ability, both between samples of particular models and with changes of temperature. Because of this the reader may come across vehemently held opinions of opposing polarity on the performance of some ceramic models. The lucky purchaser whose home is not subject to wide temperature variation may reasonably choose a ceramic cartridge if the dealer is prepared to indulge in a selection-by-demonstration process, though the economy effected by avoiding a good magnetic cartridge may well prove false in terms of relaxed musical enjoyment. In many 'budget-stereo' set-ups the price is kept down by fitting a relatively cheap pickup cartridge to the arm in a mass-produced player unit, the two together costing little more than £15, whereas these are risky points in the reproducing chain for application of a cheese-paring philosophy. Unfortunately, choice of a magnetic pickup raises also the necessary minimum cost of the amplifier, because of the need for higher sensitivity coupled with correction for the RIAA recording characteristic. Absolute rock-bottom outlay does therefore dictate a ceramic pickup, though we shall see later that the purchaser willing to tackle some home cabinet work can effect economies making it possible to keep the total hi-fi outlay down to about £90 yet still have a good turntable and magnetic cartridge, the latter in any case starting at under £6 even though the very best will cost six times this price (**fig. 61**).

As already noted, personal taste will come into choice of even the best pickups, and the reader may find this reflected in the recommendations of various dealers or enthusiasts – even magazine editors! I at one time had three highly-regarded cartridges, employing moving-magnet, variable reluctance and moving-coil principles. With prolonged listening to a wide range of music they became characterised, respectively, by quality of effortlessness, clarity and smoothness, each feature desirable and each just sufficiently more noticeable on one cartridge to make it seem to me the most important quality of the three for the time being. Visitors, however, could not say which pickup was in use and needed a carefully presented comparison to hear differences which to me had become fairly obvious. Because of such subtleties it is important when

Fig. 60 Pickup arms. Top, Audio Technica model AT–1005 II; below this, a more modest arm by Neat (G.30). Next, Connoisseur SAU–2 equipped with lowering device. Below this, Ortofon RS 212; and finally the SME, an arm of extreme flexibility. Prices span £12 to £32.

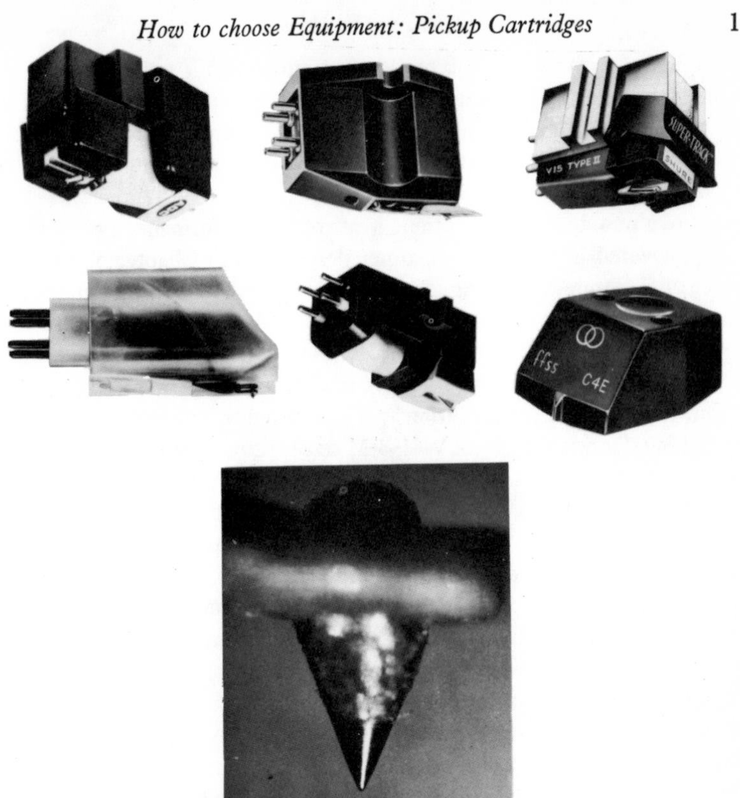

Fig. 61 Pickup cartridges using various transducer principles. Top row, left to right: ADC 10E induced-magnet, Ortofon SL15 moving-coil, Shure V15/II moving-magnet. Second row: Decca Deram ceramic, Goldring 800 'free field' (induced-magnet), Decca C4E moving-iron. These range in price from £5 to £40. Bottom, magnified view of stylus tip of the sort fitted to above cartridges.

judging pickups to note the loudspeakers in use, as an assertive over-bright speaker quality may complement a 'smooth' pickup but sound a little spiky with a 'clear' one, and vice versa. Listen to a wide variety of music when judging pickups, particularly the soprano voice which can so easily sound 'edgy'.

Another point on magnetic cartridges is the output level in relation to amplifier input sensitivity and S/N performance. Most high-grade stereo pickups give 1–2 mV for each cm/sec of groove modulation, resulting in outputs of 5–10 mV from average discs. An amplifier input rating of 4–6 mV is therefore common, normally working at an impedance of 40–100 kilohms (K). However, there is a trend towards pickups with rather lower outputs and amplifiers with higher sensitivity (0·5–1·0 mV/cm/sec and 1–3 mV respectively) and if one of the new cartridges is used with an older amplifier it may happen that when the volume control is set for fairly loud orchestral music the preamplifier background noise is audible during silences. It probably won't happen, but it is worth checking.

We move now to radio reception. Most of the performance points to look for were covered in the course of tuner description in Chapter 5, and in the absence of elaborate measuring techniques or extended investigations of long-distance VHF reception, the purchaser can only note adequacy of performance on local stations. In Britain this means the BBC's FM transmissions, and the good hi-fi dealer will obviously have reception of these in mind for demonstration purposes if local conditions permit. When a retailer's topographical situation in relation to the local VHF/FM transmitter is more-or-less the same as that at the customer's home, it is worth studying the sort of reception he gets. If good, consistent, noise-free quality is obtained when receiving multiplex stereo programmes via an aerial of the type and in the sort of position possible at home, the most important problems have already been solved, it remaining then simply to select a tuner offering adequate performance in the shop with fair confidence (though not absolute certainty!) that results will be as good at home. Be wary if the background noise increases audibly when switching the tuner from mono to stereo during a multiplex transmission, especially if the dealer's aerial is erected prominently on an external pole, as this may mean that local VHF reception conditions are marginal. If a stereo radio service is not yet operating in your area, and elaborate external aerials are needed for satisfactory mono reception even with very expensive tuners, it would be unwise to assume that stereo will be fully usable when it comes. Unless expense is no object the best course in such a case is to buy a good mono FM tuner covered by a firm statement from the manufacturer that a suitable decoder can be added later. Then, as the multiplex service is extended, local experience will reveal whether or not further expenditure is justified.

Assuming no problems of this sort, and assuming also that the user is not an FM reception enthusiast who scans the VHF band to intercept broadcasts from distant places, choice of a tuner will turn on points of practical convenience. It must be decided whether the tuner is to be self-powered from the mains or take its supplies from the associated amplifier, the latter only possible if there are HT and LT outlets with voltage and current capacity sufficient for this purpose. Ease of tuning should be checked, noting definitiveness or otherwise of the tuning indicator, and checking, with the aerial cable removed and a mere piece of wire or screwdriver in its place, that correct tuning on the indicator coincides with minimum background noise and an undistorted signal from a local station. If reception of more distant FM stations is planned it would be wise to check that signals from the local transmitter do not interfere with or distort those from the wanted station when this is not far removed in frequency. Some FM tuners designed just for the British market have three switched positions instead of a continuously variable tuning dial – a simplicity that may appeal to some, though likely to prove frustrating eventually with the expansion of local FM broadcasting.

If tape recording of received FM programmes is intended there is one small possible snag best prevented rather than cured. Whether or not a receiver is fitted with a multiplex decoder, some of the 19 kHz pilot tone accompanying stereo transmissions will be lurking around the audio output signal, possibly accompanied in the case of stereo tuners by some reconstituted 38 kHz sub-carrier. These tones may *beat* with a tape recorder's HF bias signal to produce *difference* tones, and if the latter happen to be within the audio band they are recorded on the tape and reproduce as annoying continuous whistles. Stereo tuners usually incorporate circuits giving drastic attenuation of the interfering tones, but as an extra precaution suitable filter units are available for connection at the input to some tape recorders. If it is possible to record material from the proposed tuner on the proposed recorder in the shop while stereo is being received, this will settle the matter before cheques are signed.

No mention has been made here of AM reception for reasons given in earlier chapters, though of course many tuners combine AM and FM. Prices range widely, from about £20 upwards for a mono FM tuner and £30 up for AM and FM combined; stereo decoding facilities can add from £5 to nearly £20. For reliable FM reception in stereo it is wise to look upon £35–40 as a minimum outlay, remembering, should this shock, that one can pay £400 for an FM tuner! Some representative units are shown in **fig. 62**, a couple of these being tuner-amplifiers, which serve as a reminder that purchasers wanting all the electronic 'works' in one piece will find plenty of good models to choose from. These notes on tuners and forthcoming comments on amplifiers should of course be amalgamated when choosing tuner-amplifiers – sometimes known simply as 'receivers', but not to be confused with traditional radio receivers equipped with built-in loudspeakers.

Although this chapter is concerned with the practicalities of choice rather than the niceties of functional logic, it has for convenience adopted a sequential approach, dealing first with those components designed to deliver tape, disc and radio signals to the amplifier. However, the purchaser making his first foray into hi-fi usually treats the amplifier as the core of his system – as indeed

Fig. 62 Some tuners and tuner-amplifiers. Left, top to bottom: Goodmans Stereomax AM/FM tuner; Philips RH 690 stereo AM/FM tuner; Fisher 160–T FM tuner-amplifier with pre-set station selection; Armstrong 523 AM/FM tuner. Right, top to bottom: Sansui 3000A FM tuner-amplifier; Rogers Ravensbourne II FM tuner; Quad Stereo FM tuner; Leak Stereofetic FM tuner. Price range is £39 to £187.

it is – deciding on this before thinking seriously about other components. With an eye on the technical features described in Chapter 5, there are three main points to consider when buying an amplifier: quality, facilities and power output, all of which affect price. As a general rule a high standard of performance in relation to distortion, noise and frequency response goes hand in hand with increased flexibility and power output, though there are a few stereo amplifiers with excellent technical qualities but offering a low price for the 'budget' man by omitting HF filters and elaborate switching facilities, and with a low power output. While nearly all amplifiers have RADIO and AUX or TAPE inputs, some models in this budget class are designed, on the GRAM side, to take signals from ceramic pickups only. Such inputs usually have a sensitivity of 50–100 mV at an impedance of 1–2 megohms (M), with a flat frequency response suited to a type of pickup which, as we have seen, provides its own (rather approximate) correction for the recording characteristic. Connecting a magnetic pickup to such an input will give a weak, over-bright, bassless sound; weak because such a pickup requires about ten times the amplifier sensitivity, and unbalanced because one is listening to the recorded RIAA response of **fig. 28** (page 77) without replay equalisation. The purchaser should therefore be wary on the matter of pickup input matching, for while it may seem reasonable to avoid unused facilities in the initial hi-fi budget, if the user wishes eventually to upgrade his installation at the extremities – pickup and speakers – this may not be possible at the input without additional expenditure on an extra 'front-end' preamplifier for matching a magnetic cartridge to an amplifier designed for ceramics. A few record player deck units make provision for small plug-in preamplifier units – another approach to later improvements.

Apart from the matter of power output and associated distortion, to be considered shortly, there is little in amplifier sound quality that the purchaser can judge aurally during a brief stay in a shop. When working well within their power capacity, hi-fi amplifiers nearly all sound very much the same on most musical material until one starts using filters and other facilities, though prolonged listening to widely varying music in relaxed home conditions does eventually label some models as more easy-to-live-with than others. A simple test for crossover distortion in Class-B amplifiers is to listen to a man's voice reproduced at *very low volume*; it should not sound 'gravelly'. As with other components, thoroughness of supporting literature, attention to detail in construction and finish, the manufacturer's reputation for reliability and service – all these should be considered. Also, it is reasonable to expect the dealer to demonstrate the amplifier to which one is attracted using the intended pickup and speakers, and the customer should take particular note that the system is free of hum and noise at the volume setting used for loud orchestral music. It is very easy to whip down the volume control between extracts from records, but the purchaser should insist on hearing what happens if the pickup is simply raised with the controls left untouched. Background noise in these circumstances should be absent, or practically so, not with one's ear on the loudspeaker, but as observed from a reasonable listening position. If the dealer cannot satisfy you that any apparent shortcomings here are due to ancillary equipment or interconnection problems, and not to the amplifier itself, then hear another sample of the model in question, and if still not satisfied look for an alternative.

In cheaper stereo amplifiers manufacturers sometimes fail to maintain an accurate balance between the two channels as the volume control setting is changed. This depends on the characteristics of a particular ganged potentiometer, so it must be judged on the actual sample being purchased. Ask the dealer to play a *mono* disc with everything set as if for a *stereo* record. It should then be possible to find a listening position approximately on the centre-line between the loudspeakers where the sound image seems both narrow and central – the double-mono situation discussed in the previous chapter. Rotating the volume control over the sort of range likely to be used in practice should not affect this apparent image position; and if it does there is an unacceptable tracking error on the control. Similar problems can occur with poorly ganged tone controls, particularly at high frequencies, the effect being a vague broadening of the image in a manner varying with programme material. As similar effects can arise with loudspeakers it is very important to remain in that exact 'stereo seat' position when assessing these points in relation to amplifiers.

Various preamplifier facilities may be judged desirable or superfluous according to taste and/or price, though it would be wise not to regard an HF filter as an unnecessary refinement if disc records are to be a major source of programme material. When using modest loudspeakers with a relatively poor performance at high audio frequencies a filter may indeed seldom be needed, as the filtering job is then done willy nilly by the speakers; but if the user should ever decide to invest in better speakers the improved overall response would be likely on occasion to reveal records that need 'taming' in a manner not really possible with a treble control response of the type plotted in **fig. 40** (page 114). Where the budget necessarily rules out a more sophisticated amplifier there is a case for the 'wrong' type of tone control response shown in **fig. 39** (page 113), at least for the treble-cut function. Here, the highest frequencies are cut first, and the −2 setting would usefully attenuate any edgy distortion above 7 kHz without affecting musical brilliance below 4 kHz. In fact the response is more like that of a filter than a true tone control. If in a modestly priced system the combined responses of pickup and speakers produce a tonally balanced sound from most records, without creating an urge to reduce musical brilliance because of a hard and assertive type of quality, then the theoretically inferior treble control may be useful enough: if it is not needed to control brilliance it can at least do a little filtering. In some amplifiers the reverse situation occurs, with a supposed HF filter which fails to give a steep roll-off above a specific frequency and acts, instead, like an additional treble cut control.

When no response curves are printed in the manufacturer's literature it is difficult to judge the effects of these differences without fairly prolonged listening, though if a dealer will permit the customer to operate the controls on a first-class amplifier equipped with comprehensive, calibrated filters, the differences should soon become apparent. An HF filter will reduce the cutting, metallic quality of cymbals, remove the transient fizz from wire brushes, and mellow the subtle feathery edginess of violins playing in their upper register; yet the overall balance, clarity and general brightness of orchestral sound will hardly be affected. Conversely, a genuine treble control will most definitely affect balance and brilliance, boost adding presence and apparent clarity, cut introducing an overall dimness of texture.

Apart from matters of appearance and the mode of accommodation, to be settled according to choice of cabinet housing or free-standing arrangements as noted earlier, this leaves the question of power output and speaker impedance matching. 'How much power will I really need?' is a perennial question from newcomers to hi-fi; an honest answer, unfortunately, begins with the famous tag 'It all depends . . .'. It depends on three things – size and furnishings of room, efficiency of loudspeakers, and personal taste in reproduced volume levels. The easiest way to cover all these variables is to opt for an output of thirty watts (30 W) or more per channel, though this is obviously unreasonable in otherwise modest installations. A capacity of 10 W is sensible for average speakers in average homes, though the user with a listening room in the 30 ft. × 15 ft. category, with fitted carpet, heavy curtains, and speakers of below average efficiency, may at times wish to call on powers in excess of 20W. For the sound powers needed in all but the most exceptional domestic surroundings the electrical power required from the amplifiers depends mainly on speaker efficiency. The quantities of actual acoustic energy transmitted into the air in a normal sized living room when reproducing orchestral music at realistic loudness are absolutely minute, a power level of about half of one acoustic watt representing the very maximum needed during peaks of the

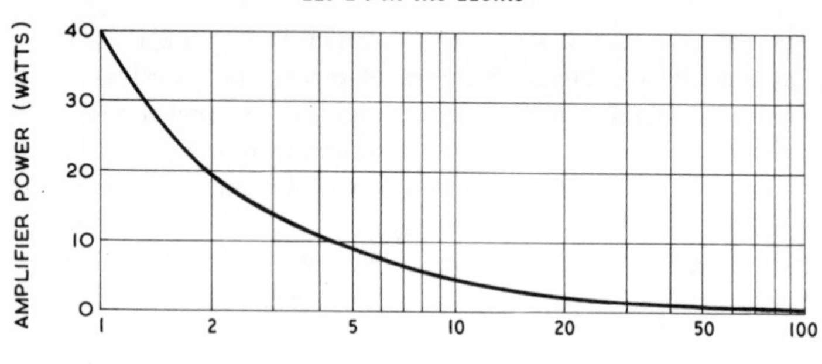

Fig. 63 Approximate working relationship between electro-acoustic efficiency of loudspeaker and necessary amplifier output power for realistic reproduction of orchestral music under domestic conditions.

most enormous climaxes. But most loudspeakers are so inefficient that production of this ½ W of sound output requires about 10 W of electrical input, corresponding to an electro-acoustic efficiency of 5 per cent. A rough graphical guide to the situation, based on an idea by Rex Baldock,* is given in **fig. 63**. Speakers generally sit in the 3–6 per cent efficiency region, though a few horn-loaded types with very high flux drive units have efficiencies of over 30 per cent. At the other extreme, tiny totally enclosed IB systems are often exceptionally inefficient, perhaps descending below 1 per cent. In theory, then, the range of output powers required is very large if any chosen speaker is to use something approaching the amplifier's full-power capacity, though in practice most hi-fi systems employ amplifiers capable of 10–20 W per channel, users automatically adopting volume control settings appropriate to speaker efficiency.

From the power point of view the most important thing to remember when choosing an amplifier is that there must be adequate output to achieve the required sound-level from the proposed loudspeakers, without ever encroaching on to the overload region where distortion becomes audible. That is, without forcing the amplifier into the region of rapidly rising distortion shown, for instance, in **fig. 43** (page 121). Listen to full orchestral music reproduced as loudly as you are ever likely to want – perhaps a little louder – and make sure that during the noisiest climaxes there is no additional roughness of the sort that disappears with a very slight lowering of the volume setting. Also, find some musical passages in which practically the whole orchestra is letting rip molto-fortissimo, preferably with a chorus doing likewise, where a cymbal

* 'Selecting a Speaker', by Rex N. Baldock, *Hi-Fi Year Book*, 1964.

clash and bass drum then add their crowning impact. The sustained sounds behind the cymbal and drum should continue unimpeded, and if there is a momentary blanketing or roughening of the musical backcloth during this additional transient, try reducing the volume just a trifle. If the effect goes it is probably due to the amplifier – a point easily checked by hearing the same passage on the same speakers via a model of higher power rating – but if it remains independently of volume it is probably a recording limitation. When highly efficient horn-loaded speakers are to be used there will be no problem with amplifier overload, though it is a worthwhile precaution to listen to the system in quiet surroundings with the volume control turned right *down*. This will leave just the inherent power amplifier background noise – usually a gentle hiss only audible with the ear fairly near to the speaker, but sometimes verging on obtrusiveness.

Closely related to power output is impedance, amplifiers being designed to deliver their signals into loads satisfying certain electrical requirements. Hi-fi models usually specify a loudspeaker load somewhere between 16 ohms and 4 ohms (the symbol for the ohm is Ω), valve circuits often including several outputs or facilities for altering the matching impedance. Loudspeakers have nominal impedance values (often *very* nominal – see **fig. 45** (page 127)) which should correspond approximately with the preferred amplifier load if the full potential power output is to be available at minimum distortion. With valve amplifiers, full power will only be obtained near the stated load value, and if the speakers have some other impedance it is better if this is higher rather than lower than the nominally correct figure. With transistor models, available power rises approximately in proportion as the load impedance falls, the manufacturer usually stating the tolerable lower load limit. These points should be borne in mind when choosing either amplifiers or speakers, especially if the latter are of rather old manufacture, when they will probably have a nominal impedance of 15 Ω and will therefore not draw full power from transistor amplifiers designed for the 4–8 Ω region. However, developments in transistor circuitry may eventually establish a trend back to higher load values.

A final point on power output in relation to speakers concerns a common misunderstanding* about power ratings. A question often asked is: 'will a 10 watt speaker match a 20 watt amplifier', or something similar. The answer is yes, because a power rating applied to a speaker simply indicates the maximum continuous wattage that the device will safely handle, and while it would be *possible* to overload a '10 W' speaker in such a case, the user wouldn't in practice succeed in damaging the speaker on musical material. In the reverse

* 'Some Common Misunderstandings', by John Crabbe, *Hi-Fi News*, April 1964.

case, where the speaker will handle more than the amplifier can deliver, there will obviously be no risk; but a very high power rating sometimes indicates – euphemistically – that a speaker is particularly inefficient, so it would be wise in such a case to apply the amplifier overload test mentioned earlier.

We are now almost on to choice of speakers as such, but first a brief look at amplifier prices. For the rock bottom stereo budget, with 8 W or less per channel and input facilities for a ceramic cartridge only, a few amplifiers are available at about £20. For under £30 one may obtain 10 W per channel, good overall quality and finish from established manufacturers, and the beginnings of control refinement by addition of a switchable rumble filter to cope with 'budget' turntables; but still usually for ceramics. A move up to £35 brings the sensitivity and equalisation needed for magnetic cartridges, with a switchable HF filter, various input/output facilities, etc. From here on one is in the region of good, genuine high fidelity, with steadily improving performance in relation to distortion, S/N ratio, control facilities and power output. The pattern of amplifier prices will change as models designed to cope with 4-channel stereo come into the picture, but for a few years the selection shown in **fig. 64** may be regarded as typical, spanning £21 to £160.

Now on to loudspeakers (**fig. 65**), the most idiosyncratic of all the elements coming between listener and music, and consequently in greatest need of calm, relaxed demonstration on a wide variety of musical material. Ideally, speakers should be judged in the actual room in which they will be used, preferably over a period of several days. The lucky purchaser may find a dealer willing to operate an approval scheme permitting this, but if such an arrangement is not possible at least choose a retailer with a quiet, domestic-like listening room who will let you hear some opera, full orchestra, chorus, small classical orchestra, string quartet, piano, organ, solo voice, and jazz band, and who is happy for you to come back again several times for similar sessions on the same speakers or in comparison with others. This 'fresh approach' technique is useful and important as the ear quickly adapts to modest coloration; but to come across the sounds from a speaker after a week on holiday, or after hearing a quite different speaker, can be something of a revelation. If a speaker system always has a particular type of character when heard in this way, and even if a few minutes of music makes one forget it – as can happen if the tonal balance is reasonable despite the coloration – be wary. Limitations to which the ear adapts in the short run can sometimes tire over a longer period. But when sampling the sound qualities of various speakers in succession, do insist that the volume control is operated to compensate for differing efficiencies, as a sudden increase or decrease in loudness can affect one's judgement yet be

Fig. 64 Stereo amplifiers with various facilities and power outputs, prices ranging from £21 to £160. Proceeding from top to bottom at left of page: Leak Stereo 30 Plus, Radford SC22 control unit, Armstrong 521, Heathkit TSA 12 (available also as a kit), Grundig SV40M, Rogers Ravensbourne. Top centre, Goodmans Maxamp 30, then down right-hand side: Quad 303 power amplifier and 33 control unit, Trio KA6000, Peak Sound SA/8–8 (also as a kit), Sony TA–1120, Sansui AU777.

Fig. 65 Selection of loudspeakers covering a wide range of sizes and prices. Giant on the left is Acoustech X electrostatic system, supplied in pairs with built-in power amplifiers and associated control unit. To its right is a BBC Monitor, also with its own amplifier, and below this is the Quad domestic electrostatic speaker. At bottom-left, Radford Monitor, and above BBC model is Celestion Ditton 10. Top-right, Goodmans Maxim (with hands to show size), then odd shaped Rectavox Omni. Next, KEF Concord, then Wharfedale Airedale. Price span: £20 to £1,000.

strictly irrelevant. Also, do ensure that programme material used for judging speakers is of the highest quality, as some very good speakers might otherwise sound worse than relatively inferior ones simply because their wide frequency range and freedom from coloration exposes limitations in the signals.

Appearance and finish must accord with personal taste and room furnishing, though size will generally increase with overall quality of performance, as will price, ranging from £15 to about £500! Weight is also a clue to quality, because both large, high flux magnets and tough, well-made cabinets are bound to be heavy. Avoid speakers which sound boomy or thick-textured, excessive resonance often causing all low frequency tones to sound much the same – 'one note bass'. Listen for a clearly defined bass line in orchestral music, crisp transients, and freedom from 'boxiness', nasal qualities or 'tizzy' sounds. Tonal balance should be natural, without either an emphatically bright, over-forward quality or a dim and distant sound. If one has a feeling on orchestral music that a volume setting which is right for average musical texture is somewhat high when upper strings or trumpets dominate the score, with an associated assertive brightness, this probably indicates some peakiness in the 2–4 kHz region, where the ear is most sensitive. It is unlikely that such a speaker would be easy to live with unless fed with consistently under-bright programme material, though such might be the case with some ceramic pickups having a slightly lowered output above 1 kHz.

The extreme high frequencies should be present, contributing a subtle tracery of inner detail to quiet orchestral music and the genuine metal-against-metal sound of cymbals and triangle. But the upper regions should not seem detached or artificially disembodied, a fault more easily detected when listening to a single speaker on mono. A speaker should sound like a single sound source on all types of music and when heard from any likely direction; various types of boxy coloration are also more apparent when speakers are heard without benefit of stereophony. Listen on piano music for separation of the notes: each should be clearly defined, except perhaps in some fast passages with running scales and much pedal work. Normally, as the pianist's fingers go up the keyboard so should the listener's ears, note by note; if it seems that the artist has an 'off day' discover whether this impression is due to his performance or to the loudspeaker. Above all, avoid speakers that impart a similar sound quality to all musical material, regardless of recording company. If after going through a wide selection of records one is not very much aware of differences in recorded balance and ambience, everything sounding equally acceptable but nothing more, then move on to the next pair of speakers.

If the eventual listening room will be fairly large, say over 25 ft. × 12 ft., but choice is restricted by finance to a pair of miniature IB speakers, ensure that these will handle the sound volumes required without distress, buzzes or rattles, especially when operating with some bass lift on the amplifier tone controls, as is often necessary with small speakers employing the acoustic suspension principle. At the other extreme, very small rooms can seldom accommodate large speakers, even though the latter may still give an improved deep bass performance despite a widely-held view that this is not possible in a confined space. One problem with very small rooms – particularly when the main dimensions are all similar – is that the major natural acoustic resonances, called *eigentones,* are high enough in frequency and close enough together to colour bass reproduction almost regardless of speaker positions. For instance, a room measuring 10 × 9 × 8 ft. will have eigentones at 56, 63 and 70 Hz (see **fig. 68**, next chapter), tending to enhance the effective sound level over the whole band from just above 50 Hz up to about 75 Hz. Large full-range speakers in such a room would probably succeed in extending the useful range to 40 Hz or lower, but the 50–75 Hz region would be exaggerated and the sound probably judged to be rather overbearing or 'forced' in the mid-bass region. This is the situation where the small IB comes into its own, for it would in any case normally require some bass boost below 100 Hz, so that with luck – and some experiment with speaker positioning in relation to walls and corners – a tolerably balanced sound may be produced with adequate bass on most music for most of the time. The matter of optimum room dimensions will be covered in the next chapter.

There remains the matter of stereo performance, though judgement of speakers in this respect is extremely difficult without full freedom to experiment with positions and angling, preferably in the actual room to be used for listening. In the next chapter we shall deal with practical setting up for stereo, and as advice on choice of speakers from this angle would necessarily duplicate much of what is said there, the reader is referred forward for detailed notes. One basic decision should be made at an early stage: whether to use (a) speakers in which all the radiating diaphragms face forward into the room, or (b) models in which either all the sound is directed back towards a wall for subsequent reflection, or some of the HF sound at least is diffused and reflected before reaching the listener. A large majority of commercial speakers come into the first group, and with these the only stereo problem concerns how nearly, and with how few aural anomalies, they may approximate in directional behaviour to the ideal discussed in the previous chapter (**fig. 49**, page 144). Speakers in the second group can never give really accurate and unambiguous stereo because there is no distinct pattern of arrival-times to actuate the aural mechanism. On the other hand, models relying on some degree of reflected sound will usually give a broad stereo spread over a much larger fraction of the listening room, with a pleasing quality on choral music or lush types of orchestral sound but an inability, for instance, to place consistently and convincingly the members of a string quartet. Another difficulty with reflected sound is that high-frequency performance depends very much on the surrounding furnishings, both generally in the effect on tonal balance and differentially between left and right speakers. This criticism is partly met in some designs by provision of separate volume controls at the rear of speaker cabinets, giving a range of signal level adjustment on the treble units. This facility is also included on some conventional forward-facing speakers and should be taken into account when judging overall tonal balance. Some otherwise excellent models sound distinctly over-bright when these controls are turned up fully.

In an attempt to escape the need for two speakers there are a few designs with both left- and right-hand units mounted in one cabinet. These depend very much on reflected sound, middle and high frequencies being projected laterally for reflection from nearby walls or a room corner. Even when used in completely symmetrical surroundings there are problems due to changes of apparent sound-stage width with frequency. When assessing such a speaker it is important not to be misled by the very surprising breadth of sound that can sometimes be created, for true stereo has more than undifferentiated breadth.

Another escape from loudspeakers is offered by headphones, which have improved vastly in recent years and are now available in various shapes and sizes for serious private listening (**fig. 66**). Headphone stereo differs from loudspeaker stereo, as the sound-stage seems to pass through one's head and moves around with the head. There is no stable external reference as with speakers or real music, though special circuits have been devised to counteract some of the more curious subjective effects. Some amplifiers are fitted with output sockets for headphones and instructions are usually issued with headsets regarding suitable attenuation of signal level if they are used on amplifier outputs intended for speakers. Many people enjoy listening to stereo in this oddly shut-off fashion, though for others it is a taste difficult to acquire. It is certainly quite different from listening to live music – necessarily so – and in this sense could be regarded as an interesting blind alley from the hi-fi point of view, assuming the search for fidelity to imply faithfulness to something with a real existence. But for those quite unable to reproduce music via loudspeakers for domestic reasons the headset idea is worth a try; it is also economical, tip-top private listening costing no more than a very modest conven-

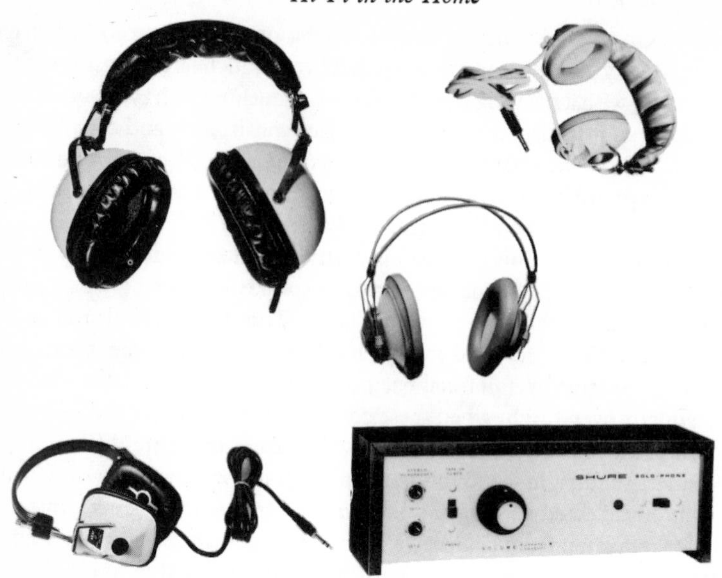

Fig. 66 Headphones. Top left, Pioneer SE–30; below these, Akai ASE–9S. Middle, Amplivox Astralite JL 26. Top right, Lafayette F–767 headset. At bottom right is the Shure Solo-phone SA–2E amplifier designed to feed headphones for private listening. Prices of headphones shown here range from £5 to £15.

tional system. One tiny hint: do not judge turntable or tape recorder wow when listening via headphones, as our ears are more sensitive to pitch changes if some of the sound arrives after reverberant delay, as it does from loudspeakers in a room.

Thus far we have examined separate components only, though many retailers – and some manufacturers – offer 'package deals' at various prices, and these of course may be judged for their overall performance, bearing in mind all the foregoing cautionary remarks. One type of semi-package worth consideration by the non-handyman is an integrated tuner-amplifier and player unit, designed for use with separate speakers of one's own choice. The dealer who has bothered to assemble particular combinations of equipment to suit three or four budgets, with appropriate cabinets and all the cables and connections worked out, is worthy of respect, even if his overall stock of other units is limited. To many customers this is a more reasonable service than that offered by vast electronic comparators where the buyer may, at the touch of a button, connect together any of several hundred thousand possible combinations. There are also hi-fi exhibitions, such as the annual *Audio Fair* and

Sonex shows held in or near London, where equipment may be judged; though here as elsewhere one must beware of slick showmanship and the easy impression of quality created by well chosen records of gimmicky music. But such events are most certainly worth attending, especially as the more sophisticated and musical of manufacturers tend to shine out as bright beacons.

To give some idea of the sort of balance to strike between cost and quality of components when compiling hi-fi systems, five possible combinations are listed in **fig. 67**, all based on known and priced equipment at the time of writing (summer 1970). It was suggested earlier in this chapter that reasonably good hi-fi starts becoming possible at about £80; indeed, by selecting components carefully and 'shopping around' various discount suppliers it is possible to acquire a surprisingly good set of stereo record playing equipment for under £100. First column in the table lists such a selection at £97, followed by a rather better 'budget' set at £140. Then come good, very good and superb systems corresponding to the price categories labelled in **fig. 57**

The 'discount' budget encompasses a modest moving-magnet pickup cartridge fitted with a 0.7 mil stylus, an adequate arm and simple but good two-speed motor/turntable unit. These have to be assembled on a wooden plinth fitted with a Perspex lid, thus making a properly protected self-contained player unit. The amplifier has a power capacity of 10 W per channel, matches a magnetic pickup despite its modest price, and incorporates proper bass and treble controls giving both boost and cut. Styling is for free-standing use on a shelf. Two small totally enclosed IB speakers complete the scheme and offer a very creditable performance at £30 for the pair.

Working in this lowest price region one has to keep a close watch on the components available, and of the five selections listed in **fig. 67**, this first would be the most subject to change – almost from month to month. The important point is that with careful selection of component parts, very pleasing stereo quality can be obtained for this sort of outlay. There is also the second-hand market to consider, where, by studying the classified advertisements in hi-fi magazines or in patronising shops which handle good used equipment, one can usually get excellent value for money. Here, though, the non-technical buyer is best advised by a friend who is familiar with the items in question. Also, the handyman willing to tackle some carpentry could devise a simple lidded plinth of his own to save most of the £7·50 allowed in the table, while an even bigger economy could be effected by purchasing separate loudspeaker drive units in the form of manufacturers' kits, to be assembled in home-made enclosures. As an example of this approach, an established high-quality miniature IB system normally selling for £16 is available also as a drive

Component	'Discount' Budget (free-standing)	Better Budget	Good Hi-Fi (free-standing)	Very Good Hi-Fi	Superb Hi-Fi
TURNTABLE	£15·00	£27·25 (player unit)	£45·00 (player unit on plinth)	£36·50 (player unit as in previous col., but without plinth)	69·50
PICKUP ARM	£10·50				£31·25
PICKUP CARTRIDGE	£5·15	£13·00	£22·00	£22·00	£40·75
CONTROL UNIT	£29·00 (integrated amplifier in case)	£42.50 (integrated amplifier)	£99·00 (integrated FM tuner-amplifier in case)	£59·50 (integrated amplifier	£43·00
POWER AMPLIFIER					£55·00
RADIO TUNER	—	—		£62·00 (FM)	£82·50 (AM/FM)
TAPE UNIT	—	—	—	—	£150·00
TWO LOUDSPEAKERS	£30·00	£43·00	£87·00	£107·00	£204·80
FURNITURE	£7·50 (plinth for turntable)	£15·00	—	£25·00	£50·00
COMPLETE SET-UP	£97·15	£140·75	£253·00	£312·00	£726·80

Fig. 67 Price breakdown of some possible stereo installations.

unit module at £10·40. Two of these mounted in small enclosures constructed in accordance with the manufacturer's suggestions – perhaps even using odd structural alcoves in the house to obtain the necessary enclosed air volume – would save a further £9. Allowing, say, £5 to cover materials for plinth and enclosures, this brings the 'discount' budget down to just under £86.

The £140 'better' budget in **fig. 67** is based on normal retail prices and assumes use of a modest equipment cabinet rather than a free-standing arrangement. Most of the £15 allowed for this is in fact saved elsewhere, as no separate plinth is needed for the turntable and the amplifier may be purchased in chassis form rather than cased in its own small cabinet. Turntable and pickup arm are purchased as an assembled entity in this case, with a rather more sophisticated magnetic pickup cartridge fitted with a 0·5 mil stylus. The amplifier is rated at 15 W per channel and features switched high- and low-pass filters in addition to normal tone controls. Loudspeakers are slightly bigger and better IBs than those chosen for the first budget – 'bookcase' rather than ultra-miniature types.

The third column brings 'good' hi-fi for £253. Assuming a free-standing layout, this employs an integrated turntable/arm system mounted on its own plinth, and a high quality cartridge fitted with an elliptical stylus and capable of tracking even difficult discs at 1½ gms. Stereo FM radio reception becomes possible in this price region, amplifying and receiving features being combined in a tuner-amplifier of excellent all-round performance. Speakers are again more ambitious, being floor-standing models of higher efficiency and better bass performance.

On to 'very good' hi-fi for £312, this list features the same player in its (cheaper) cabinet-mounting version, and the same cartridge, but amplifier and tuner are now separate components, each more refined and ambitious than the corresponding parts of the previous list's tuner-amplifier. A cabinet is included for housing, and speakers are yet again bigger and better – this time employing three drive units each to ensure uncoloured reproduction of the vital mid frequency range.

The last sample includes a high-quality tape unit and is housed in an appropriately larger cabinet. The turntable is a finely engineered device with a massive platter and electronic speed control – it is almost totally silent in operation. A precision pickup arm with lowering device has every facility for exact setting up and accurate adjustment of playing weight and side-thrust compensation; the cartridge is a top quality moving-magnet type fitted with elliptical stylus and exhibiting exceptionally high 'trackability'. The amplifier system, though transistorised, has separate control unit and power sections;

this is partly to satisfy professional usage, and also to make the control unit of sensible size and weight, which it cannot be if coupled to a very large stereo power amplifier. The high price of these two items (£98) procures 45 W per channel at almost immeasurably small distortion, with filter and tone control facilities of the most elaborate kind and exceptionally low levels of background noise. An AM/FM stereo multiplex tuner has automatic switching between mono and stereo signals. Finally, the speakers are domestic equivalents of professional monitoring models. Low frequency loading is of the transmission line type, each system using four drive units and complex electrical crossovers to give exceptionally smooth and uncoloured reproduction over a very wide frequency range.

Although based on actual components, these five lists should not be taken as a literal guide to purchase of equipment, as there are obviously endless permutations and possible shifts of emphasis and price. No names have been given for this reason, the exercise simply indicating the sort of priorities to adopt when confronted with the problem of dividing expenditure between various links in the chain.

Finally, a few words about judging the sound quality of complete hi-fi systems. Ultimately, this is all that the non-technical purchaser can do, for until the ear is well accustomed to the characteristics of all the supporting equipment it is difficult to pinpoint the more subtle deficiencies in individual components. A complete installation, of course, is what the user must live with, and it could be said that aural 'liveablewithness' is the supreme virtue in hi-fi – at least for the fastidious listener. People can live with the sounds generated by small transistor radios, but we do not normally apply hi-fi standards of judgement to such devices. It is when one starts to listen for the sounds heard in the concert hall that the interposing layers of coloration and distortion become apparent.

For the beginner in hi-fi there are some temptations to overcome, easier for those familiar with the sounds of live music. A small radio lacks output at low and high frequencies, and when real bass and real extreme treble are heard for the first time they can impress mightily. They can sometimes impress even more mightily if exaggerated rather than left to speak for themselves, a fact used on occasion by unmusical demonstrators of equipment. Equipment should sound impressive, if at all, either because the music being played is the sort of music that would impress in real life, or because the reproduction is so startlingly real that closing one's eyes makes it difficult to believe that the performers are not out there in front of one. 'Startlingly real' must be interpreted with care, for some sounds can be startling without much reference to reality. Startlingly 'forward' or 'present' may indicate a particular type of recording, but if *everything* has great presence it is probably a matter of treble peakiness. At a suitable volume setting it should be possible to close eyes and feel that nothing very significant or obvious is intruding on natural sound. With good reproduction it should never seem ludicrous to suggest that the sounds heard might, with only a little imagination, be coming from real musical instruments. Any limitation noticed should seem acceptable rather than irritating. An indication that one is beginning to apply the right standards is the reaction, on hearing again the portable radio, not that bass and treble are missing (which they are) but that the sound is coloured, boxy, confined, muzzy or shrill, lacking in detail, distorted. It couldn't possibly be confused with the real thing, whereas well chosen hi-fi can – on occasion at least.

8. INSTALLATION

ALTHOUGH a grasp of basic hi-fi principles is extremely helpful when choosing equipment, we have seen how choice can be complicated by differences in the height of fi between various audio ingredients, and further confused by the influence of personal taste in such matters as loudspeaker performance. Likewise, when equipment has been chosen, purchased and delivered, its assembly and siting in the home is bound to vary because of differences in connections and housing, and according to the shape, size and acoustics of listening rooms – the more so if the very best musical performance is to be obtained from any given pair of loudspeakers. Apart from the actual positioning of cabinets, components have to be connected together in the most convenient and efficient manner, and if the user is doing his own assembly of parts in an equipment console, there are possible snags with wiring and layout which will not occur to the beginner. The new hi-fi owner may find himself in any position between the purchaser of a 'package deal' with all wiring finished and foolproof printed instructions provided, and the eclectic who picks separate items to fit into a home-made equipment cabinet, perhaps even with the intention of building his own speaker enclosures. There are also likely to be considerable variations in the attitudes of dealers when it comes to a request for help: equipment paid for and at home poses a different psychological problem to that waiting in the shop to be sold! The best retailers will give advice and help at the installation stage, but as service takes time and effort, the most co-operative in this respect are unlikely to offer large price discounts. This chapter will attempt to cover installation in sufficient detail to be of use to the man intent on doing most things himself, though it will be assumed that anyone proposing to make up his own interconnecting cables is at least familiar with simple electrical soldering techniques. The reader will no doubt skip those parts not applicable in his case, though I hope there is at least something of use or interest to anyone concerned with hi-fi in the home.

The nature of that home – or at least the nature of the room to be used for listening to music – will already have determined basic things like the choice of free-standing or cabinet-mounted units, high or low power amplifiers, large floor-standing or small bookshelf speakers, and so on. Finance also comes into this, as it may well happen that there is room for a large system but insufficient money for more than a 'budget' set-up. It is a moot point whether this is more frustrating than being wealthy but forced, for other reasons, to live in a bed sitter! Anyway, let it be assumed that all such factors have led to purchase of equipment which has been delivered and awaits assembly. For most users there will be only one room in which an audio installation can be contemplated, though those moving into a fresh house may have an opportunity to include acoustics in the list of room features when planning family geography. For these and others with a choice of hi-fi accommodation it is worth considering a few general rules relating musical enjoyment to size, shape, structure and position of rooms.

Within normal domestic limits the first rule is: 'the bigger the better'. In a large room there is more space for sitting in comfort without undue restrictions arising from stereo loudspeaker geometry; also, stereo or not, there is aural virtue in being moderately well back from the speakers – 'distance lends enchantment' in sound as well as vision. Consideration of room resonances comes into this; as mentioned in the last chapter, if a room is small its major eigentones occur well within the audible spectrum and can add coloration to reproduced music. Whenever sound is generated in an enclosed space our ears respond not only to the originating wavefronts which determine the apparent direction, but also to the multitude of reflections which constitute ambience or reverberation. Where there are large parallel surfaces – walls, floor and ceiling, for instance – sound-waves are bounced back and forth many times before decaying to insignificance, and at some frequencies the reflected waves are all in phase because the surfaces are spaced by an exact multiple of half the sound wavelength. This situation is similiar to that in musical wind instruments, and a room can indeed be regarded as to some extent 'tuned' to particular frequencies. When a continuous note is played at one of these eigentones the sound level builds up above that created by tones at other frequencies, and because of the steady, resonant pattern of reflections the listener may find very noticeable concentrations of loudness when moving about the room. This is called a *standing-wave*.

As the lowest and most prominent eigentones occur at those frequencies where room dimensions equal just half a wavelength, it follows that in a very large room the major effects are out of harm's way at or below the lowest bass frequencies (see **fig. 68**) and incidentally make it somewhat easier to reproduce the lowest notes. It may strike the reader that even in the most palatial homes ceilings are seldom over 12 ft. high, and that there is thus bound to be at least one main room resonance above 50 Hz. This is true, though ceilings and floors are usually constructed of plaster and boards on a supporting framework with a space above or beneath, making them less efficient as sound reflectors at low frequencies. This means that those eigentones potentially most troublesome from the frequency point of view are in practice usually damped down to insignificance. An exception is in all-concrete buildings, where how-

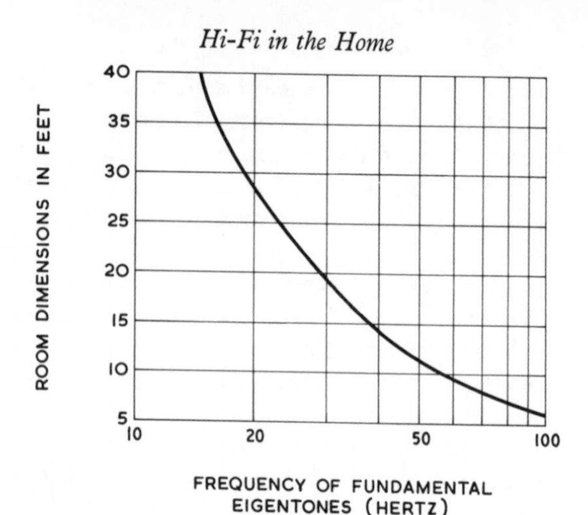

Fig. 68 Relationship of room dimensions and frequencies of associated fundamental eigentones (half-wavelength resonances between parallel surfaces).

ever much absorbant material is present to reduce 'liveness' at higher frequencies, rooms can on occasion give trouble with excessive LF resonance.

The main eigentones corresponding to the three dimensions of a rectangular room are accompanied by a multitude of others occurring at harmonics of the fundamental resonances and at frequencies determined by diagonal reflections, the resulting complex web of sound-waves providing an overall acoustic character peculiar to that particular room. This we take into account unconsciously, though when listening to reproduced sound – particularly in stereo, where the music can have a convincing studio ambience of its own – it is generally desirable to minimise local reverberation and ensure that any unavoidable room resonances are evenly distributed in frequency. Ideally, as already noted, the basic eigentones should occur at frequencies too low to matter, say below about 30 Hz, but as this involves a room width of 19 ft. or more the ideal cannot be attained in most homes. Also, even if a room were 20 ft. square the coincidence of resonances resulting from equal width and length would give a coincidence of harmonic eigentones which might reinforce each other sufficiently to accentuate some bass notes well above the basic frequency. The latter – 28 Hz in the case of a 20 ft. dimension – would not be reproduced by the vast majority of domestic loudspeakers even though occasional pure A_4 organ pedal notes may manage to get on to recordings, but its octave at A_3 and the next harmonic a fifth further up (E_2) are both well within the normal orchestral bass region. This means that whether the listening room be large or small, it is better if its major dimensions are staggered somewhat, both to avoid

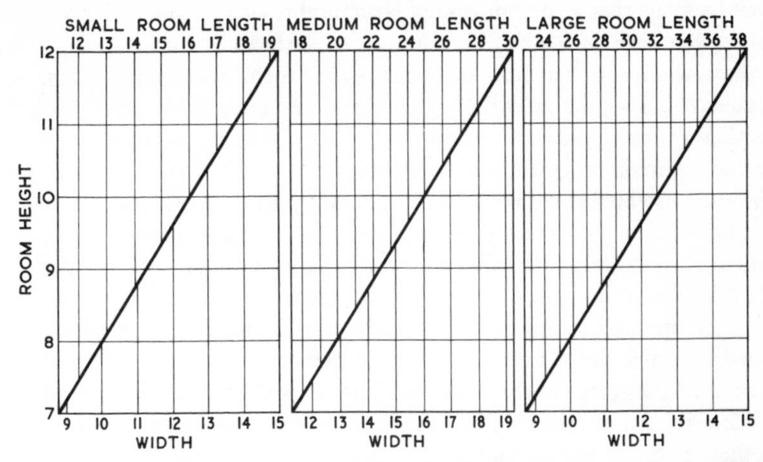

Fig. 69 Optimum dimensions in feet for best distribution of eigentones in rectangular rooms. Where height meets diagonal line, project up and down for other dimensions.

duplication of the main resonances and to randomise distribution of the higher eigentones. A uniform distribution of relatively minor resonances is preferable to a situation where more prominent eigentones are grouped in narrow bands with gaps in between. With these factors in mind, ideal height/width/length ratios have been computed for domestic rooms, giving least alteration of low frequency balance for sounds created or recreated therein. Taking the height as unity, ratios for small, medium and large rooms respectively are 1/1·25/1·6, 1/1·6/2·5 and 1/1·25/3·2. These are plotted for various room heights in **fig. 69**, from which it should be possible to see almost at a glance how nearly any particular room approaches the ideal.

Many rooms, of course, are not purely rectangular, and unless speaker placement for a reasonable listening area becomes too difficult, some irregularity is to be welcomed as this ensures even greater randomisation of eigentones. Recesses, alcoves, window bays and changes of ceiling height all tend to modify the main resonance pattern, and can even improve a basically poor room (one with width and length equal or in a two-to-one ratio) to the point where a 'perfect' oblong would be little better. Mollification also arises through absorption of acoustic energy by the room's boundary surfaces. The reduced LF reflection from floors and ceilings has already been mentioned, and when walls are of the plaster partition variety rather than solid brick or concrete they will have their own structural resonances which, unlike acoustic effects within the room itself, result in dips rather than peaks in a room's frequency 'response'. It often happens that a combination of structural resonances in walls, floor and ceiling, together with some breaking up of surfaces by furniture and a few recesses, will make even quite small, near-cubic rooms relatively 'dead' at bass frequencies. Also, the extent to which eigentones are excited depends on the positions of loudspeakers in relation to a room's boundaries – a point to which we shall return later. Qualifications of this sort mean that while a perfectly proportioned listening room is obviously desirable, it may in practice be reasonable to choose a theoretically inferior one if other factors make it more convenient, possibly experimenting with the equipment (or at least the speakers) first in one room then the other – family arrangements permitting. Apart from LF eigentones, all rooms influence reproduced sound quality at middle and high frequencies in a manner and to an extent dependent on furnishings, carpets, curtains, etc; but as these – unlike room dimensions – are controllable by the occupant and have some bearing on loudspeaker arrangements, this is another matter for later attention.

There are a few more points for consideration by those with a choice of music listening rooms, the first of which may apply to some readers who had not regarded themselves as having any choice. In many British houses the largest room is the main bedroom on the first floor, often spanning the hallway and front sitting-room down below. There is frequently no reason other than convention for using this room as a bedroom, and when family circumstances permit it is at least worth thinking about its use as a sitting room equipped with hi-fi. If the house is not detached and one's neighbours are usually late to bed, hi-fi upstairs during the day or evening is rather less inhibiting with regard to the volume control setting than in the more common ground-floor situation where the sound of next-door's television comes ominously through the wall, to warn that one's hi-fi will do likewise in the reverse direction at a higher phon rating. Incidentally, this latter situation is less likely to arise in houses which are attached on the hall side, so when house-hunting look for property with adjoining front doors. Transmission of sound from room to room within one's own household should also be kept in mind when deciding on a listening room, especially as very considerable structural alterations are usually necessary to reduce mechanically conveyed rather than air-borne sound. Incoming noises can also be troublesome, and if one's house is situated on a busy road a decision to put hi-fi in a front room will almost certainly be regretted sooner or later; double glazing will help, but remember that traffic *always* gets worse! The roar of a central-heating boiler, the chug-chug of a gasmeter and the emptyings and rushing noises associated with plumbing – all can irritate the serious music listener. Small points, of course, but worth taking into account when installing hi-fi for long-term musical pleasure in the home.

Returning now to the average reader for whom all this talk of choosing rooms is only a hypothetical exercise, how shall things be arranged in the space that is to be used? The answer to this revolves around speaker positions and the convenience or comfort of possible listening areas. We shall see later that speakers generally give of their best in the bass when near to room corners, and the use of two similar corners for a pair of stereo speakers automatically settles matters in many cases. In a normal oblong room the speakers are usually best positioned at the ends of a short wall, permitting the listener to sit well back from the sound sources without being forced into a 'backs to the wall' position at the far end. A speaker/seating layout that places the 'optimum' seat approximately as far from the speaker wall as the speakers are from each other is ideal, as this gives a viewing angle of 50–60 degrees, which most people find very satisfactory for stereo listening. There is of course no objection to a wider stereo picture if the listener prefers it, though the closer one gets to the speakers the less room there is for manoeuvre away from the stereo centre-line. Also, in medium sized rooms there is a tendency for deep bass reproduction to be rather thin for listeners out in the middle of the floor area, due to standing-wave effects whereby sound pressure falls to a minimum mid-way between the end walls around the lowest eigentone frequency. There could be other factors to consider: furniture, fireplace, doorways, etc, may make speaker positions depend on seating rather than vice versa: small IB speakers might dictate stereo geometry by their location in bookshelves; and odd room shapes or curious acoustics could demand unconventional layouts. Points such as these – and the need to follow a fairly careful loudspeaker setting-up procedure if the best is to be made of a room – mean that one cannot really make a final, detailed decision on equipment layout until everything is connected and working. Despite all theoretical predictions, and a likelihood that the 'obvious' positions for speakers will in any case give quite acceptable results, there is nothing to compare with an actual musical signal when making final adjustments: the proof of the stereo pudding is in the listening – to music. So we shall leave speakers for a few pages to deal with some of the points and pitfalls in mounting and connecting other units in the chain.

Whether amplifier, tuner, turntable, etc, are to be free-standing or housed in a cabinet, it is wise whenever possible to position them within easy access of the proposed listening area. Just behind or to the side of one's seat is sensible; between the speakers, where it is difficult to adjust controls to proper effect without constantly bobbing up and down, is nonsensical. In some situations a little nonsense may be inevitable; with shelf units, for instance, it might easily happen that the only surfaces available either for speakers or electronics are on the one wall, as in **fig. 56** (page 158), but generally this particular inconvenience should be avoided if at all possible. Assuming such avoidance, a very definite advantage in using wall-mounted shelves for the input end of a system is that the turntable/pickup assembly is firmly supported well away from floor vibrations. Despite every precaution and care in the design and suspension of player units, a thump on the floor or a minor collision with an equipment cabinet may still cause a pickup stylus to leave the groove, a fate more easily avoided when turntable plinths rest on shelves firmly screwed to brick walls. If a floor-standing equipment console is used, risk of mechanical disturbance is best minimised by situating the cabinet at a point where there is no 'spring' in the floor, preferably against a structural wall as floor-bearing joists are usually terminated solidly at a room's boundaries.

Apart from style and finish to suit personal taste or existing furnishing schemes, the cabinet itself should satisfy ordinary common-sense standards of woodworkmanship and be rigid enough to avoid the least suspicion of swaying motions when knocked or pushed – some otherwise excellent designs suffer from spindly legs. Hi-fi furniture ranges from the simplest lidded box with space for turntable, amplifier and tuner, up to comparatively massive cabinets designed to take the above units plus tape recorder and a record collection. Some furniture at and between these extremes is shown in **fig. 70**.

There are a few functional and technical points to remember when choosing a cabinet. If the user would normally wish to handle equipment and records from a sitting position it is better to have a long low structure, as while a tall cabinet can house the electronic items at a convenient height on a front panel, the turntable and pickup are usually too high to reach without standing up. There is an opposite danger in some very low cabinets, as front-mounted units in these can be too low down for a clear view of the controls unless one sits on the floor. Assembly of equipment is in some respects easier in a tall narrow cabinet, as items are simply mounted one above the other and interconnecting cables follow short, direct routes. However, heat generated by any and every component causes warm air to rise and gather beneath the turntable, eventually raising its temperature to the possible detriment of records. This is unlikely to be a serious problem with transistor equipment, but large valve power amplifiers generate considerable quantities of heat which, when all convection paths lead to the turntable, can give trouble. In such a case it would probably be wise to introduce deflecting sheets of wood or metal to direct the warm air flow away from vulnerable regions. Related to this, any equipment

Fig. 70 A selection of equipment cabinets. Top-left, the Hampstead console by Hampstead High Fidelity, with hinged hopper to reveal control unit and tuner. Below this, a striking vertical system in the Scan range mounted on Hi-Raks, with the Lowflex cabinet to its right, both by Record Housing. Bottom-left, the Largs Caithness, with record storage on its left; also by Largs in the Lanark coffee table. Bottom-right, the simple Viking cabinet by Howland-West standing on a plinth, and at the top, an unusual stereo reflector system by A. Davies, who make many loudspeaker cabinets.

cabinet should have liberal openings to encourage a cooling air flow over enclosed units, with an inlet slot or holes in the base or at the bottom of the rear panel, and an outlet at the top of that panel to minimise any pocketing of air beneath the turntable board. Similar considerations apply to a tape recorder that is to be housed in part of a cabinet: if the recorder has an air intake grille on its base and would normally stand up slightly on feet to allow an inward air flow, remember that when lowered into a cabinet this normal cooling will be impaired. A few holes cut through the board on which it stands will help here, and it would be wise to see whether the proposed console will permit this.

Ensure that a cabinet offers reasonably easy access to the rear of the units mounted therein, either by means of removable back panels or via cooling holes large enough for passage of hand and arm – it is extremely irritating if a whole front panel assembly has to be removed simply to change a plug or pickup adaptor. Likewise, in a long low console where units will be spaced laterally, check that leads may pass easily from section to section; if there are no holes for this purpose through dividing panels, make sure that drilling some will be straightforward. Make sure also that there is sufficient space to get the power amplifier (or at least the mains transformer corner or end of it) well away from the pickup head: 18 in. should be regarded as a minimum distance, and over 2 ft. is preferable if hum induction into a magnetic cartridge is to be avoided. Check that smoothly operating lid-stays are fitted, as a jerky hinging action can transmit vibrations to the pickup; the types depending on frictional grip or pneumatic lowering are preferable to those with slides and rotating stops which require extra lid movements for releasing purposes.

Finally, take care that the space for accommodating turntable and pickup really is large enough for the proposed units, not forgetting the rear of the pickup arm with its possible assemblage of counterweighting devices which swing round as a record is traversed. One tends to accumulate odds and ends such as record and stylus cleaning devices which take up room around the turntable, and as handling of records is generally more relaxed in uncramped situations it is wise to opt for a generous rather than strict ration of space around the player assembly. However, do not extend this generosity to the vertical plane, as there is no virtue or sense in scrabbling around in a deep well when handling delicate instruments. Fortunately this is a matter easily dealt with in most equipment cabinets, as the turntable mounting board usually sits on runners or corner pieces and may easily be raised above these by spacers if necessary. The ideal turntable height in relation to a cabinet is that which gives safe clearance between the underside of closed lid and uppermost point on the pickup arm (say $\frac{1}{4}$ in.) while at the same time leaving a completely clear view of the record surface when viewed horizontally, the latter being very useful when setting up the pickup cartridge.

Most dealers supplying a cabinet will, if it is part of a complete hi-fi purchase, cut the necessary holes, fit the units and connect up. For those needing or preferring to do the job themselves, all that is required for the cutting and fitting operations is reasonable competence at elementary woodwork and simple mechanical assembly. Manufacturers usually supply paper templates for panel cutting and fairly complete instructions for basic fitting, and provided the various earlier remarks about operating convenience, spacing and heat dissipation are intelligently applied, there should be no significant difficulty with non-electrical installation matters. A keen carpenter may even tackle the whole cabinet making job, which should be fairly straightforward for an experienced amateur woodworker once a list of constructional ideas has been compiled following a close look at some commercial cabinets. The reader may even have items of furniture which could be adapted for housing hi-fi equipment – a rug chest, for instance, lends itself to this and many have been converted with great success. People have even been known to employ the wall cavity left after removal of a fireplace, building in a shelf or shelves flush with the main wall to provide an extremely substantial support for equipment without taking up any room space. Others, with aesthetic objections to free-standing amplifiers and tuners but wanting the benefits of wall-mounting for the player unit, have bracketed a turntable plinth to the wall and built a mock cabinet around it. There are endless possibilities for the hardened general do-it-yourself man with a newly acquired interest in hi-fi.

Moving on now to more specific matters, we shall look at: (i) fixing and adjustment of turntables and initial setting up of pickups; (ii) cables and connections both inside and outside the cabinet, including aerial fitting; (iii) impedances, sensitivities and pre-set adjustments; (iv) possible snags such as hum, rumble and acoustic feedback; (v) setting up loudspeakers for optimum tonal and stereo results; and (vi) final pickup adjustments when the system is working. Although these points follow an approximately logical sequence, there will be some need when installing equipment to flit back and forth, as it were, between pickups and loudspeakers; should there be any readers in the extraordinary position of using this book as their sole source of hi-fi advice, please don't take the rest of this chapter as chronologically sacrosanct.

It is customary for equipment cabinets to have an easily removed horizontal panel on which to mount the turntable and pickup, and turntable units are normally supplied with fitting data on this assumption. In most cases a large hole has to be cut in the panel to accommodate motor and drive mechanics which project downwards. Once this is done there is unlikely to be much room for manoeuvre, and to save subsequent trouble it is a wise precaution to shape a sheet of paper corresponding to the panel, position the turntable template on this and ensure that the pickup arm will have sufficient room both to sit on its rest well off a 12 in. disc and run right in to meet the label of a 7 in. disc ($1\frac{3}{4}$ in. from centre spindle) without overhanging the panel at any point. Turntable position thus decided, the panel may be cut and unit fitted following the maker's instructions. This panel, incidentally, should be tough and thick: $\frac{1}{2}$ in. plywood represents about the minimum tolerable material.

Next comes the pickup arm, which, unless incorporated with the turntable as part of a player unit, involves further work on the main baseboard. Arms are supplied with templates to facilitate correct positioning in relation to the turntable centre spindle, and in the case of a complete pickup as opposed to the more usual combination of cartridge and adjustable arm, there is one – and only one – position if geometrical tracking error is to be minimised. As templates for pickup fitting have to allow for a difference in height between baseboard and turntable, it is often rather difficult to arrive at the exactly correct position for an arm base. Therefore, in cases where fixing involves no more than drilling a hole for leads to pass through and then securing the pivot with three or four wood screws, the lead hole should be made rather larger than the instructions suggest, final screwing down depending on accurate positioning for minimum tracking error, to be described shortly. In the case of a universal arm with a built-in adjustable pivot position, this precaution should not be necessary. Ensure that the baseboard is flat, as a warp can make the 'horizontal' reference plane of the pickup different from that of the record surface, and check that when the cabinet is in its intended permanent position the turntable platter itself is truly horizontal. Ensure also that the turntable unit is screwed down firmly to the baseboard, ignoring the maker's fitting instructions in this one respect if they involve suspension of the unit on springs or rubber. While compliant mounting of a complete player may help with isolation from mechanical jarring, it is important that this applies to pickup and turntable as a rigid entity; when the arm pivot is fixed firmly to a separate baseboard the turntable should be similarly fixed. Some luxury turntables (see **fig. 59**, page 168) include a special panel at the side for pickup mounting, giving rigid coupling between PU and TT and yet permitting the whole assembly to ride on resilient rubber mounts.

Next step is to fit the cartridge. This will have a pair of holes or slots with $\frac{1}{2}$ in. spacing for fixing by nuts and bolts to the head shell. The latter, normally detachable from the arm by simple unplugging or release of a clamping ring, has corresponding holes or slots. Cartridges are supplied with appropriate screws, nuts and spacing washers, and in many cases careful adherence to the instructions accompanying arm and cartridge will lead to a successful fitting operation. Some complete pickups have plug-in heads which combine the functions of a separate shell and cartridge, and installation of these is obviously rather less complex. However, to cope with all cases, and to cover the odd omission from instructions, the following notes give an idealised step-by-step procedure. Having unpacked the cartridge, first safeguard delicate parts either by removing the slide-out stylus assembly – normally employed in moving-magnet and induced-magnet pickups – or by ensuring that any guard or shield supplied for protective purposes remains firmly in position for the time being. Check that there is no obvious incompatibility between shell and cartridge, then fix one to the other with the 'hardware' provided for this purpose. In some cases it may be necessary to space the cartridge downwards away from the shell to achieve adequate clearance between shell/arm and record when the arm is finally adjusted, and at least one spacing washer should in any case be included if it appears that ribs or other mouldings on shell underside or cartridge top surface will prevent an unstressed fit. Make sure that the bottoms of the bolts do not protrude beyond the cartridge base.

The normal cartridge/shell set-up is shown in **fig. 71**, from which it is important to note two geometrical points: the cartridge should be lined up accurately with the shell, the two components' imaginary centre-lines being coincident when viewed from above or below; and the natural horizontal surfaces of the two should be parallel unless there are instructions to the contrary. The first point is very important, as eventual arm alignment is made with reference to the head shell on an assumption that its contour reflects the cartridge beneath it. Most cartridges have parallel sides over part of their length, and it is usually possible to arrange some pieces of paper or card against these in order to judge whether the shell lines up accurately when viewed from above. Always check this alignment finally after the screws have been tightened.

Electrical connections must then be made between cartridge and shell. Most cartridges have four projecting pins at the rear, two for the left channel and two for the right, and there are usually corresponding pins and/or leads within the shell for feeding signals through to a plug which fits a socket on the arm. The short insulated wires used for cartridges and shell are supplied with small lugs soldered to their ends; these push on to the pins as shown in **fig. 71**. In some cases small spiral springs are supplied for this purpose, but whatever the mode of connection, the object is to make a firm, trouble-free electrical

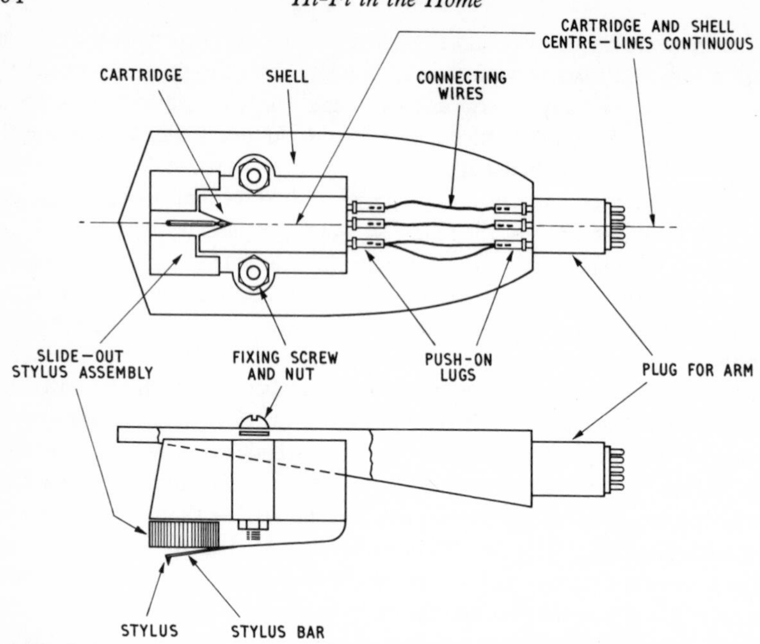

Fig. 71 Pickup cartridge mounted in head shell. Centre-line of cartridge must line up with centre of shell and plug; cut-away side view shows cartridge top parallel to shell.

contact between each cartridge pin and its corresponding shell pin – necessarily relying on a friction contact at the cartridge, as heat from a soldering iron would damage delicate internal connections. Great care should be exercised when fitting the push-on lugs, using a pair of tweezers or fine-nosed pliers for proper control and easing open the lug tubes slightly if brute force would otherwise be needed to get them on the pins. Try to avoid fracturing the soldered joints between wires and lugs.

Apart from mechanical care, it is important that these connections are made correctly from the viewpoint of signal conveyance, as a reversed pair of leads could ruin stereo and introduce hum. In any audio system an electrical signal is carried from point to point by *two* conducting paths, as currents from A cannot travel through B without a return path to keep the flow going, and voltages across one component or circuit cannot appear across another unless both sides of each are connected. Taking, say, the left-hand signal generated in one coil winding in a magnetic pickup cartridge, both ends of the coil must somehow be connected to the LH preamplifier input even though one side of that input may be reached via a simple metal screen surrounding a

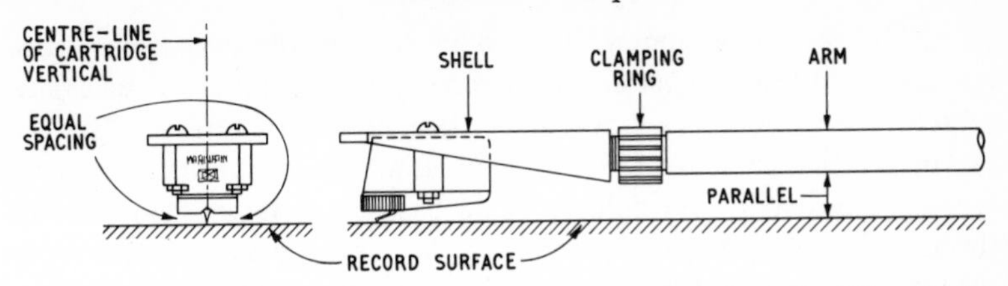

Fig. 72 Angling cartridge and arm in relation to record surface. Cartridge should be vertical when viewed from front, and arm parallel to disc when viewed from side.

socket and attached to the chassis. This chassis side of circuits – often shared by several inputs in a preamplifier – is known commonly as *earth* or *ground*, and for convenience the 'earthy' side of each stereo cartridge output is frequently marked with a 'G' to indicate its destination in the preamplifier and to distinguish it from the non-grounded or 'live' connection. The four pins on a cartridge are therefore commonly labelled: L, R, LG and RG, meaning left, right, left-ground and right-ground. When a similar coding is not used on the head shell there will either be appropriate instructions to cover the point or colour-coded connecting wires. In the latter case the convention is: white, left-live; blue, left-ground; red, right-live; green, right-ground. Sometimes arms and shells use a three-wire system, with the two earthy paths sharing a single conductor. To meet this contingency, cartridge 'hardware' normally includes a shorting link which is pushed over the two earthy pins before connection of a single common earth wire. Whatever the precise arrangement, when connections have been made ensure that push-on lugs are touching neither each other nor the shell, and tuck the wires gently up into the shell where they are screened by its metalwork.

Cartridge mounting complete, re-insert the stylus assembly and fix the head shell to arm. Following the arm maker's instructions, carry out any balancing operations that may be necessary; then, with the arm adjusted to swing freely with cartridge just clearing the turntable surface, ensure that no strong inward or outward bias is applied either by twisted leads at the pivot or by non-vertical mounting of the pivot assembly. Free the leads or level the arm base if necessary, then use whatever procedure is laid down to apply a correct nominal playing weight or force to the cartridge. Place a record on the turntable – preferably a blank disc* as this eases the next adjustments – and lower the pickup on to it. Viewed from the front, the cartridge or pickup head should be approximately vertical in relation to the record surface (see **fig. 72**), a point more easily judged when a clear reflection can be seen in the record,

* Available in the U.K. from Wilson Stereo Library.

hence the blank disc. Assuming the head not to be wildly away from vertical, check next that the arm is parallel to the record surface, or is at whatever angle the instruction manual says is desirable, when viewed from the side (**fig. 72**). An error here is very likely when using separate arms and turntables, as the latter vary considerably in height above the baseboard. Raise or lower the arm pivot assembly as appropriate, mounting the pickup base up on a wooden block if necessary should the range of adjustment be insufficient to cope with an exceptionally high turntable. Having thus paralleled arm and record, examine the cartridge again from the front with a view to accurate vertical alignment. Many arms permit a small degree of rotation at the head to facilitate this, either by movement against some friction when the shell clamping ring is relaxed, or through freeing the internal socket by means of a small screw. If there is an error but no adjustment is possible and it seems that the shell's top surface is parallel to the record as viewed from the front, any skew must be attributed to the cartridge or its mounting and further attention should be paid to this, possibly by addition of an extra spacer or some packing on one side; remember to retain longitudinal alignment of cartridge and shell as discussed previously. On the other hand, if shell and cartridge lean over together and cannot be rotated relative to the arm, it may be necessary to place some packing beneath one side of the arm base because of a mounting baseboard not in the same plane as the turntable surface. In all cases the end result should be that the cartridge sits correctly on the record surface as shown in **fig. 72** at any point across the normal playing area when set to its nominal tracking weight. Take care during all these operations to avoid sudden dropping or knocking of the pickup, using the built-in raising and lowering device whenever possible and clamping the arm back in its rest position during adjustments.

The last important adjustment when installing a pickup is alignment for optimum lateral tracking. It will be recalled that arm geometry is carefully arranged to keep tracking error within modest limits despite the curved path followed by a pickup stylus when using a pivoted arm. This care is wasted, however, unless the pivot is placed at a particular distance from the turntable centre spindle. With a given arm, correctly designed, this precise position is most easily achieved indirectly by adjustment for zero lateral tracking error at the innermost recorded groove radius of $2\frac{3}{8}$ in. As linear groove speed falls towards the record centre, thus cramping recorded waveforms into shorter distances, any angular error in stylus motion represents more distortion. It is therefore customary to design pickups for zero error at the inner grooves, and provided an arm is positioned for exact alignment of cartridge and groove at a

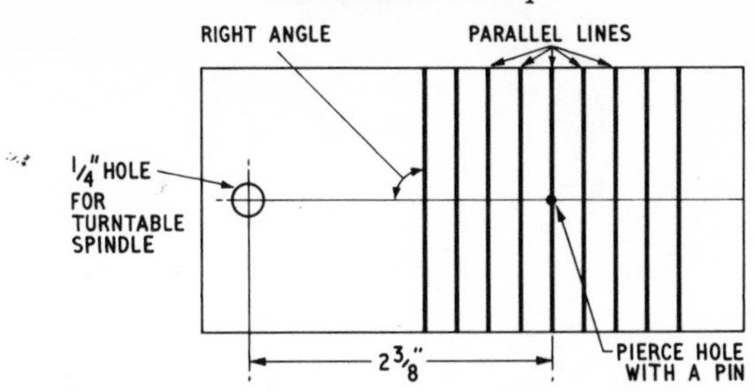

Fig. 73 Simple pickup alignment protractor. This layout is used by SME Ltd. (to whom thanks are due), on protractors issued with their pickup arms.

distance of $2\frac{3}{8}$ in. from the centre spindle, other geometrical requirements are satisfied automatically. A simple home-made *alignment protractor* eases the task of setting up by permitting the relationship of arm and turntable to be seen at a glance. A suggested layout is shown in **fig. 73**; this may be drawn accurately on white card, but if the circular hole for the turntable spindle cannot be cut cleanly it would be wise first to force a smaller rough-cut hole over the spindle to give a reference point from which to draw the protractor lines (a true $\frac{1}{4}$ in. hole will give a tight push-fit, as the spindle diameter is 0·285 in.). It is very important that the centre-line on the finished protractor projects truly towards the turntable centre when fitted in position, with the series of parallel lines accurately at right-angles to this.

Alignment procedure using a protractor is as follows. Place a record on the turntable so that the protractor card lies flat rather than across ribs on the turntable mat, fit the card and lower pickup. With the eye down at TT level, aided by a bright light, rotate the protractor around the spindle and move the arm across (ever so gently) until the stylus engages the tiny hole pierced in the card at a distance of $2\frac{3}{8}$ in. from the spindle. This is shown in **fig. 74**, where it will be noticed that the pickup headshell is exactly parallel with the guidelines, indicating correct alignment for zero error. If the shell does not line up in this fashion, lift pickup from protractor and shift the pivot position a little towards or away from the turntable as appropriate, then check again. It should be possible to find the correct position after a few repetitions of this, when the pivot adjustor on a universal arm should be locked according to the maker's instructions, or the base on a non-adjustable model screwed down permanently. On some arms this longitudinal adjustment is made at the head, with slots instead of simple holes in the shell so that the cartridge may be moved rather than the complete arm. In such cases the cartridge must be accurately re-aligned with the shell after each movement, as the whole exercise would be

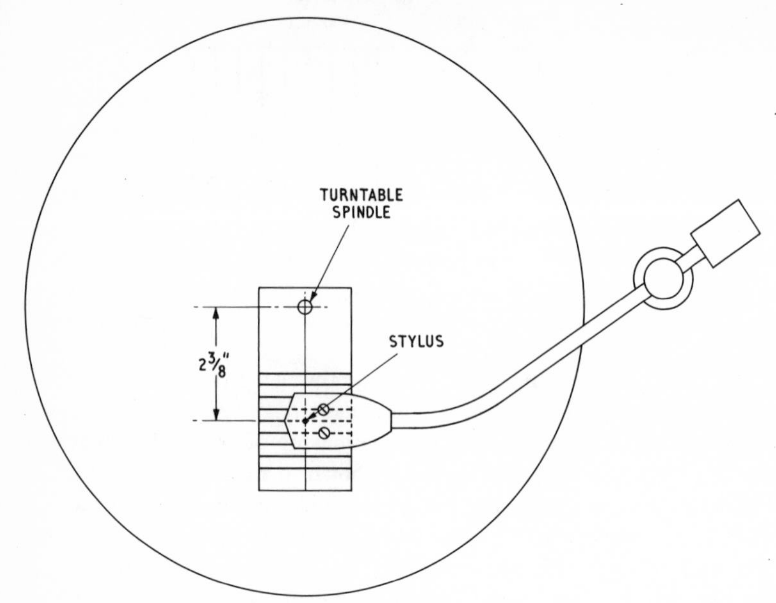

Fig. 74 Use of an alignment protractor for setting lateral tracking error to zero at inner record grooves.

self-defeating if the shell were finally lined up with the protractor but did not represent the cartridge on its underside.

Apart from final setting of playing weight and bias compensation, to be covered towards the end of this chapter, that completes the mechanical and geometrical parts of pickup and turntable assembly, leading us on to cables and interconnections. In nearly all domestic set-ups the most critical signal feeds and connections are those between pickup and amplifier, for here are found the weakest audio signals and the highest sensitivity, with a correspondingly high risk of hum induction. The better pickup arms have sockets at the pivot base, with made-up cables carrying appropriate plugs for connection to arm at one end and amplifier input at the other. Screened leads are used within such cables to shield signal-carrying wires from stray hum fields. Such leads comprise an inner conductor or conductors, insulated from a surrounding braided wire sheath, which may in turn be insulated on its outer surface. The inner or 'live' wire conveys the signal between two units or components in safe seclusion from outside interference, while the outer conductor or braiding completes the circuit by connecting together the two earthy sides. If the pickup arm and its internal wiring could be omitted, the labelled pins on a cartridge

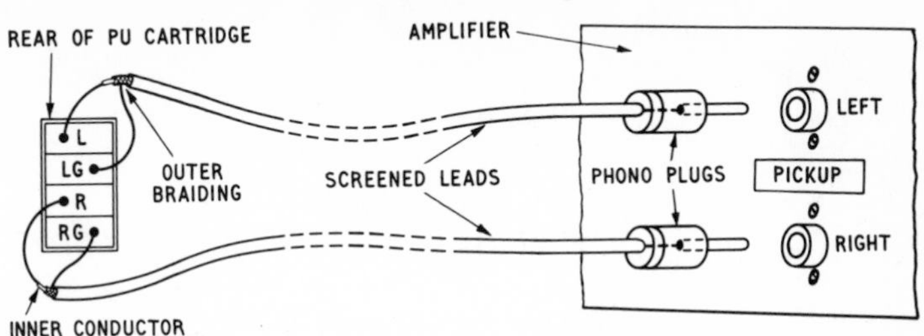

Fig. 75 Idealized arrangement of connections between stereo pickup cartridge and pre-amplifier input panel. In practice, a pickup arm intervenes, with its own internal leads and terminations.

would be connected to an amplifier input rather as in **fig. 75**. The cable provided with a practical pickup might well contain two screened leads fitted with plugs for the amplifier end, as in the illustration, but with a proper termination at the other end via a multi-pin plug for coupling to the arm base. In some cases, as noted earlier, there will be only three conductors, the two earths being combined in a dual screening. With a cheap arm, where leads simply protrude from the base through the mounting board with no further connection facilities, the best course is to fix a tag-strip or terminal board beneath the arm base and solder or screw the PU wires to this, each with the appropriate braiding or inner conductor from a pair of screened leads (**fig. 76**). The wires from the PU can then be positioned for minimum drag down through the base and will not be affected by movements or weight of the connecting cables. It is assumed here that signal wires from the pickup are not screened and are simply coded for appropriate termination by the user; but in some cases screened leads will already be in use within the arm, perhaps with braiding

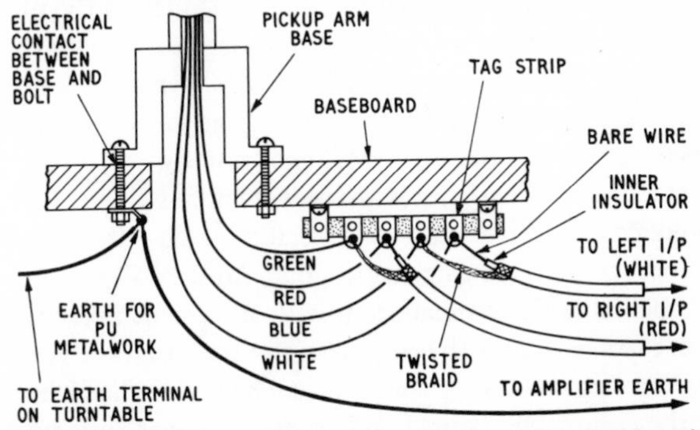

Fig. 76 Suggested scheme to cope with a simple arm not equipped with multi-pin socket at its base. In this case the PU wires are not screened, but a similar arrangement could apply with screened flex, taking live to live and braiding to braiding.

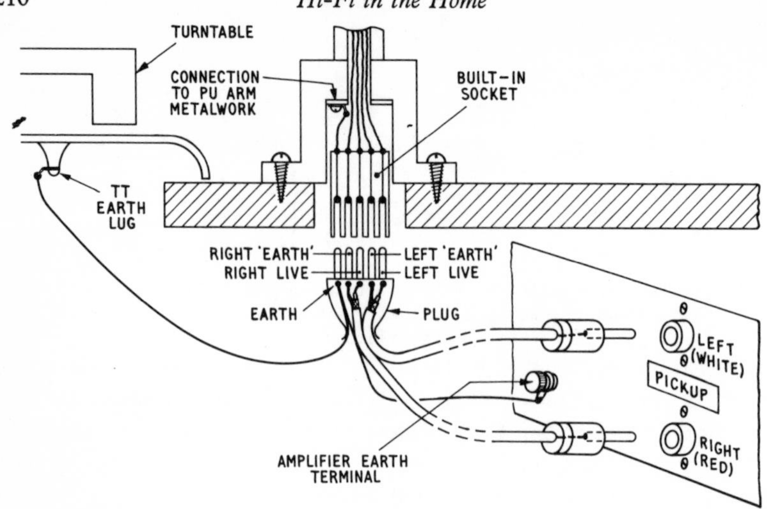

Fig. 77 Complete lead system for pickup signal connections and earthing of arm and turntable; such a scheme is used on expensive arms. On combined TT/PU player units there is only one metalwork earth.

contacting the arm's metalwork, thus making the extra earth wire shown in the illustration superfluous as the arm will be earthed via the signal leads.

When installing screened leads between pickup and amplifier it is important to use cable of low capacitance (which a good dealer will supply), especially if a run of several feet is involved, otherwise an electrical resonance can occur with magnetic cartridges, giving a false boost to high audio frequencies. The most sophisticated and foolproof pickup cable systems include another earth wire separated from the signal earths and bonded to the pickup's metal structure at one end and for connection to a separate amplifier earth terminal at the other. A wire from the tag normally fitted to turntables for earthing purposes would in this case join that connected to the pickup's non-signal earth. A system incorporating all these points is shown in **fig. 77,** where the separate earth wires would be ordinary flex with the insulation stripped back at the ends either for soldering to suitable tags or for clamping directly beneath terminals or nuts-and-washers at the earthing points. An amplifier not fitted with a specifically labelled earth terminal will usually have some covering note or suggestion in its manual, but in the complete absence of any advice, metal parts not used for carrying audio signals may be connected to the mains earth where this enters the equipment cabinet. When mains power systems do not carry a third socket for earthing via three-pin plugs (the three-way system is unknown in the U.S.A.) it may in extreme cases be necessary to provide a separ-

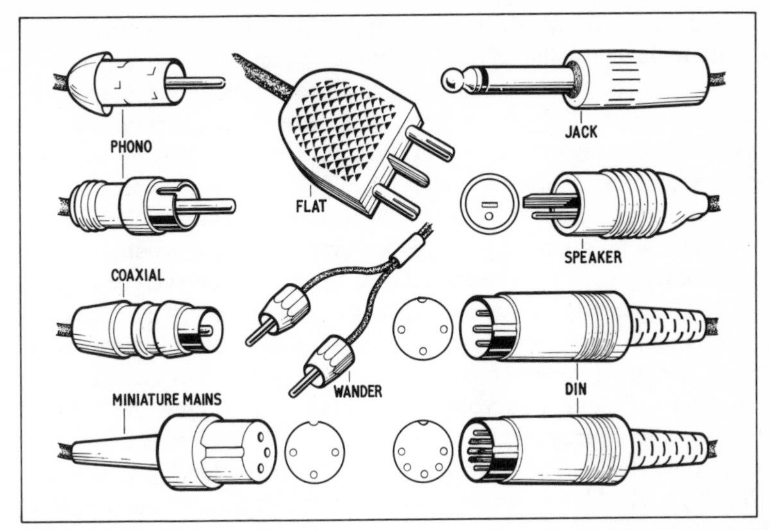

Fig. 78 Some types of plug used for interconnecting hi-fi equipment. Phonos are general-purpose connectors for conveying a single screened conductor, coaxials are used for aerial feeds, and jacks (in standard and miniature versions) very commonly for microphones. DINs may carry double or quadruple screened wires from pickups or to and from tape recorders, and speakers may be connected by wander plugs, flat plug or flat-and-round. Any and every combination, and some other types of plug and socket, will be found in practice.

ate earth from the water supply pipe system or from the Earth itself. This is all tied up with hum prevention, to which we shall return shortly after considering some other aspects of interconnection.

In recent illustrations small plugs have been shown on the ends of screened leads, with corresponding sockets on the amplifier. These are known as *phono plugs,* widely used on audio equipment and a simple means of conveying a screened signal, the cable's inner conductor being soldered to a central pin, with the braiding separated from the inner to be either soldered or clamped to the outer shield. Other types of plug and socket are also in use, particularly those conforming to the DIN standard which is rapidly becoming universal. The DINs incorporate several pins and are employed for multiple functions such as: carrying left, right and common earth for stereo pickups; likewise for tape recorders; two stereo pairs for both recording and replay when coupling a recorder to a hi-fi system; special non-reversible speaker plugs; and various other applications. Several types of plug and socket are shown in **fig. 78.** Whenever multi-pin DIN plugs are used great care should be taken to avoid confusing left/right and in/out connections, bearing in mind that the recommended patterns are not always followed by manufacturers. Again, close attention to installation manuals and panel markings is advisable.

Apart from the special case of a pickup arm without a made-up cable for coupling to the amplifier, the business of interconnecting hi-fi units is usually fairly straightforward, as dealer or manufacturer will provide appropriate leads for couplings between tuners, amplifiers and tape recorders, leaving simply external connections to aerials, loudspeakers and the mains supply. In cases where appropriate signal leads are not supplied, some specialist firms produce a range of cables terminating with any and every type of plug or socket*. The user who wishes nevertheless to make up his own inter-unit connections should always take the greatest care with soldered joints, avoiding any risk of short circuits or intermittent contacts, and never forgetting to put the rear parts of plugs on to cables before connecting the pins. One of the most frustrating and infuriating experiences in electronic hobbies is to discover an unfitted plug shield *after* expending much effort on soldering a cable's ends to the plug itself. Always push plugs firmly home into sockets, and when split pins are used prise them open a little to ensure a good contact; intermittent signals arising from sloppy connections can involve hours of investigation. Follow the advice of manufacturers on the type of wire or screened cable to use; make wires short, compatible with avoidance of mechanical strain or heat from valves and the need to keep signal leads away from transformers, electric motors and AC power wiring; allow for ease of access in the event of alterations or breakdowns; avoid contact between separate screened leads when their metal braiding has no outer insulation; and make sure that such leads cannot dangle near exposed mains connections.

Mains power wiring deserves a separate paragraph. Assuming the equipment to be positioned conveniently in relation to a mains socket, a normal two- or three-core flex will convey power from a plug to the equipment cabinet; the British user is well advised to make use of a proper three-pin earthed supply. With a ring-main system employing fused plugs, the fuse used should match the wiring and possible current consumption of the equipment rather than be left at 13 A, as is commonly but quite wrongly done. A 5 A fuse should cover all domestic hi-fi possibilities. Many amplifiers provide mains outlet points for feeding other equipment, and with such models the power lead can come straight to the amplifier, twin-flex feeds for turntable, tuner (if self-powered) and tape recorder being taken from the socket or terminal board provided. All such connections should be made with care, as befits high voltages; beware of stray wire ends, sometimes left out when twisting flex. With the

* Tape Recorder Spares Ltd. have an enormous range.

main power feed in, take care to follow the colour code on a three-wire system: green/yellow earth (top pin), brown live (RH pin), blue neutral (LH pin); or with older leads: green earth, red live, black neutral. When equipment has an adjustable mains tapping, make sure that this is set to suit the local supply voltage. When AC outlet points are not fitted to the amplifier it may be useful to mount a terminal board just inside the cabinet (but inaccessible to small fingers), connecting the input mains feed to this and using it as a distribution point for power to all units. Earth from a three-wire mains supply would go from here to the amplifier's earth terminal, which from then on should be used as the earthing point for items such as turntables and pickup arms, as already noted. Any mains wiring within the cabinet must be kept away from signal-carrying cables, especially at the turntable, where flex feeding power to the motor should not come near to pickup leads. In short, treat the mains power supply wiring as something quite separate from the signal feeds between units, only digressing from this rule if power amplifier, control unit and tuner are supplied with interconnecting cables which perform both tasks together. Note that these comments all refer to *mains* power and not to HT and LT which are sometimes fed separately from an amplifier to control unit and/or tuner when the latter are not self-powered. Once again, the manufacturer's manual should cover such matters.

Loudspeakers are fed from separate sockets or terminals on the power amplifier, connections usually being quite straightforward and obvious. Ordinary twin 5-amp lighting flex is suitable, though it might help when *phasing* the speakers for correct stereo performance (see later) if the two wires within a pair carry different colour insulations – black and red twin flex is very common. It is sometimes tempting to use fine bell-wire or telephone wire for speakers, but this should be avoided, as unless the runs are very short the wire's own electrical resistance may amount to a significant fraction of the speaker impedance, causing wastage of amplifier power. Once speaker positions have been decided the leads can be tucked permanently out of sight – under carpets, or perhaps even beneath the floor if this has to come up for any other job. In the latter case, a neat finish is achieved by fitting flush sockets on the skirting board near the equipment cabinet and by the speakers; but use a type of two-pin socket which cannot be mistaken for a power point.

The remaining 'extra mural' wiring arises from an aerial for radio reception. VHF/FM radio on Band II uses, like television, frequencies giving rise to curious local reception problems, the surest way to avoid difficulty being erection of an elevated external *dipole* or more complex array. In some areas reception of local FM broadcasts will be adequate with a mere 'picture-rail' dipole, and if there is a wall in the listening room at right angles to a line pointing towards the transmitter, such an arrangement is worth trying. A simple dipole is shown in **fig. 79**, where the two limbs can be aluminium tube of about $\frac{3}{8}$ in. diameter and the lead used for feeding radio signal to the tuner is a type of screened wire called *coaxial cable.* This is designed specially for RF applications at a particular matching impedance – usually 70–80 ohms – and is used also for TV aerials. Most FM tuners made in Britain have aerial input sockets to take the standard type of coaxial plug used to terminate such cable (see **fig. 78**), though many imported models have terminals or sockets for a flat type of twin-feeder. The latter often has an impedance of around 300 ohms and is designed to feed what is known as a *balanced* circuit, whereas conventional 70–80 ohm coaxial feeder is employed for unbalanced circuits. American readers will normally use a balanced arrangement, and British users of American or Continental VHF tuners may need to use a small transformer called a *balun* to match a 70-ohm aerial and coaxial feeder (unbalanced) to a 300-ohm balanced input. In some such cases of nominal mismatch a possible simple solution is to connect the coaxial cable's inner conductor to one of the twin input sockets and its outer braiding to any nearby screw or earth point on the chassis. Returning to the simple case, if a plain indoor dipole of the **fig. 79** type (mounted horizontally in Britain for BBC transmissions) gives completely noise-free reception on local stereo broadcasts, then proceed no further unless wishing to experiment with Continental reception in Britain or inter-city reception in the States. If multiplex stereo is not yet available locally but mono reception is excellent, one can again stick to the picture-rail device for the time being, hoping that stereo will be satisfactory when it comes but with the possibility of aerial improvement if signal-to-noise ratio is degraded.

Where a simple downstairs dipole fails, one can have a more elaborate aerial array of the type shown in **fig. 80**, properly erected on a chimney stack – a job for experts. But for the impecunious or keen do-it-yourselfer there is a compromise likely to be successful in most U.K. localities: a home-made dipole with reflector mounted in the loft. A possible design is shown in **fig. 81**. Here, the active dipole element is supported in a wooden bar, which can be drilled from the sides to gives a tight fit for the tubes, with electrical contacts made by means of self-tapping screws which couple soldering tags on the cable to the two tube ends inside the wood. As before, the inner ends must be separated, though the reflector is one continuous rod. Such an array, mounted horizontally in a loft well away from metal and masonry and pointing towards the local FM transmitter as indicated, should provide adequate signals for a sensitive tuner, but if it fails the matter really does become one for experts.

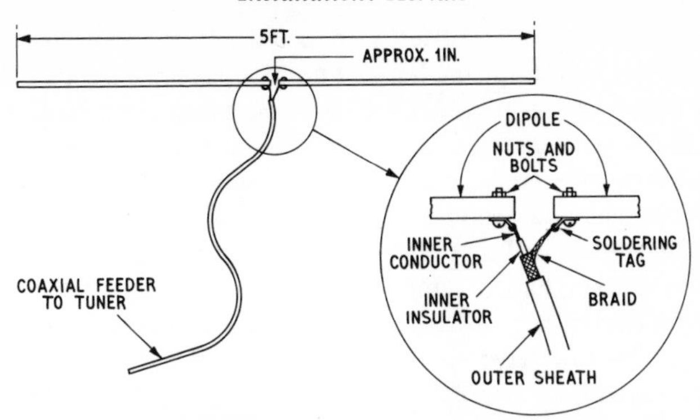

Fig. 79 Simple dipole for Band II. Where reception conditions are very good, this could be rested on a picture rail.

Fig. 80 An elaborate commercial VHF dipole suitable for multiplex reception in difficult localities. Manufactured by J–Beam.

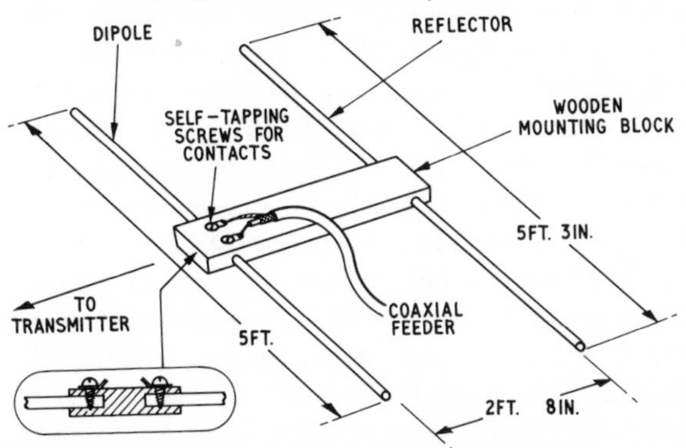

Fig. 81 Home-made dipole with reflector, for loft use.

The coaxial feeder from aerial to tuner should take the shortest convenient path, which may be outside the house if necessary; remember that many upstairs rooms have plaster partition walls which can sometimes be used to carry a cable out of sight from the loft down to ground-floor ceiling level. The screening effect of its outer conductor gives coaxial cable an advantage over twin-feeder in that it may run near metal objects on its way to the tuner, while a balanced feeder must be positioned more carefully. The simple types of aerial described here have a natural impedance of 75 ohms, hence the direct connection of coaxial cable; there are, however, commercial types to suit either a 75-ohm or 300-ohm feeder system.

Having dealt with the external connections – mains, speakers and FM aerial – there are three points on pre-set adjustments and impedance matching to be mentioned before we pass on to troubles likely to arise if things are mounted or connected wrongly. Following advice given in the last chapter, the application of some general rules regarding impedances and signal levels should have led to a satisfactory choice of units; but there will in some cases still be room for error or misunderstanding.

When amplifiers offer a choice of loudspeaker matching impedances, whether a range of values between upper and lower limits or several specific impedances at separate output terminals, make sure that the speakers satisfy any load conditions specified. Many speakers are quoted as having a particular nominal impedance, but some data sheets will say: 'impedance, 8–16 ohms', in which case the *lowest* figures should be taken as most relevant. Provided the speaker impedance is not less than about two thirds of the stated load value for a particular pair of amplifier output terminals (or sockets), connections to that output should be satisfactory. However, it is generally better from the distortion point of view if the speaker offers an impedance above rather than below the amplifier's nominal load value. In a marginal situation where, for instance, a valve amplifier has alternative outputs labelled 3–5 ohms and 12–16 ohms, and the speaker is stated to be 8 ohms, the safest course would be to use the lower output. If this permits reproduction of the loudest musical climaxes at an adequate volume level via the speakers in question, leave it that way. Connecting to the higher impedance output will give greater loudness at a given setting of the volume control, but the peak output power available before onset of audible distortion will not necessarily be greater. Some amplifiers have one speaker output only for each stereo channel, but provide for various matching impedances by means of internal connection changes or alternative links between terminals on the output transformer. Appropriate adjustments should be made at the time of installation in accordance with the above notes.

One final point on amplifier outputs: never switch on a power amplifier with the speaker disconnected, or unplug the speaker while an amplifier is in use, or short-circuit the amplifier output terminals. These could in some circumstances cause damage, loudspeaker wiring shorts being particularly prone to blow fuses or damage components in transistor amplifiers. When making alterations to speaker connections, play safe by switching off in the case of a transistor amplifier and at least turning the volume control to zero on a valve model.

The second matching adjustment that could be necessary when installing equipment is at the preamplifier input. There may, for instance, be a choice of sensitivities for magnetic pickups, either via separate input sockets or by changing a plug-in adaptor, or the socket panel might have pre-set controls needing attention. These points will usually be covered fully in the amplifier installation manual, but such adjustments or choices as are available should generally be made so that most programme material from all sources requires reasonable settings of the volume control for normal use. If the signal level from pickup, tuner or tape recorder is so high for the relevant preamplifier input that the volume control is always confined to the first few degrees of its total rotation, this carries not only the inconvenience of accommodating large changes of reproduced loudness within tiny movements of the knob, but indicates a possibility that signal peaks will exceed the overload margin of stages before the volume control. In the case of efficient horn-loaded speakers fed from amplifiers of unnecessarily high power rating there may be a similar effect, the volume control being used not to hold down a high input, but to prevent stimulating the power stages to a high output. However, with speakers of average efficiency this should not occur, and it is reasonable to use any pre-set sensitivity adjustment facilities to give normal volume control settings around the half-way mark – say between 5 and 6 on a knob calibrated 0–10. This is by no means crucial, and it does not matter if on occasion the volume has to be turned right up towards maximum, provided the signal-to-noise ratio is satisfactory at this setting. If pre-set controls seem to affect the S/N ratio, the settings finally adopted should give the best compromise between background noise and volume control flexibility. Remember that with disc reproduction via a magnetic pickup the volume setting needed for healthily loud (not over-loud) reproduction of orchestral music should give no hum and no more than the gentlest suspicion of amplifier hiss as judged from the listening area in a quiet room when the pickup arm is lifted with everything else left untouched. Remember also that an increase of background noise when the volume control is turned above this setting is of no consequence for practical musical purposes.

The last small matching matter concerns pickups and preamplifier input impedances. Magnetic cartridges are usually designed to work with an electrical load impedance in the region 50–100 K, the precise value within this range generally being unimportant. However, it will be recalled that ceramic pickups deliver their self-equalised (or corrected) outputs at a higher voltage level into an amplifier impedance of one or two megohms, and it could happen that for reasons of economy a pickup of this type must be used for a while with an unsuitable amplifier input. A common situation of this type involves coupling a ceramic cartridge to a magnetic input, the effect of a 50 K load being to modify the frequency response of the pickup such that it begins to resemble the velocity-sensitive characteristic of a magnetic model. At the same time the cartridge's output voltage is lowered towards that obtained from a magnetic type, the net effect being that the signal becomes more suitable for feeding to the sensitive, equalised, magnetic pickup input in the very act of being connected thereto. In practice it is necessary to use a few extra components if this matching and attenuating task is to be carried out in a reasonably accurate manner; some ceramic cartridges are supplied with a suggested circuit for the purpose. The reader unconversant with circuits or the handling of resistors and capacitors should leave this problem to his dealer, and of course such an approach should not be attempted if the amplifier has a high impedance input of sufficient sensitivity for ceramic pickups. In the reverse situation where an amplifier has such an input, but cannot cope with a magnetic pickup which the owner wishes to use, it is necessary – as noted before – to employ an extra front-end preamplifier unit with built-in equalisation and high sensitivity. Such devices are designed specially for the user wishing to graduate to a magnetic PU but owning a 'budget' amplifier suitable only for ceramics. Installation should be straightforward, following the maker's instructions and the general notes on interconnection and earthing in this chapter.

Apart from final placing of loudspeakers, the installation task is now complete, and it should be possible to switch on and get things working. But errors in connection, earthing or positioning of components can lead to various snags, which will now be tidied up before we move on. Hum is the most common irritant in hi-fi, and is inclined to turn up at the slightest provocation and in the most expensive set-ups simply because certain elementary precautions have not been taken. Audio equipment normally operates from the mains supply, and in consequence AC voltages, currents and magnetic fields at mains frequency exist in and around wiring and components. When such agitations succeed in inducing small signals into the audio circuits, these are heard as a continuous low background note of definite pitch – hum. Unwanted hum signals occurring within complete amplifiers are kept down to negligible levels in good designs, though the occasional poor earth contact or faulty component – often a smoothing capacitor – can give trouble. However, most hum problems arise from the disposition and interconnection of separate units, which may be covered under four interrelated headings: screening, magnetic induction, wiring loops and earth paths.

Whenever voltages occur, the wires or circuits in question are surrounded by an electric field, and in the case of AC mains this field fluctuates at 50 Hz or 60 Hz and is fairly intense due to the high voltage. This electrostatic field will affect nearby wiring due to capacitance existing across the intervening space, and if a conductor in the vicinity happens to be carrying audio signals some unwanted hum could be added to them. The higher the associated circuit impedance and the lower the audio signal level, the greater the likelihood that a conductor will suffer from hum induction of this type, so some means must be found whereby vulnerable wires are shielded from hum fields. Screened lead is the answer, used in the matter already described for pickup connections, with the outer conductor connected to the earthy side of signal feeds and so arranged within terminating plugs that live conductors are always surrounded by some earthed metalwork. Pickup leads are particularly sensitive, and if, for instance, hum were experienced with the connection arrangement shown in **fig. 76**, it would be worth mounting a tin-plate shield over the tag strip, not touching any of the tags and with a separate connection to the earth wire. Unless manufacturers' instructions advise differently, or made-up interconnecting leads are provided with equipment, all cables carrying signals between units should employ screened wire – live via inner to live, earth via outer to earth – and the braiding should itself be covered with an insulating sleeve to avoid contact with other components which can introduce alternative earth paths: another cause of hum, as we shall see shortly. The exception is speaker leads, where impedance and signal level are too low and high respectively for electrostatic hum induction to be troublesome. As noted earlier, signal cables should be as short as is consistent with reasonable distance from mains leads, remembering that electric fields are coupled via capacitance, and this is reduced as the conductors are moved further apart. Contrariwise, the mains wiring itself should always employ twisted or closely paralleled twin flex, as this minimises external fields by a process of mutual cancellation, voltages and currents in the two wires being in anti-phase.

Magnetic hum induction comes from devices in which mains current passes through coil windings to produce, deliberately, powerful AC magnetic fields. Transformers and electric motors come under this heading, where the fields are used respectively to generate suitable supply voltages (in other windings) and rotary movements. Careful design limits the external effects, but some field always gets out and can cause trouble by inducing hum into coil windings used in signal-carrying or generating components. One difficulty is that magnetic hum fields are not deterred by ordinary electrical screening, so that special magnetic shields made of materials such as mu-metal must be employed. Pickup cartridges, tape-heads, filter coils in preamplifiers, miniature transformers used with some moving-coil pickups – all have appropriate magnetic screens. But close proximity to a mains transformer or motor can still give hum trouble with some of these, a factor to consider when planning the layout of an equipment cabinet. As usual, the pickup is the most likely component to need special consideration, particularly as it must automatically be near to the turntable motor. Good TT motors do not give trouble, though pickups differ in their sensitivity to external hum fields, variable reluctance types generally being most troublesome in this respect. If magnetic hum induction from the motor is suspected, try switching the motor on and off with the pickup positioned at various points across the turntable, and if the suspicion is confirmed look into possible alternative dispositions of pickup arm to keep the head further from the motor. If hum varies with movement of pickup even when the motor is off, try repositioning the amplifier in case its mains transformer is the culprit. If distance cannot be increased, a simple change of angle will sometimes improve matters. Tape recorders have their own appropriate internal layout to minimise hum induction into the tape-head, but it can happen that when a recorder is fitted into a cabinet with other units the head finds itself rather too near an extra mains transformer; this again is a matter for individual experiment.

Just as alternating magnetic fields will induce hum into exposed coil windings and their cores, so they may also find their way into unintentional 'coils' existing as loops in the signal wiring. Anxious that everything should be earthed thoroughly, the user might link various units in his hi-fi system with a separate wire, thinking perhaps that he is following the sort of arrangement shown for pickups in **fig. 77**. However, in the pickup case illustrated the metalwork is isolated from the signal leads, whereas in nearly all electronic units the earthy sides of signal circuits are connected to chassis. In the latter case an extra earthing wire (**fig. 82**) duplicates the function of the screened signal cable's outer conductor, at the same time creating a sort of loop or single-turn coil that could be situated within a magnetic hum field. In **fig. 82** some field from the amplifier mains transformer is seen passing through the *earth-loop* formed by the wiring, the effect being to induce a tiny circulating current

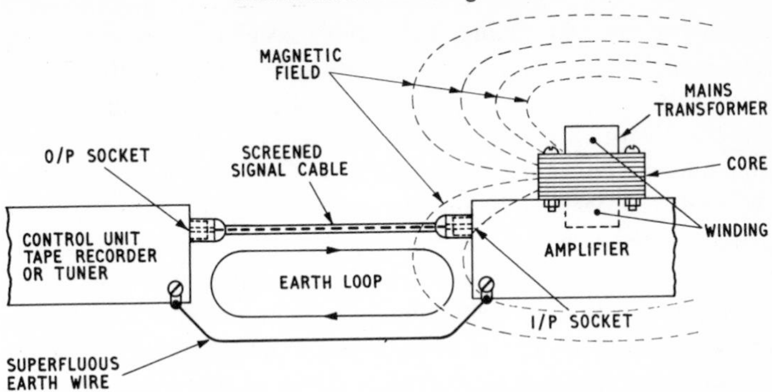

Fig. 82 This shows how connection of an extra earthing wire between chassis, 'just to be sure', can create a loop into which hum might be induced.

into the loop which, because of the resulting minute AC voltage generated across the earthy side of the signal feed, causes hum. Because of this common effect, earth-loops or similar arrangements are known also as *hum-loops.* Ideally, interconnecting cables in an audio system should be arranged to convey signals by one path only between any two points, that path comprising one live and one earthy conductor in the form of a screened lead. Such leads should provide earth connections between amplifier, tuner, recorder, etc, and also the HT earth return in the case of units obtaining their power supply from the main amplifier, separate earth wires only being used in the case of items such as turntables and most pickup arms, which have metalwork isolated from signal circuits. Separate earthing on such components – particularly turntables – is not only permissible, but desirable, as spurious static charges or tiny mains leakage currents are best conducted directly to the system's main earthing point rather than via signal earths, where they could generate small interfering voltages.

Any extra metalwork in an equipment cabinet might usefully contribute to screening if connected to the main amplifier earth terminal, but only if not already in contact with an equipment chassis, otherwise an earth-loop would be introduced. Remember that the type of pickup arm in which screened leads come straight out of the base without plug and socket fittings may depend on the signal lead braiding for earthing of the metalwork, in which case an extra wire for the arm itself might create a hum loop. With this sort of arm it would be better if the turntable earth wire went straight to the main earth terminal. A rather puzzling type of earth-loop can arise with VHF/FM tuners, where an unsuspected break in the outer insulating sleeve around the coaxial aerial feeder causes the braiding to rub against a gutter or similar object that is vaguely earthed via the building. This connects the tuner chassis to an external earth, while the signal feed from tuner to amplifier also couples the chassis to mains earth, thus giving a rather roundabout hum-loop. Similarly, a tape recorder fitted with a three-core mains lead will provide a double earth path if the third wire is left intact when signals are fed to and from a hi-fi installation. When a recorder is to be installed permanently its mains earth connection should therefore simply be omitted, all earthing coming via the signal cables. A recorder that is occasionally used separately may have its mains plug fully connected, but a corresponding three-pin socket in the equipment cabinet can be wired without an earth connection, so that in this location only the recorder is earthed not via the mains but to the other equipment by the signal lead earths.

Finally on earthing – or rather its absence – if the mains supply is of the plain two-wire type, all the procedures mentioned here should be followed except that there will be no mains earth to connect to the central amplifier earthing point. In this situation the equipment may well work quite satisfactorily, in which case leave well alone; but if hum is obtrusive try reversing the mains plug. If this fails to improve matters, or if hum comes and goes according to one's movements or what one touches, it may be necessary to install a proper external earth connection. A lead water pipe that clearly disappears into the ground is ideal, and small clamps are available from electrical shops for making earth connections to these. Alternatively, a metal plate, pipe or old saucepan may be buried where the ground is normally moist, with stout wire firmly attached in good contact and fed to the equipment cabinet for connection to the earth terminal.

So far we have considered electrical hum which comes through the amplifier and is turned into sound at the loudspeakers. There is also mechanical hum – not common, but sometimes very puzzling because it is difficult to locate until diagnosed, and often seeming to vary with cabinet position. This usually arises from the amplifier mains transformer, its laminated iron core tending to buzz slightly at mains frequency, thereby setting up sympathetic vibrations in the equipment cabinet. Very often this is easily cured by tightening the bolts used to clamp the transformer core to the chassis, or to its own supporting frame (see **fig. 82**). In extreme cases it may be necessary to stand the whole amplifier chassis on a piece of foam rubber to isolate its vibrations from the cabinet.

These are would-be outgoing vibrations to be kept in, but the technique used to contain them may in some cases be necessary to keep incoming vibrations of various sorts away from the pickup. Casual jarring of the equipment cabinet or, in some difficult rooms, transmission of substantial sound vibrations from loudspeakers via the floor, can cause the pickup stylus to jump, judder or even oscillate continuously in the record groove, with corresponding signals going back through the amplifier to produce further noises from the speakers which may lead to continuous acoustic or mechanical feedback. As noted earlier, the pickup arm should be secured firmly to the turntable baseboard, and if incoming vibrations are to be barred the whole TT/PU assembly must be isolated or decoupled from the cabinet. Some player units are fitted with springs or rubber pads for this purpose, but when separate components have been mounted on a common baseboard, it is this that must be resiliently suspended. In most equipment cabinets this board sits on wooden runners or corner pieces, and if its edges are shaved to give clearance on all four sides it is then a simple matter to lay felt or foam strips on the supports and rest the baseboard on these. Whatever the detailed arrangements, the aim is to ensure that (a) the turntable and pickup base are coupled together firmly, either as integrated parts of a player unit or on a common baseboard, and (b), that if there is trouble from acoustic feedback or excessive sensitivity to external vibrations, this assembly must be isolated from the cabinet to the extent that all contact is via some spongy or compliant material. If, despite these precautions, vibrations transmitted from the speakers produce unpleasant effects by feedback to the pickup, it may be necessary to experiment with positioning of equipment, perhaps in the last resort standing the speaker cabinets on thick felt pads or even – as has happened in very exceptional cases – placing the player unit on a slab of concrete, the slab in turn being isolated from its supporting cabinet or table by substantial blocks of foam rubber.*

From such dire and unlikely extremes we must now move on to the more general subject of loudspeaker positioning. The aim here should be to achieve the best tonal balance and stereo presentation consistent with points of convenience discussed earlier. In all likelihood there will not be much room for manoeuvre, and in most rooms the wall against which the speakers stand has been decided at the outset, the problem reducing itself to one of height and angling within, or distance from, a pair of corners. Corners certainly enhance loudspeaker performance in the bass, as sound radiated in this frequency region would, if a speaker were suspended in mid-air, move out equally in all directions, whereas situating the speaker cabinet first on a floor, then against a wall, and then in a corner, progressively reduces the 'solid angle' into which a given amount of sound energy is radiated. This increases the sound power available for projection forward into the room, relatively more at low fre-

* See *Stereo for Beginners*, by B. J. Webb, page 88.

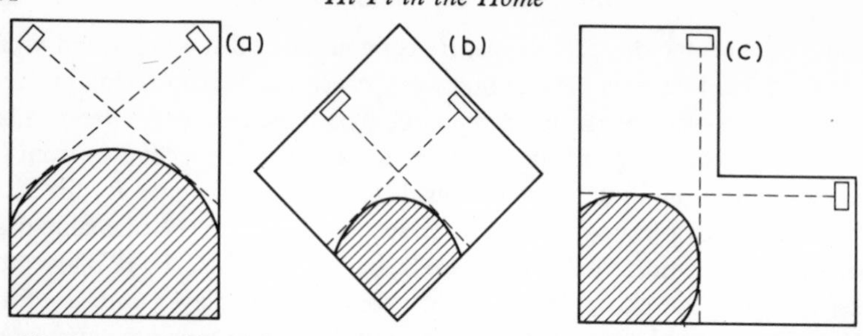

Fig. 83 Some possible stereo speaker layouts. Conventional scheme in a rectangular room (a) usually gives the largest listening area (shaded). In a square room with troublesome resonances the speakers might provoke least coloration if arranged as in (b). Various layouts are possible in L-shaped rooms, (c) offering best symmetry.

quencies because there would in any case be less diffraction around the cabinet further up the scale, and also because of an acoustic loading effect on the speaker. Folded horn enclosures, particularly, depend on corner mounting for proper performance, though some large non-horn-loaded systems in the professional monitoring class are designed to have a balanced acoustic frequency response when stood well away from walls – an unusual feature in domestic speakers. However, one must be careful with this corner mounting business, as improved overall bass performance is frequently accompanied by noticeable coloration due to LF room resonances; simply against a wall but away from a corner is often a useful compromise. In a room with dimensions near to one of the 'ideals' in **fig. 69** it should be possible to make full use of corner mounting for a conventional stereo layout (**fig. 83(a)**) without undue trouble from eigentones; but in a poorly proportioned room a corner mounted speaker might 'excite' the main resonances rather too easily, and in such a case it would be worth experimenting with various positions modestly away from the corners. It could happen that one corner provokes room resonances and the other doesn't, suggesting some asymmetry in stereo positioning. In the extreme case of an impossibly boomy near-cubic room it might be necessary to locate the speakers so far from the corners as to suggest the unusual layout of **fig. 83(b)**. If the room and other furnishings preclude a listening position well back from the speakers, as could happen if the only available space for speakers is on a long wall, it might be necessary to bring the enclosures in from the corners so that the gap between them is not significantly greater than the listening distance. Non-rectangular rooms offer much scope for stereo experiment, one suggestion for an L-shape being **fig. 83(c)**.

Apart from the effects of room boundaries on bass performance, the overall acoustic character of a room can influence apparent tonal balance very noticeably. A room with few soft furnishings and no carpets usually has a bright, lively acoustic and may become tiresome and rather overbearing when used for extended listening to large-scale music. The degree of liveliness will also influence speaker angles and positions chosen to give the best stereo performance, and it can be said that in general stereo requires a 'dead' rather than 'live' room for full enjoyment. If heavy curtains or thick carpets are ruled out, a few rugs and some soft chairs will help, and if all else fails a persistently over-bright room will benefit from a wall or ceiling of acoustic tiles, which are available in many attractive patterns and may be emulsion painted. There is a common misunderstanding about acoustic tiles or boards: they do *not* give a very significant improvement in sound insulation between rooms – their functions is to reduce reverberation within rooms. Whatever is done in the way of treatment or decoration, it is desirable to maintain reasonable symmetry between the surroundings of left- and right-hand speakers. In the simple layout of **fig. 83(a)**, for instance, it would be wrong for one side wall to be hard painted plaster with the other covered in heavy curtains, particularly near the speakers. A large absorbent area is best situated on the speaker wall or behind the listeners, though when speakers are of the type depending to some extent on reflection for high frequency distribution, it might be wise to leave some bare surfaces in the crucial regions – again with an eye on stereo symmetry. Full-range electrostatic speakers of the open, flat (doublet) type present special problems because of radiation to the rear at low frequencies, and while these also benefit from absorbent surfaces, behaviour in relation to walls and corners can be rather unpredictable and may not conform to earlier remarks. This type of speaker should certainly be tested at home before purchase.

Before discussing final stereo positioning, there is a question of height to consider. Large commercial speakers are normally designed for floor-standing use, tweeter units being positioned within cabinets at around the average head height of a seated person. When no other furnishings are tall enough to obstruct the path between tweeters and ears, everything is satisfactory. Medium sized speakers are not so easy to use on the floor: not only do chairs tend to block the aural view, but people also get in the way on group listening occasions. The really miniature IB speakers must of course be mounted well above floor level, and in most homes will be placed on some other items of furniture or in the bookshelves from which some models take their names. When the stereo sound-stage is well presented laterally it is inclined to seem slightly more realistic and pleasing if elevated just a little above ear level. An impression is conveyed that the more distant performers – apparent distance being set by recorded reverberation – are tiered up as if on a concert platform; **fig.**

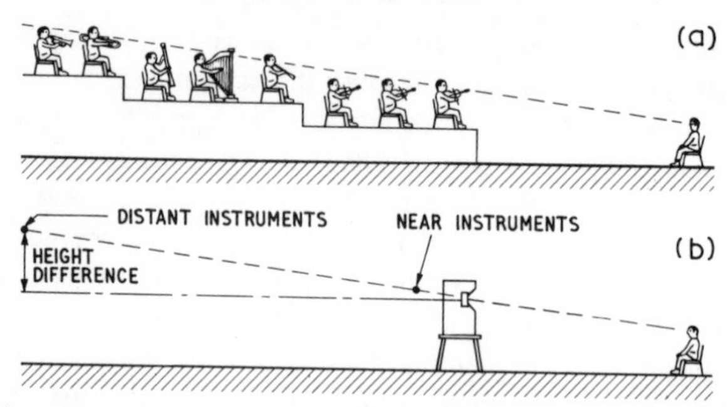

Fig. 84 Orchestras usually perform on a tiered platform (a), and some aural impression of this arrangement may be created by mounting speakers slightly above ear level (b).

84 shows the real and imaginary situations. A small point, perhaps, but worth bearing in mind when planning an installation. When hi-fi equipment has to be housed in a tightly packed living room there are also practical advantages to some modest speaker elevation, as a couple of judiciously placed shelves can leave space beneath for other furniture (**fig. 85**). In the future, 4-channel reproduction will require two further speakers, and these will very probably need to be mounted quite high up – perhaps in the rear ceiling corners.

To give optimum stereo performance over the largest listening area speakers must be angled to make use of their directional properties as explained in connection with **fig. 49** (page 144). This is done by a process of trial and error, the first step being the phasing process mentioned earlier. If all the fitting and connecting has been done correctly, the signals for left- and right-hand speakers will arrive at the amplifier output terminals or sockets with the correct phase relationship, and if the connections from each output to each speaker follow the same pattern this will preserve correctness right through to sounds in the air. As noted before, colour coded cable helps in this. To make sure that everything is in order, place the speakers in any convenient position permitting easy listening with exact left/right symmetry, in the manner of the central listener in **fig. 48** (page 140), switch the amplifier for mono operation on both channels (double-mono), and reproduce a signal from any convenient source – disc, tape or radio. If phasing is correct the sound should seem to come from a point mid-way between the two speakers and will be exactly central and distinct when listening precisely on the centre-line. If a sharp, detached, central image cannot be found, but there is instead a vague, broad sound inclined to leap from side to side with only small head movements, this indicates wrong phasing. Reversing the wires to *one speaker only* should effect a cure. With amplifier and speaker connections thus confirmed, switch to stereo and play a

Fig. 85 The typical sitting/living room cannot accommodate large floor-standing speakers, but corner shelves for a pair of modest units lift the sound-stage above other furniture to give good stereo in ordinary surroundings.

mono record. As we learnt in an earlier section, a mono disc will produce identical outputs from the two channels of a correctly wired stereo pickup, and apart from a possible need to shift the amplifier balance control to cope with a decibel or so of left/right sensitivity difference in the cartridge, there should still be a firm and clearly defined central image as heard from the 'stereo seat'. Switching between mono and stereo functions while playing a mono disc should therefore produce no appreciable shifting or broadening of the central sound image, as in both conditions an essentially double-mono signal is being fed to the speakers. If a mono disc played via a stereo pickup gives a thin, distorted and very weak sound with the amplifier switched to the mono function and will not produce a central sound image when switched to stereo, this indicates a reversed pair of connections on one cartridge channel.

Phasing checked, the stereo angling process may now begin, again using a mono record. Place the speakers in their proposed positions and arrange their axes to cross just in front of the intended listening areas as in **fig. 49** (page 144). Playing a mono record with the system switched for stereo operation, check that a narrow, central sound image is heard along the left/right centre-line. If necessary, compensate for any one-sided bias – caused by reflecting surfaces or room asymmetry – by altering the angle of one speaker only. Now move off the centre-line within the proposed listening area and make a mental note of the amount by which the sound image appears to shift and broaden. If by some acoustic miracle it remains central and narrow despite large lateral movements, leave the speakers where they are. Normally there will be considerable changes, and the object of speaker angling is to minimise them. Try facing the speakers more nearly towards each other and listen again, checking first that a central image may still be heard on the centre-line. If there seems to be either more or less shifting and broadening of the image, alter the speaker positions accordingly, repeating the procedure several times until the best compromise is reached. In the case of speakers using reflected sound the central image will be less precise at the outset, but though broader it may seem to shift less as one moves to the side. Every speaker and room combination is different, in one case giving a reasonably stable double-mono image straight away, and in another requiring much experiment and only yielding consistent results if the wall surfaces between and around the speakers are made absorbent with curtains or acoustic tiles. Remember that this whole exercise makes use of the tendency for speakers to focus their sound forward to some extent, and as no practical speaker is likely to have exactly the right radiation pattern at all frequencies there is bound to be compromise and some ambiguity in what is heard.

It follows from the nature of stereo perception that if an identical signal fed to both speakers produces an impression of a more-or-less central sound source at most points in the listening area, then a stereo signal fed to the same speakers will be well presented over the same area. In fact the situation is better than this, for a double-mono signal is no more than its name implies, there being no subtleties of surrounding ambience to give the ears additional clues regarding the sound's 'correct' direction. So, if that mono record disappoints by giving a sound that refuses to remain central and always broadens when heard away from the 'stereo seat', it may still transpire that a nominally similar sound on a stereo record will be much more stable. Having achieved the best compromise with a mono recording, play some stereo records which feature a soloist against an orchestral backcloth, when the 'imprisoning' effect of the surrounding orchestra and recorded ambience may surprise in its effectiveness. With some care and patience it is possible with most commercial speakers in nearly any room to obtain effective and satisfying stereo to the extent that foreground soloists and the main instrumental sections of an orchestra remain clearly separated and reasonably spread in proper order between the speakers for any listener in an area large enough to contain a dozen or more people.

So far we have considered only the disposition of complete commercial

Fig. 86 Some speaker drive unit assemblies for those who like to make their own cabinets. Top-left, KEF preassembled baffle type K2; left, one of a range of Peerless kits shown mounted on a baffle; top-right, Lowther PM2 unit designed to cover the whole frequency range when mounted in a horn type enclosure; below this, Module by Richard Allan for use in miniature IB cabinet; centre-bottom, packaged Unit–3 kit by Wharfedale.

loudspeakers, though the keener handyman may wish to save money and exercise his skill by building cabinets to suit separate drive units. Many speaker manufacturers offer a full range of units and publish details of suitable cabinets for the home carpenter to tackle, and in some rooms there will be odd corners, recesses, alcoves, etc, across or into which panels could be built to give appropriate enclosed air volumes. A long bay window, for instance, may benefit from installation of a window seat, into the ends of which could be built some speaker enclosures. Generally, the non-technical home constructor would find it safer and easier to build cabinets for single full-range drive units, though some multi-unit assemblies are available ready mounted on panels with crossover circuits all wired into place, while other manufacturers offer kits of separate units with crossover networks and detailed assembly instructions. Some examples are shown in **fig. 86**, covering large and small assembled systems, large and small kits, and a twin-cone unit for housing in a horn-type enclosure – only to be tackled by the ultra-enthusiast. Acoustically suitable decorative grille cloth materials are available for disguising the front of home-made speaker enclosures, and indeed for hiding the whole stereo speaker system should room decor happen to lend itself to such treatment (**fig. 87**). There are distinct psychological advantages in hiding the speakers when listening to music,* especially if the disguise permits one to imagine a concert hall extending back behind what would normally be an all-too-solid looking wall. Also, stereo perception benefits, as removal of immediate awareness of the sound sources makes it easier to accept the idea of musicians spread right across the space beyond the speakers and not just concentrated around them.

We have now come right through the hi-fi chain from the installation angle, though one or two pickup matters were left incomplete on the way because of the need to get the whole system working before judging finer points of performance. Pickup tracking weight and side-thrust compensation must be set, and both adjustments are most readily made with the help of special test-discs issued to explain technical record playing problems in an easy manner.† Using the most heavily modulated bands on one or more of these records, the playing weight should be set to avoid mistracking and then checked with a pair of tiny pickup scales (available with various other disc and tape accessories from several manufacturers) to ensure that it is not above the upper limit recommended by the cartridge manufacturer. If an altogether excessive weight is needed – bearing in mind the findings published in magazine reviews as well as claims in the maker's literature – make sure that the pickup arm is not imposing extra drag because of poor lead positioning or some similar factor. Assuming weight to be set and of acceptable magnitude, side-thrust correction should be adjusted using a test-disc with identical heavily modulated passages on alternate left and right channels, the correct lateral bias being that giving an equal propensity to mistrack or distort on either groove wall at a given playing weight. The various test-discs referred to in the bibliography carry other setting-up instructions which help in assessing overall quality of an installation and in revealing deficiencies great and small.

One apparent deficiency which may plague the owner of older mono records is excessive surface noise. Stereo pickup cartridges fitted with elliptical styli usually avoid this trouble as the stylus tip sits well up the groove walls, but models with 0·5 mil spherical tips sometimes explore parts of the groove into which dirt has been pushed by mono pickups with larger styli. Any user with a

* See *Aural–Visual Discord*, by Rex Baldock, *Audio Annual* 1966.
† See bibliography.

Fig. 87 End of the author's sitting room, showing curtain used to hide loudspeakers. A dark grey fibrous material (type 'T–100 Charcoal' by BFF) performs the main task of reducing light transmission while remaining transparent to sound, a frontal open-work coloured net completing the disguise.

substantial collection of earlier discs would be wise to opt either for an elliptical tip or a 0·7 mil spherical stylus, the latter available as an additional slide-in assembly on some moving-magnet cartridges. Also, and despite the stereo setting-up technique described in recent pages, mono discs should normally be played with the amplifier switched for mono operation, as such switching connects the two pickup channels together, thereby cancelling the cartridge's sensitivity to vertical modulation and reducing surface noise and various minor types of distortion. Noise due to dust is kept down by use of appliances mentioned in the next chapter, though static electrical charges on records make the situation very difficult at times, and one helpful item suitably covered under an installation heading is an anti-static turntable mat. If records seem particularly prone to dust attraction and exhibit electrical cracklings when moved to and from the turntable, it might be worth replacing the existing TT mat with a special anti-static one.

Stylus wear worries many hi-fi equipment users, a situation not helped by a scarcity of retailers with suitable microscopes. A diamond tip working in dust-free conditions at a playing weight of $1\frac{1}{2}$ grams or less in a modern cartridge should cover two thousand or more 12 in. record sides before appreciable wear takes place, and if one keeps a rough tally on playing time it soon becomes apparent how many months or years may pass before an examination is necessary. When the time comes, the cartridge or stylus assembly is best returned to the maker, who will advise on replacement or how much further use is permissible. Tape users have similar worries with tape-heads, though a more likely short-term cause of deterioration in quality here is contamination of the head with oxide coating from the tape, a matter for modestly regular attention by fastidious owners. Low-speed cassette players are particularly prone to troubles arising from tape-head contamination, as quite small specks of dust can upset channel balance and degrade HF performance. But now we drift to maintenance rather than installation, a sign that it is time to get on with some listening.

9. ADVENTURES IN LISTENING

THE EQUIPMENT is installed and in working order – what shall we play? This is our first venture into stereo and high fidelity – which recordings will start the new collection? Disc records are delicate, precision products – how should they be stored and handled? Good reproduction brings home listening nearer to the concert hall or opera house – how can this be shared with other music lovers? These are the sort of questions to be answered in this chapter.

The practical business of record storage and handling forms a natural post-script to installation covered in the previous chapter, and is a fitting prelude to enjoyment of the sounds of music. In the tiny grooves may be deposited blazing masterpieces or shimmering subtleties of sound, and from sheer love of music it is sensible to show respect for these mundane pieces of flat black plastic. Flatness is important and is most easily retained by storing records vertically side by side, packed with sufficient tightness to prevent individual discs from leaning, but not so tight that removal is difficult. Cases, cabinets and racks are available for record collections large and small (**fig. 88**), but whatever the accommodation it is wise to face the open ends of record sleeves away from the cabinet front to minimise exposure to dust. Some equipment consoles have storage space for a modest record collection with access from above. Here it is most important not to have sleeve openings facing upwards – a direct invitation to falling dust, grit, cigarette ash and a dozen other household contaminants. When records do not quite fill a storage compartment, an easy way to avoid the risk of warps inherent in a leaning position is to pack the spare space lightly with a roll of corrugated cardboard or piece of foam plastic. A reasonably constant temperature is desirable for record storage, and an ultra-dry atmosphere should be avoided if possible; though this is a counsel of perfection perhaps best taken to mean simply that extremes should be eschewed. An atmosphere of dust, fumes and general grime is likely to trouble record grooves in the long run.

When buying records, patronise establishments giving an absolute guarantee that discs are unplayed; before handing over your money always examine the surfaces for scratches or other deformations and check that there are no obvious warps or ripples, and that discs are not *dished*. The latter refers to a saucer-like effect causing a record to bow up towards the centre-hole or towards the edges, depending on which side is uppermost. Another possible manufacturing fault – a rare one – will not be noticeable until the disc is at home, rotating with pickup stylus in the groove. If the record is a *swinger* the pickup will move from side to side once per revolution because the centre-

Fig. 88 Record storage may take various forms. At top-left is the Foldaway cabinet by Record Housing, with sliding doors that fold back; beneath this, handy Paddock Tidy by Power Judd. Top-right, an unusual system by Recordaway which may be fixed to the undersides of shelves, with slide-out transparent containers for discs. Bottom-right, the Criterion by Pheonix holds up to 320 discs.

hole is not truly central in relation to the groove. Normal tolerances permit a maximum error of five thousandths of an inch (5 mil), just about visible to the naked eye in terms of pickup arm movement, but a definite swinger will be obvious at a glance and may introduce audible wow on some types of music. The record dealer should change such a disc.

Assuming that perfect records are perfectly stored, and that they are played at $1\frac{1}{2}$ grams or less with a high quality pickup fitted with a well-shaped, polished and unworn diamond stylus, it is still possible for discs to suffer through simple mishandling. The need to avoid scratches is obvious, but some of their backdoor causes are not so evident until discovered by bitter experience. Firstly, always use a properly designed lowering device for putting the pickup stylus in the groove. This is not because a hand cannot lower a pickup

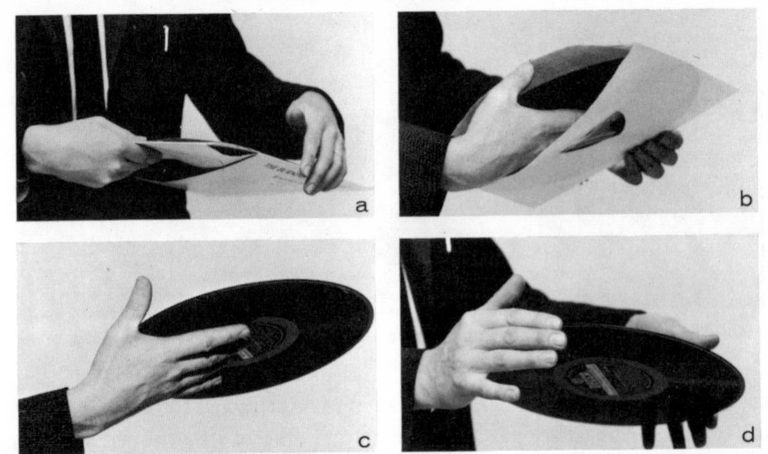

Fig. 89 Removing a record from its sleeves: (a) bow the outer sleeve to ease extraction of inner; (b) support disc without touching grooved surfaces and hold inner sleeve apart; (c) remove, still without fingers on the grooves; (d) hold record at edges only, ready to lower on to turntable.

gently enough, but because only the slightest disturbance or involuntary nervous impulse is needed to convert a simple downward action into a heartrending lateral graunch as the pickup skids across the record. With care, a pickup may on occasion be lowered by hand on to the run-in groove, but the process of searching for a particular passage by hand half-way across the disc will sooner or later end in disaster.

Next comes the literal handling of records as they are moved in and out of sleeves or on and off turntables. The danger here is that fingers will touch the grooved surfaces, leaving deposits of grease and acid which trap dust and other grime – possibly to contaminate the pickup stylus – and to which tiny gritty particles may adhere, eventually to be rubbed across the grooves as records are returned to their inner sleeves. Fortunately the habit of holding records to avoid this is easily acquired. **Fig. 89** illustrates the technique: (i) bow the outer sleeve slightly to ease removal of the inner; (ii) open inner sleeve by parting it with thumb and fingers, extending hand beneath disc to support underside lable with second and third fingers, the record edge then pressing into the fleshy region between base of thumb and forefinger; (iii) with disc thus firmly supported without any part of the hand on grooved surface remove the inner sleeve; (iv) support far edge of disc with the other hand, remove fingers from underside to grip the first edge, then lower on to turntable. The procedure is simply reversed when returning the record to its sleeve, but

Fig. 90 The Watts 'Dust Bug' in use on a record; a simple suction pad fixes the pivot in position.

take care with polythene inners which tend to be attracted to discs by electrostatic charges. Any gritty particles present may thereby be rubbed across the surfaces.

With records thus safely transported to and from the turntable there comes the need to keep them clean during playing. As mentioned in the previous chapter, a cleaning device of some sort should be permanently installed for this, the most popular, effective and widely used product being known as a Dust Bug (**fig. 90**). This 'plays' the record at the same time as the pickup, nylon bristles exploring the grooves to loosen dust and other material which is taken up by a small plush pad. The pad is freed of collected matter between sides and may be moistened occasionally with a fluid supplied for the purpose and which helps to reduce electrostatic charges. Employment of this or a similar device is mandatory for high quality disc reproduction if records are to be kept in good condition and background noises minimised.

A related accessory for use in a constant battle against the snap, crackle and pop caused by dust is the Parostatik Preener (**fig. 91**). The Preener is a big brother to the Dust Bug's plush pad, spanning the full width of a record's grooved section between rim and label. It employs a moistening anti-static principle and is applied to the rotating record to remove all superficial dirt in a second or so. With new discs, which often carry an excess of packaging dust, it is most useful, while the really fastidious user 'preens' each record before playing in addition to employing a Dust Bug. When playing one side only of a disc it is wise practice to preen the other side before returning it to its sleeve, as any dust picked up since the record was last used is then removed and the two sides are equally protected against scratches from gritty particles carried into the sleeve. This applies particularly to records likely to be used one-sidedly fairly frequently, for it is often found in such cases that the unplayed

Fig. 91 The Watts 'Preener' removes superficial contamination from the whole grooved surface with a couple of record revolutions.

side becomes almost unplayable due to repeated exposure to the air and the turntable without any corresponding cleaning. It helps if the turntable mat is cleaned from time to time, an extra Preener reserved for this task alone not being an undue extravagance.

I sadly recall learning these particular lessons with great embarrassment, when some friends came particularly to hear a performance of Berlioz's 'Harold in Italy', having been much impressed on an earlier occasion by a short extract from the last movement. Alas, that last movement – The Brigands' Orgy – had been used so many times for demonstrations of equipment that the other side of the record was quite dreadfully noisy.

One final point on dust and contamination: despite every care, some particles will find their way on to the disc surface between sweeping by Dust Bug and tracing by pickup stylus, and some of these will cling to the stylus. Tiny brushes are available for keeping styli free of clutter, an occasional gentle *forward* stroking action with one of these normally sufficing to keep things in order. In extreme cases where records have been exposed to a particularly grimy atmosphere – such as regular envelopment in tobacco smoke during handling – a thin layer of 'goo' can accumulate in the grooves and tends to build up a rather more resistant deposit on the stylus tip. Again, stylus cleansing kits are available to cope with such problems.

Now we are in a position to discuss building a library of stereo gramophone records to do justice, musically and technically, to the principles, techniques and equipment covered in the preceding five chapters. Every music lover has his favourites and it is only natural that the first records purchased will follow a well-trodden personal path. But be on guard: do not assume that the particular versions which appeal to you on paper, perhaps because of the artists, are going to be the best sonically. Read plenty of record reviews – and read between the lines. The musical sympathies and idiosyncrasies of critics will gradually become familiar, as will some inconsistencies in judgement of recorded quality. Some publications are more helpful than others in this respect, but as standards of domestic reproduction have risen over the years critics have been forced to listen more and more to the actual *sounds* of music in addition to the scores sometimes apparently printed on their ear drums. It is important to remember that hi-fi equipment not only does justice to the best records, but is also rather more revealing of deficiencies in the worst. We shall see shortly that the filters and tone controls on good preamplifiers can be very helpful with difficult discs, compensating to some extent for the ability of wide-range loudspeakers to reveal high frequency distortion, though careful purchase can usually ensure that most records in one's collection will not put either equipment or musical pride to shame.

One way of judging records before purchase, without invoking the iniquitous retail practice of playing discs in the shop with pickups of doubtful quality, is to borrow them from commercial libraries. The gramophone departments of public libraries are often efficient and well stocked, in many cases with a good quota of stereo discs, but quality of record surfaces is often dreadful due to multiple mishandling by users, and musical waveforms can become distorted in the grooves when borrowers are permitted to play records with non-hi-fi pickups. A few first-rate commercial libraries* require that patrons use reproducing equipment of a certain minimum quality, keeping a close check on records as they pass in and out. Borrowing from such organisations is not cheap, but many people find that a steady flow of recordings by post keeps them musically on the move, a list of 'musts' for personal acquisition soon germinating and growing as the months go by. Thus one can both hear a wider range of music than could be afforded by outright purchase, and know for certain which recordings are worth adding to one's own collection.

It is wise to give the collection a broad musical base, starting by all means with those personal well-loved gems, but with a daring excursion here and there to encompass music previously regarded as rather austere, too 'early', perhaps too florid or by a composer whose name frightens through simple unfamiliarity. Do not be put off by the previously unknown, for it may then

* See bibliography for addresses.

never be known and a whole world of musical experience is lost. An acquaintance learning to play the Spanish guitar was once attracted to my home by a gramophone concert because the programme included Rodrigo's Guitar Concerto. The last item was Shostakovich's 5th Symphony, which he decided to sit through without any expectation of enjoyment. To his delighted amazement this turned out to be the event of the evening, and he has now joined the ranks of those who regard this as one of the most important symphonic works composed in the mid-twentieth century.

There are some who will insist on listening only to one very limited type of music because it happens to show off their equipment in its best light, considerable sums being expended in high fidelity for the sake of 'sounding brass and tinkling cymbal' and little else. The situation was well expressed in a letter which arrived in the *Hi-Fi News* editorial office just before this book went into production: 'I've heard it said that hi-fi enthusiasts very often aren't musical at all; they are only interested in obtaining "perfect" sound . . . I know someone like this, he has two *x* recorders, two *y* turntables, two *z* pickups, two tuners and three amplifiers, and all he ever listens to are military marches. All that outlay just for military marches – its absurd!'

Well, there is nothing wrong with military marches, but of course they form but a drop in the musical ocean. In the hope that readers will make some attempt to span stretches of that ocean, or at least try an occasional trip away from the immediate home waters of musical habit, there follows a short list of good stereo recordings stretching from Bach to Bartok and from string quartets to full orchestra with chorus. As there are now very many hundreds of good stereo issues this list cannot be regarded as anything more than a personal selection, but neither in standards of performance nor quality of recording should any of these disappoint. For up-to-date fully tabulated references compiled on reliable but less personal lines, various indexes are available.†

BACH	Art of Fugue (organ)	HMV CSD 3666–7
BACH	Brandenburg Concertos	Decca SET 410–11
*BACH	St. John Passion	Telefunken SK 19 (1–3)
BACH	'Switched on Bach' (various electronic realisations)	CBS S 63501
BARTOK	Bluebeard's Castle	Decca SET 311
*BARTOK	Music for Strings, Percussion and Celeste; Divertimento	Argo ZRG 657

† *The Stereo Index*, edited by W. J. Wilson, and *The Art of Record Buying* (an Annual).

BEETHOVEN	Symphony No. 5	DGG SLPM 138804
*BEETHOVEN	Symphony No. 7, etc.	Decca SXL 6447
*BEETHOVEN	String Quartets Op. 127/135	Philips SAL 3703
BEETHOVEN	String Quartet Op. 131	Philips SAL 3790
*BEETHOVEN	Hammerklavier Sonata	Decca SXL 6335
BERLIOZ	Requiem	Philips 6700 019
*BERLIOZ	The Trojans	Philips 6709 002
BLOCH	Violin Concerto	HMV ASD 584
BRAHMS	Piano Concerto No. 2	Decca SXL 6322
BRAHMS	Piano Trios	Philips SAL 3627–8
*BRITTEN	Curlew River	Decca SET 301
*BRITTEN	Midsummer Night's Dream	Decca SET 338–40
CHABRIER	Orchestral Pieces	Decca SXL 6168
*DVOŘÁK	Symphony No. 3, etc.	Decca SXL 6290
ELGAR	Cello Concerto, etc.	HMV ASD 655
*ENGLISH MUSIC FOR STRINGS	Pieces by Purcell, Elgar, Britten, Delius and Bridge	Decca SXL 6405
*ENGLISH SONGS	Janet Baker recital	HMV HQS 1091
*HAYDN	Harmoniemesse	Argo ZRG 515
*HAYDN	Horn Concertos	Argo ZRG 5498
HAYDN	String Quartets Op. 3/33/76	Decca SXL 6093
*HAYDN/J. C. BACH	Harpsichord Concertos, etc.	Decca SXL 6385
*HOLST	The Planets	HMV ASD 2301
JAZZ	Blues in Thirds (Hines)	Fontana FJL 902
JAZZ	Communications (Scott)	Columbia SCX 6149
JAZZ	Far East Suite (Ellington)	RCA SF 7894
JAZZ	Popular Duke Ellington	RCA SF 7835
JAZZ	Swinging Era (Hines, *et al.*)	Fontana DTL 200
KODALY	Hary Janos	Decca SET 399–400
LISZT	Organ Works	Argo ZRG 503
MAHLER	Symphony No. 2	Decca SET 325–6
MAHLER	Symphony No. 5, etc.	HMV ASD 2518–9
*MAHLER	Des Knaben Wunderhorn	HMV SAN 218
*MOZART	Symphonies No. 13–16	Argo ZRG 594
*MOZART	Symphonies No. 32, 35, 38	HMV ASD 2327
MOZART	Piano Concertos No. 14, 15	HMV ASD 2434
MOZART	Six 'Haydn' Quartets	Philips SAL 3632–4
ORFF	Carmina Burana	DGG SLPM 139362
ORGAN RECITAL	Westminster Cathedral	HMV CSD 3648

OVERTURES OF OLD VIENNA	By Johann Strauss, Nicolai, *et al.*	Decca SXL 6383
POP	Abbey Road (Beatles)	Parlophone PCS 7088
POP	Blood, Sweat and Tears	CBS S 63504
POP	Bridge over Troubled Water (Simon and Garfunkle)	CBS S 63699
POP	Sergeant Pepper's Lonely Hearts Club Band (Beatles)	Parlophone PCS 7027
PROKOFIEV	Symphony No. 5	DGG SLPM 139 040
ROMANTIC RUSSIA	Borodin, Glinka, Mussorgsky	Decca SXL 6263
*ROSSINI	String Sonatas	Argo ZRG 506
*ROUSSEL	Spider's Banquet, Bacchus and Ariadne	Columbia SAX 2562
SCHUBERT	Quintet in C major	WRC ST 992
*SCHUBERT	Piano recital	Decca SXL 6260
SHOSTAKOVICH	Symphony No. 5	RCA SB-6651
*SHOSTAKOVICH	Symphony No. 10	DGG SLPM 139020
*SIBELIUS	Symphonies No. 3 and 6	Decca SXL 6364
SIBELIUS	Symphony No. 4, etc.	DGG SLPM 138974
SOUSA	Thirteen Marches	Columbia TWO 113
STRAUSS FAMILY	New Year Concert	Decca SXL 6256
STRAUSS (RICHARD)	Der Rosenkavalier	Decca SET 418–421
STRAUSS (RICHARD)	Also Sprach Zarathustra	Decca SXL 6379
STRAVINSKY	Pulcinella; Apollon Musagête	Argo ZRG 575
STRAVINSKY	Rite of Spring	DGG SLPM 138920
TCHAIKOVSKY	Symphony No. 4	DGG SLPM 139 017
TCHAIKOVSKY	Ballet Suites	Decca SXL 6187
*VAUGHAN WILLIAMS	Symphony No. 4, etc.	HMV ASD 2375
*VAUGHAN WILLIAMS	Symphony No. 6, etc.	HMV ASD 2329
*VAUGHAN WILLIAMS	Five Tudor Portraits	HMV ASD 2489
*VERDI	La Traviata	Decca SET 401–2
VERDI	Nabucco	Decca SET 298–300
VERDI	Otello	HMV SAN 252–4
*VERDI	Requiem	Decca SET 374–5
WAGNER	Götterdämmerung	Decca SET 292–7

As the collection begins to expand the user will become aware of subtle differences in recorded balance adopted by the various record companies for different styles of music. There will be recordings in which the performers sound close and almost without supporting ambience, and others which manage to combine closeness with reverberant warmth and yet remain convincing. More distant sounding performers will sometimes be accompanied by foggy imprecision, yet others at an equal apparent distance will seem clear and unconfused because of a helpful spatial disposition. Some records will be generally over-bright, others – made with less skilfully placed microphones – may exaggerate the higher overtones of some instrumental sections only. Some recording studios sound spacious and airy; others seem too small, imparting some unpleasing thickness of texture. Bass may be light or heavy. The soloist in a piano concerto may blend well with the supporting orchestra – even though perhaps brought further forward aurally than in the concert hall – or there could be an impression that the piano is set in a slightly different acoustic. In opera recordings one company will place the singers well in front, producing the sort of sound one might hear if sitting on the stage, with the orchestra at the far side of an empty opera house; a rival set of the same opera puts the singers in their place and lets the orchestra almost drown them at times, as happens in real life. A string quartet can be placed well back, stereo being used not so much to separate the four instruments from each other as to place them as a small group within an acoustic setting. Alternatively, it may be spread right across the space between the loudspeakers and recorded with the sort of tonal balance requiring an 'almost in the room' replay level for a natural effect.

Many of these differences arise from the personal preferences of recording producers and are not necessarily right or wrong in any absolute sense; but others call for action at the listener's control unit if hi-fi equipment is to be used to give maximum musical enjoyment. Apart from a correct volume control setting, of which more shortly, many of the best recordings sound just right without further adjustment when reproduced via a high quality system. A few discs in this class are asterisked in the foregoing list and may be used for assessing the tonal balance of an installation, as mentioned in the previous chapter. Assuming that by such means – preferably supplemented by visits to live concerts to keep one's aural mechanism in touch with musical reality – it has been established that tonally balanced programme material reproduces most naturally with the controls set for a flat response, it is then possible to make intelligent use of preamplifier facilities in coping with minor errors of recorded balance. 'Errors' is perhaps not quite the right word here, for a recording that seems tonally over-bright may in fact be a true representation of the sound patterns picked up by microphones in the studio, any emphasis of high frequencies arising from microphones being placed nearer to the instruments than average domestic listeners would choose to be. In this sense it could be argued that apparent falsities of balance should be ignored, though the whole situation when recording with several microphones dotted around the orchestra is so artificial that the ear must be the final arbiter. In short, if a recording *sounds* over- or under-bright, then it *is*.

Many otherwise first-class records suffer from slight over-brilliance, and when such an excess of treble is not accompanied by edgy distortion – the sound being 'clean' in hi-fi parlance – matters are put right by some simple treble cut. This assumes, of course, use of the preferred type of tone control response discussed in connection with **fig. 40** (page 114). It also assumes an appropriate overall loudness, as the feeling of greater presence resulting from boosted treble may be confused with loudness, the volume control being set unwittingly at a level too low to produce a natural perspective from the stereo ambience as recorded. A record may be put on and the volume turned up to what might be the correct sort of level, but the sound seems a little overbearing so the volume is lowered. There is then a feeling that things are not quite right, the sound-stage seeming too distant for its width, the orchestra too bright for its apparent distance, and so on; had the treble been turned down a little at the outset the feeling that the sound was too loud might have been avoided.

The matter of volume level is therefore worth some experiment, for with any good stereo recording there will be a natural replay loudness appropriate to a given listening room and loudspeaker spacing. The spatial disposition of orchestra or operatic stage, both laterally and in depth, should 'click' into position and seem most correct at a particular volume control setting. This is because the ear senses the reverberation pattern of the recording studio and relates this to the relative loudnesses of near and distant instruments or voices on the sound-stage. There can be only one overall loudness at which these various aural signals are free from anomalies, and while the brain does not throw all its calculations into consciousness, it does make us aware that some volume settings are better than others. This applies only to well managed stereophony, for when multi-microphone techniques are misused, recorded ambience fails to complement the main signal and the ear is at a loss to tell what the final loudness should be. The massive Mahler recording session depicted in **fig. 52** (page 149) illustrates this well, as the resulting records offer a sound with almost complete lack of ambient integration, everything coming and going in a most confusing fashion making it impossible to find a volume setting that is convincing and natural for more than a few minutes. Fortunately most stereo recordings by the major European companies keep up a higher standard than this.

With mono recordings the situation is different, there being no obviously correct replay loudness or natural relationship between music and ambience. Due to imperfections in stereo speaker systems – whereby there is some broadening and shifting of a double-mono sound image as one moves away from the 'stereo seat' – some listeners prefer single-channel recordings of solo instruments or solo voices to be reproduced from one loudspeaker only. When the performer has in addition been recorded in a fairly 'dead' acoustic the replay loudness is then most naturally set to place the instrument or voice apparently just behind the loudspeaker.

Returning to stereo, and having found a volume setting that is right from the perspective/ambience/loudness point of view, tonal balance should be corrected if it seems false. It is unwise to make prior assumptions about the use of tone controls: the 'flat' positions are not sacrosanct, and boost or cut should never be a matter of unthinking habit at either end of the audio band. Let tone control serve musical ends, counteracting a dim or lifeless recording with treble lift, and over-brilliance with treble cut; adding missing fullness or body with bass lift or correcting muddy, over-resonant low frequencies with bass cut. Remember also that if a record seems to need some treble boost to give added clarity or definition, it might benefit instead from a slightly higher volume setting, or perhaps a modest mixture of the two.

These are all matters of balance, of using the tonal information in recordings to best musical effect. However, many discs suffer from uneasiness at very high frequencies due to distortions introduced during the original cutting process or by pickup resonances and stylus tracing errors. Such uneasiness may come from a hard edginess on fortissimo brass or the soprano voice, an unnatural fizz added to the natural guttiness of strings, exaggerated sibilants in vocal music, or a generally inflated extreme treble of the sort that adds a wiry, metallic quality to music but which is not easily cured by a simple levelling down of the whole treble band. Here the HF filter should come into play, attenuating any added upper harmonics or high frequency intermodulation products without affecting genuine musical brilliance. The sort of effects to listen for were noted in Chapter 7.

If the preamplifier includes only a single switched HF filter operating around 6 kHz, it is a simple matter to try its effect when reproduction quality is in doubt. The response may be clipped at a lower frequency than is strictly necessary or its rate of cut-off may be too rapid, but generally a simple filter of this sort is designed with typical problems in mind and should be used whenever musical enjoyment is significantly degraded by distortion at high frequencies. The listener with a more elaborate control unit offering a choice of cut-off frequencies, perhaps even with a variable rate of slope, can always find the optimum response for maximum music and minimum distortion from any programme source. In such cases a little experiment nearly always pays dividends in musical quality, as one record may need a fairly steep cut at say, 7 kHz to tame some severe edge in brass passages, whereas another may benefit from a modest fall-away from 5 kHz to cope with a not very distorted but rather glaring or wiry extreme treble. Some tapes or radio transmissions may have no more than a suspicion of edginess above 10 kHz. It is always wise to use the minimum rate of slope necessary to remove the unwanted effect, and in cases where a recording manages to be both tonally dim *and* edgy there is no reason why conventional treble lift should not be applied in addition to some filtering. Despite great improvements in pickups, including reduction of tracing distortion by use of elliptical styli, some deterioration of quality towards the inner grooves of records is still inevitable, and a slight increase in the severity of HF filtering at the ends of sides carrying heavy modulation is sometimes worthwhile.

The reader should not conclude from all this that disc reproduction is one long agony of adjustments, as most of the better recordings need little attention beyond a touch of treble cut when played by good pickups. Also, once any necessary control settings have been decided they can easily be marked on the inner record sleeve for future use. A few discs (very few) are slightly unbalanced in the left/right sense. When this is suspected they should be heard from the 'stereo seat' and the channel balance control set for a proper spread – this setting can also be noted. Finally on controls, some records carry what seems to be recorded rumble, and when a switchable LF filter is available it should be used if the resulting reduction of background noise is judged to be preferable despite any associated loss of extreme bass.

Following the theme of this book, hi-fi equipment is now properly installed in the home and is used to give maximum musical satisfaction from a growing collection of well chosen and carefully tended stereo records, with the added occasional delight of outstanding radio concert broadcasts. From the aural point of view public concerts may be substantially simulated at home; despite remaining technical limitations to be discussed in the next chapter, the gramophone at its best now offers experiences cramped mainly by psychological and social shortcomings. Informal semi-musical evenings with visiting friends is one way to make up for the sense of occasion and shared experience inherent in concert and opera going but not automatically switched on by the lowering of a pickup stylus. Here, snippets from the latest discs are played, partly to delight the visitors and partly to create a sense of well-being in the host, who would not be human if his hi-fi equipment failed to stimulate a sense of pride when reproducing an exceptionally good recording. At other times the user may feel that the art of music is not well served in this way, perhaps preferring to sit down in privacy and concentrate, bar by bar, on a full length work. One curious thing about public concerts is that each individual may concentrate on the music privately yet somehow sense and contribute to the corporate reaction. A feeling that this valuable aspect of musical experience is lost in home listening has probably contributed to the growth of Gramophone Societies, where people meet for planned musical evenings either with set concerts or selections of music illustrating particular themes. Some smaller Societies hold their regular meetings in members' homes in rotation, others hiring suitable halls. The British reader wishing to make contact with gramophone oriented music lovers in his locality – perhaps even with a willingness to form a local Society and use his own home and equipment for some of its meetings – should write to the National Federation of Gramophone Societies.*

For those who prefer not to become involved with outside bodies there is another approach. My wife and I have organised evenings of recorded music in our own home for several years, and we have found that efforts to reproduce something of the atmosphere of a public concert have attracted ordinary music lovers rather than gramophone enthusiasts as such. The degree of realism has extended to a charge for seats, the money raised being donated to Oxfam. The reader wishing to do something on these lines, at the same time helping a charity of his choice, can start with modest advertising of the activity in local shops or among neighbours and acquaintances. After a few months a regular clientele is built up and further advertising may not be necessary: people tell their friends or bring wives and husbands and a sort of concert-going club materialises. It may help if secretaries of musical societies, librarians and other custodians of local culture are notified of the activity. Those lucky enough to have large listening rooms may thus share the joys of high fidelity with many extra people, providing satisfying musical evenings and raising money for a good cause at the same time. The smaller room has its problems, but it should be possible to pack half a dozen or so visitors into a reasonable listening area in most homes, and such a number can still ensure very worthwhile events.

Assuming an adequate audience, the best guarantee of success is proper planning and preparation. First, a legal point: whether or not a charge is made for a charity, a record concert advertised to strangers is a 'public' event, and as

* See bibliography for address.

such it needs to be covered by a license (in the U.K.) from the Performing Right Society in respect of any copyright music which may be performed, and from Phonographic Performance Ltd.* in respect of the records used. This sounds formidable, though in practice small domestic concerts of this sort organised to raise money 'for the alleviation of sickness and distress' are looked upon with a kindly eye by both bodies. It is found that the fee normally payable will either be waived or the royalty charged will be extremely nominal; though permission must nevertheless be obtained in writing to remain above legal board.

Next come the logistics, concerning accommodation and refreshments. Seating should be worked out carefully, a long settee forming a convenient starting point if speaker placement permits this to go at the back-centre of a practical stereo listening area. Small stacking chairs are useful for packing a large number into a small space, though a search around most homes will gather together a motley collection of smallish seats which can be used at a push with added cushions. Whatever the resources, a maximum audience must be agreed upon and not exceeded on any account; this is more easily ensured if bookings are taken by telephone or postcard on the understanding that no one just turns up on the night of the concert without prior warning. The maximum decided upon will obviously depend on space, seating and stereophony, it being a wise general rule to keep the audience as far back from the speakers as is practicable and to make the front 'rows' shorter than the back ones as in **fig. 92**. Having decided an approximate layout, set the tone and volume controls for good reproduction as heard near the centre of the proposed seating area and then listen from all other points in that area in case room resonances or reflecting surfaces upset stereo perception or tonal balance at any point. Each listener should have an unobstructed aural 'view' of both speakers, and from this point of view the traditional staggering of seats in public halls is to be avoided, as a head straight ahead is better than a head blocking the path to each speaker.

No concert is complete without an interval, and if the lady of the house can be persuaded to provide coffee and biscuits, this adds a pleasant friendly note to the evening. People chat, discuss the music, arrange to give each other lifts home, feel relaxed and decide to come along to the next concert. This is also a convenient time to collect money for whichever charity is to benefit from the event, a good sign of success and general euphoria being the contribution of payments above the agreed half-crown or 3*s*. 6*d*.

Programme planning is not simple, even though it might appear so from a

* See bibliography for addresses.

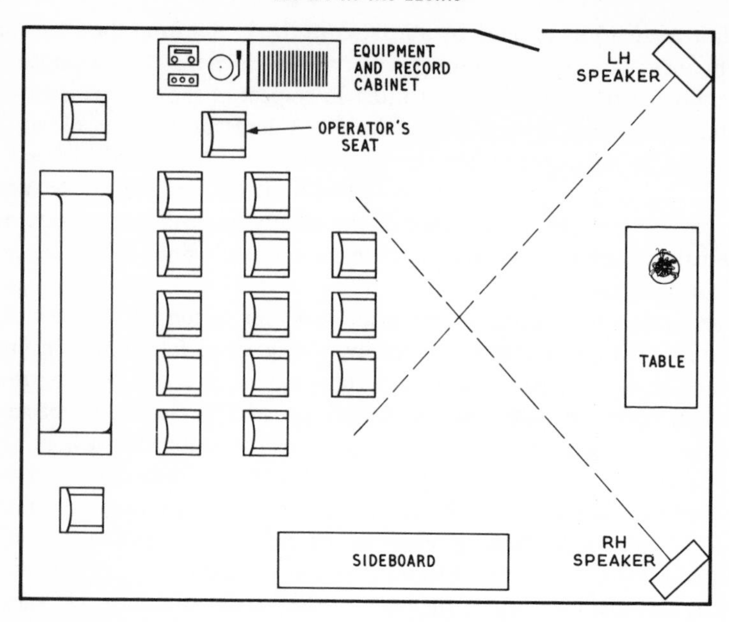

Fig. 92 Possible layout of a room to seat twenty people for a stereo concert.

glance down the jumble of superficially unlikely juxtapositions found, say, in a London Promenade Concert prospectus. We shall return to this in a moment, but assuming any particular evening's items to be agreed, it is wise for the person who will be at the controls to have a sort of dress rehearsal in order to note down the volume, tone control and filter settings needed for the particular recordings to be used; once again, as heard from a seat in the middle of the intended audience area. The loudness as judged by listeners at the front will be a trifle higher than that heard at the rear, though this should not be troublesome provided the front row is itself reasonably well back from the speakers. Also, there is a self-compensating factor here, as the wider stereo 'viewing' angle obtained when nearer to the speakers automatically suggests to the ear that one is closer to the performers, thus a higher loudness level is appropriate. Depending on the amount of acoustically absorbent material in a room initially – carpets, heavy curtains, soft furnishings, acoustic tiles – addition of a thick layer of human beings will lower the effective sound level by reducing reverberation. This means that the volume control settings found in advance may turn out to be just a little low, though by how much must be a matter of experience in that particular room as furnished: the more 'dead' it is normally, the less the effect of an audience. After two or three concerts it should be clear if there is any difference, and if there is it will usually mean increasing all volume settings by some constant amount – a quarter or half division perhaps.

Programmes will be dictated in the first place by the contents of one's record collection, though provided this arises from a reasonably catholic taste it should be possible to find an acceptable balance. A straight orchestral concert should normally start with a fairly short item – a sort of *hors d'oeuvre* with exactly the same function as an operatic overture: to allow for late comers and to settle the audience into a receptive mood. This could be followed by a piano concerto, possibly with a further short orchestral piece before the interval. The main work, a symphony, comes in the second half, and if timing seems reasonable this also might be preceded by a quite different short item – an aria perhaps. There are matters of musical mood to consider, some works ending in such triumph or sadness that they simply must come at the ends of concerts, and others best followed or preceded by works from a totally different genre – Bach against Stravinsky, Haydn against Holst. Do not follow Rachmaninov in melancholic mood with Tchaikovsky's 'Pathetique', or the 'Symphonie Fantastique' with the '1812 Overture'.

Some concerts can have a deliberate 'period' bias – Bach/Vivaldi/Corelli or Haydn/Mozart – or be split between baroque in the first half and Brahms in the second. One composer might reasonably dominate a concert, on the lines of Prom Beethoven nights, or be represented with a short item at the beginning and a major work at the end. The possibilities are endless and mistakes will be made, but the challenge is attractive. When a regular audience has been established, try a complete opera; but do make arrangements for librettos/translations to be available, perhaps one to every third person for sharing. The large booklets supplied with opera sets are ideal, though spares are hard to come by. Those with access to photo-copying equipment can be helpful here, and some libraries carry librettos.

To help set a concert atmosphere a programme should be available to each member of the audience, preferably typed but better hand-written than absent. Some examples of a possible layout follow, these being actual successful concerts which raised about £3 apiece for Oxfam.

RECORDED CONCERT—Wednesday, 8th March

Money taken will be passed to
Oxfam

DVOŘÁK (1841–1904) *Scherzo Capriccioso*

London Symphony Orchestra
conducted by Istvan Kertesz

BERLIOZ (1803–1869) *Nuits d'Été*

Régine Crespin (soprano)
with L'Orchestre de la Suisse Romande
conducted by Ernest Ansermet

CHABRIER (1841–1894) *Suite Pastorale*

L'Orchestre de la Suisse Romande
conducted by Ernest Ansermet

INTERVAL

MAHLER (1860–1911) *Symphony No. 1 in D major*

London Symphony Orchestra
conducted by Georg Solti

Next Month's Concert (*Wednesday, 12th April*)
Richard Strauss — *Don Juan* (tone poem)
Mozart — — *Symphony No. 33 in B flat major*
Sibelius — — *Swan of Tuonela*
Sibelius — — *Symphony No. 2 in D*

RECORDED CONCERT—Wednesday, 6th September

Money taken will be
passed to Oxfam

ROSSINI (1792–1868) *Overture, Semiramide*

Orchestra of the Rome Opera
conducted by Tullio Serafin

HAYDN (1732–1809) *Horn Concerto No. 1 in D*

Barry Tuckwell (horn) with the Academy of St. Martin-in-the-Fields
directed by Neville Marriner

HOLST (1874–1934) *The Planets*

New Philharmonia Orchestra
conducted by Sir Adrian Boult

INTERVAL

BEETHOVEN (1770–1827) *Symphony No. 7 in A major*

Vienna Philharmonic Orchestra
conducted by Claudio Abbado

Next Month's Concert (*Wednesday, 11th October*)
Handel — *Sonata in A minor for Recorder and Continuo*
Tippett — *Fantasia Concertante on a Theme of Corelli*
Bach — *Fourth Lute Suite* (transcribed for Guitar)
Schubert— Last thirteen songs from the *Winterreise*

The more enterprising might also attempt some programme notes to support the bare details of works and performers, possibly in the form of a brief spoken introduction to each item. Those fortunate enough to have both time and a typewriter can try their hand at written notes – which can always be cribbed from the record sleeve if inspiration runs out. In the absence of either oral or written guidance, the record sleeves themselves should certainly be available for study both before the concert and during the interval. With things organised on these lines it will probably be found that as the pickup is lowered on to the first record the audience reacts as if the conductor had raised his baton. This is an auspicious sign, for when good psychology joins good sonics it may truly be said that hi-fi in the home has justified all the technical bother by serving the art of music – especially when visitors feel impelled to clap the absent performers at the end. My most cherished memory of an appreciative hi-fi concert audience is of the occasion when, after putting on the complete Verdi opera 'Nabucco' – a three hour marathon with two intervals about which there had almost been second thoughts – everyone positively bubbled with elated enthusiasm and one visitor asked, jokingly but with meaning, whether we would be repeating it next month!

10. MUSIC IN THE HOME

THE MAN who wanted 'Nabucco' all over again paid a great tribute to the art of recording and reproduction as it stands in the early 1970s. When good hi-fi equipment is used properly to reproduce the best stereo recordings of fine musical performances, full involvement of the listener is comparatively easy, and awareness of any remaining technical limitations tends to fade. The obstacles to this 'willing suspension of disbelief for the moment'* have been progressively removed as technique has advanced, though it would be wise to remember that throughout gramophonic history there have been those who believed that reproduction had reached a stage where the differences between live and 'canned' were marginal. However, the proportion of listeners who feel this way has justifiably grown since the days of the acoustic gramophone and it can now truly be said that we are nearer to perfection than ever before. Indeed, technique has reached the position where some controlled deviations from audio accuracy might actually be preferred by many listeners as psychological compensation for absence of a real concert hall or operatic stage.

We shall look at such points later, concentrating first on those few technical limitations which, though they might be swept aside from time to time by musical involvement, are still there nagging away at hi-fi perfectionists. Such perfectionists have always been around, to feel superior towards and be taunted by those who believe that the hi-fi millenium is already upon us. They make a perpetual fuss about matters which seem unimportant at the time but which eventually become an accepted part of home music listening; their critics have usually dismissed pleas for future improvements as extravagant, while accepting without thought practices which were themselves once regarded as extravagant. This is perhaps a comment on human nature rather than high fidelity, and no doubt students of social history could offer countless parallels; but it is raised here as a warning that the time will probably come when the standard achieved by today's best domestic high fidelity stereo equipment will seem little better then than the best pre-1939 sound seems to us now. Far fetched? Perhaps, but remember that a similar statement in 1939 might have seemed equally extreme.

Where, then, is there room for improvement despite the glorious sounds with which today's techniques can capture the musical imagination? There are four areas for advance: (i) improved dynamic range and greater stereo coverage in VHF/FM broadcasting – a parochial matter for the BBC; (ii) a marked lowering of non-linear distortion in disc reproduction through general

* *Youth and Age*, by S. T. Coleridge.

use of an agreed pre-distortion technique to cancel replay stylus tracing errors; (iii) use of more signal channels to overcome certain limitations in two-channel stereophony; and (iv) a marriage of audio hi-fi and video hi-fi to take the home music lover into concert hall or opera house by eye as well as ear.

Although the BBC has covered practically the whole British population with a network of FM transmitters, there has been minimal support from the general radio industry and very little effort to publicise the advantages of VHF/FM reception (up till 1970). Increasing difficulty with interference on the MW and LW bands has nevertheless driven people to look for alternatives, and some local city stations may help by being receivable only on VHF; but any change is against a background of ignorance rather than because of a general understanding that MW/LW radio is out of date. Thus progress has been slow at the receiving end and most sound radio listening in the U.K. is still via the older medium. As the cost and complexity of duplicating signal circuits and landlines for the AM and FM transmission systems would be high, the BBC is forced to subject all audio signals to the degree of compression needed for reasonable AM reception in poor localities, thereby unnecessarily limiting the dynamic range on FM transmissions. However, as stereo radio is extended to cover more population centres this situation will probably change, for such an extension will in any case involve duplication of signal feeds. Once a network exists to provide an audio facility distinguishing FM from AM beyond simple reduction of background interference, the principle has been conceded that the VHF service may offer hi-fi luxuries otherwise unobtainable on radio. It might reasonably be hoped that as the balance of 'consumption' tips further in favour of FM, more attention will be paid to audio quality via this medium. Technical changes will help, with increasing use of super high frequency (SHF) radio links to convey audio signals between broadcasting centres and VHF transmitters.

The second area for advance concerns disc tracing errors. Despite the improvement effected by elliptical styli, there is still some degradation of reproduction quality at the inner grooves. The excruciatingly rough distortion heard at the ends of heavily modulated sides when using inferior pickups can be a thing of the past for the owner of hi-fi equipment, but the finite edge radii on elliptical tips are still unable to follow *all* the groove contortions produced by a sharp edged cutter. Further sharpening of the replay tip by reduction of the minor radius is not practical from the viewpoints of record and stylus wear, finish and setting accuracy of tip, playing weight, tracking performance and other factors, so we must look elsewhere for further improvements. A process of reshaping recorded waveforms to provide automatic compensation for replay tracing distortion was mentioned in Chapter 4, and this may eventually be the universally applied solution. The work of RCA Victor in this direction has already been mentioned, though in the late 1960s other major recording companies set about the matter in earnest, particularly the Decca group, whose Teldec cutter – evolved in collaboration with Telefunken – sets an unmatched standard of accuracy in control of cutting stylus motions at high frequencies. Much experimental work is in hand,* and already in 1970 it is possible to play the inner grooves of some heavy orchestral or choral recordings – the sort of passages normally inclined to sound rather 'rough' – with equally good quality using either spherical or elliptical styli, indicating that the system really does work.

Although this approach to the tracing distortion problem must be based on a particular stylus size and shape – very probably 0·7 mil spherical, the type in widest general use – it would be wise at present to use pickup cartridges giving the best performance with the majority of today's discs, which includes a vast library of music most unlikely to be recut. Fortunately, it seems that use of an elliptical tip to play a disc pre-distorted for optimum performance with a 0·7 mil spherical stylus still gives a better overall result than would be obtained without waveform compensation. The prospect of cutting distorted waveforms on disc records as a means of overcoming replay deficiencies has been greeted with alarm by some purists, though it could be argued that high frequency pre-emphasis adopted by everyone to improve replay signal-to-noise ratio in accordance with the RIAA convention is equally 'immoral'. It is the end result that counts, and if music in the home can be made more musical for the ear by undergoing some slightly weird processing in the groove, who are we to worry? Another possible solution to the tracing distortion problem is to abandon the sharp-edged cutting chisel in favour of a spherical embossing tool having the same radius as the replay stylus. This has been suggested many times but never implemented through lack of a suitable material to replace the lacquer presently used for direct disc cutting. If a usable material turns up this may be the way ahead.

Limitation number three is more fundamental, as its banishment would require fresh signal channels to supplement the simple left- and right-hand of conventional two-channel stereophony. Such stereo is concerned with two dimensions only – depth and width. The depth is built into the signal as reverberation that gives aural clues in the form of music-to-ambience ratio, revealing the apparent relative distances of sound sources within a particular setting. The width is provided as a horizontal segment of space bounded by the apparent far walls of the recording or transmitting studio, and by the

* See 'Replay Stylus Compensation' by Stanley Kelly, *Audio Annual*, 1969.

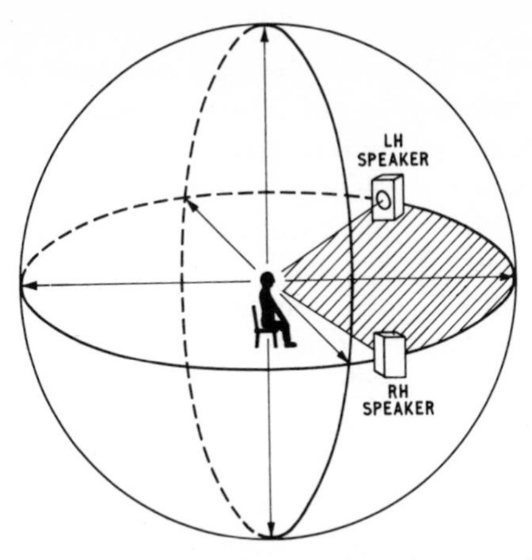

Fig. 93 Limited stereo 'vision' in a three-dimensional world.

angle subtended between speakers and listener, the latter being limited to around 90 degrees in most practical systems. **Fig. 93** shows the listener's natural three-dimensional world and the rather small piece of this encompassed by two-channel stereo. We saw in an earlier section that it is possible to gain a little of the height dimension by placing the speakers somewhat above ear level, though this is really a geometrical swindle as the apparent sound sources are still all situated in one plane. It might be objected that **fig. 93** is also a swindle, as unless one frequents the most expensive seats at the front of the stalls one rarely listens to a real orchestra occupying as much as 90 degrees of the aural sound-field. Indeed, average seats in most concert halls subtend an angle of about 30 degrees to the orchestral or operatic stage, with the better middle positions at around 60 degrees.* Since two-channel stereo can give us 60 degrees with ease and 90 degrees with a little care, and as we can regain some impression of height, what more is there to do?

We listen to live music within a 'setting'. Stereo helps us to appreciate that setting's contribution to musical sound in a more nearly natural manner than is possible with mono, but not in a *completely* natural fashion. The setting is defined by reflections from the studio walls and the subsequent reverberation, but in real life these reflections come from all directions, not just 'out in front'. It is lovely to hear a loud orchestral chord followed by a dying ambience, but two-channel stereo apparently places the bulk of the studio space behind the orchestra, whereas in practice it would be around and behind the *listener*. The unique sense of being actually in the hall with the performers – rather than

* See many plans of the world's concert halls in *Music, Acoustics and Architecture*, by Leo L. Beranek.

just looking in through a large but finite aural window – can only be created by reconstituting the complete pattern of reverberant sound in three dimensions. To do this properly requires at least another pair of stereo channels, and 1969/70 saw the launching of 4-channel stereo or *quadraphony* as a commercially viable domestic system.* At present (mid-1970) 4-track tapes are employed, with four replay loudspeakers normally situated in the listening room corners.

One important related matter from the general hi-fi point of view is that apparent limitations of dynamic range in music reproduction cannot always be explained satisfactorily by simple reference to signal levels. It seems that subjective dynamics may be greatly influenced by the quantity and quality of recorded reverberation, it being possible to produce quite different judgements regarding dynamic range by replaying various recordings of the same performance, the only differences being in the microphone arrangements used. This suggests that a more natural and complete reproduction of the total sound-field around the listener's head could be an important step towards the sort of dynamics frequently heard in the concert hall, but which at present are seldom approached subjectively despite measured levels indicating that they should be. Factors of this sort have probably added to other difficulties causing recording engineers and producers to adopt multi-microphone techniques rather than a simple coincident stereo pair in the manner of Blumlein. While a pair of human ears may find a point in space giving a perfectly satisfying balance even on quite massive musical works, a stereo microphone pair in the same position may fail because it cannot 'hear' and discriminate in three planes at once.

Pending establishment of more channels, can anything be done about multi-directional ambience with our present two-channel signals? Various coding schemes have been suggested to enable an original three- or four-channel signal to be 'hidden' within two channels,† and if these prove to be workable it may be possible to obtain a limited sort of quadraphony without great expense. It certainly seems that unused ambient information is already present in many 2-channel recordings, so that some modest electronic circuitry and a couple of extra speakers may provide a sort of 'budget' sound-in-the-round. There were also earlier experiments in 'ambi-stereophony', the idea being to

* See 'Four Channel Stereo' by Edward Tatnall Canby, *Hi-Fi News*, December 1969; also, 'The Principles of Quadraphonic Recording' by Michael Gerzon, *Studio Sound*, August and September 1970.

† See 'The J.O.K.E. System' by Michael Gerzon, *Hi-Fi News*, June 1970; 'Surround Sound from 2 channel stereo', by the same author, and 'Two Channel Quadraphony' by David Hafler, both in *Hi-Fi News*, August 1970; also, a report on the Scheiber coding system ('Listening to Four-in-One') by Edward Tatnall Canby in *Hi-Fi News*, June 1970.

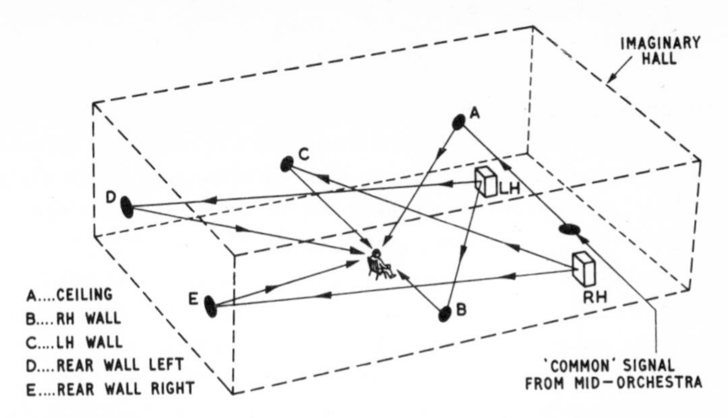

Fig. 94 Imaginary concert hall created for the listener by simulated reflections using controlled delays.

extract a suitable common signal from the two channels, subject this to some time delay and then feed it back to the listening room at a low level via obscured speakers mounted to the sides and rear of the audience.

It might be possible to simulate concert hall listening more effectively than this by producing a number of different 'reflections' from both the left- and right-hand channels, as well as the common signal, programming for different time delays according to the position of the postulated hall's boundaries that are imagined to be reflecting the sound. Referring to **fig. 94,** a combined signal could give the first reflection from the 'roof' (A); left and right signals could give reflections (B) and (C) from right and left 'walls' respectively, with appropriate longer delays; further reflections could come from the rear (D and E); then even longer time delays could be used for progressively more mixed left/right signals, as happens in a real hall when reverberation become more randomised. Thus something akin to an actual hall of definite dimensions could be created subjectively, having regard to the type of recorded ambience so as to blend the two acoustics as unobtrusively as possible. This last point is important, and is why the domestic room acoustic should be damped down in conventional stereo listening, as the short reflection times in an ordinary room can seem 'out of gear' with a large recording studio.

Returning to the simulated hall, one can imagine a standard multi-channel reverberation unit which could be programmed for different types of hall by pressing a set of buttons marked REFLECTION 1 – TOP CENTRE, REFLECTION 2 – LEFT REAR, etc, and adjusting some knobs marked DELAY. Records might come with a recommended replay acoustic to match and complement the recording studio, and if one wanted to sit further back, one could simply shorten the rear reflection times, increase the initial delays and turn down the volume! This is a fantasy, of course, but something on these lines could come one of these days unless genuine 4-channel stereo does indeed become an established and widely used medium to recreate the original environment rather than create a fresh one. When we have true multi-directional stereo, working via properly designed speakers which do not suffer from any image shifting or broadening effects as the listener moves around, it is very likely that the best of today's reproduction will be looked upon as rather crude.

Coming now to the fourth infidelity of present-day high fidelity, and moving into a controversial region beset with psycho-acoustic difficulties, it can be argued that there is no hope of reproducing the full excitement and involvement of live music while the orchestra or operatic stage remain invisible. It is certainly true that in the concert hall nearly everyone watches the orchestra fairly intently and that to some extent this focuses one's interest on the music. This applies particularly in concertos, where the soloist's part is usually easier to follow when ear is aided by eye. Checking this by means of trial periods with eyes closed, it sometimes seems – particularly during violin concertos – that the soloist's musical line becomes inextricably imbedded in the general instrumental texture, only to leap straight out when one's eyes again alight on the performer. This is one justification for using semi-close-up aural effects in recordings for domestic listening, the psychological setting being unavoidably different from that found in the concert hall. Opera is an extreme example of this, for while the value of vision is at least arguable with instrumental music, an operatic production in the theatre provides a spectacle of singular value – a unitary art-form of compelling power in which vision and sound together involve the spectator in a unique amalgam of humanity and music. That music is the main ingredient few would deny, and with skill in production the sounds of voices convey sufficient of the human element to make recorded opera very worthwhile; but as such it is a different art-form from that known in the theatre, and it may quite legitimately evolve its own rules for presentation.

Recorded opera has gone from strength to strength since the introduction of stereo, though in the long run we may find vision following sound into the home, with the full stage displayed before our eyes in colour. Already television has evolved techniques for presenting concerts and operas, complete with visual close-ups of performers – pleasing to some, irritating to others, but arising from the tiny domestic TV screen which does not permit the viewer to apply his own private selection process. Proper stereo sound reproduction cannot be allied convincingly with present-day television, for it is no use hearing an orchestra or singers spread across a large space beyond the end of a room while viewing them on a silly little screen in the middle – a sort of transistor pocket radio for vision and full hi-fi for sound. However, whether by TV, video tape recording or some as yet unknown medium, the time will undoubtedly come when 'Nabucco' or the London Symphony Orchestra, 'Götterdämmerung' or a group of jazz players, will be there almost 'in the flesh'; certainly in colour, possibly in three dimensions. This will transform the psychology of home music listening, and our present problems with replay loudness and apparent aural distance will be joined by questions of visual width and apparent depth: whether the orchestra in an operatic production should be heard in full display across a space where it is seen not to be, or tucked down in an aural pit as in the theatre; whether the singer's acoustic should change with change of scene as on some sound-only recordings, and if so what about the orchestra's acoustic; and so on. It will be an exciting world for a new breed of aural-visual hi-fi perfectionists.

In the meantime we can relax with reproduced music as pure sound, closing our eyes if this helps to produce a mental picture of the room, studio, hall or church in which the music is supposed to be played. There are those who might regret a return to the visual concert hall when listening to music in the home, feeling that we have in fact gained a new sort of intimacy with much music simply because of the way it is produced and presented for consumption at home by ear alone. No doubt the seventeenth-century satirist (and Bishop) Joseph Hall would have agreed, for in his 'Occasional Meditations' he praises the darkness of night as a setting for music: 'How sweetly doth this music sound in this dead season! In the daytime, it would not, it could not so much affect the ear. All harmonious sounds are advanced by a silent darkness.' Three centuries later Sir James Jeans put a similar view rather more forcefully:* '. . . hearing and seeing do not blend well; they rather compete – in an unequal competition in which seeing usually wins. In the opera house, many of us miss much of the music through watching the acting too intently. Only when the distraction of sight is removed can our minds give full attention to what we hear. Our appreciation of sound then becomes far keener and more critical. This is why blind people so often become exceptionally good musicians, and why people who are not blind find it well to listen to the radio with the room darkened, and to close their eyes in the concert room, resisting the temptation to watch the fingers of the pianist, or the mouth of the prima donna'.

Whether or not one agrees with Bishop Hall and Sir James Jeans, there are even now backroom boffins planning the future of domestic, wide-screen, high-definition, colour video systems, and the highest of hi-fi will probably actuate ear drums and retinas together by the end of the century. In the nearer

* *Science and Music*, by Sir James Jeans.

future there are other possible and probable developments worth noting. The most probable, indeed certain, is a steady encroachment into domestic electronic equipment of miniature integrated circuits. It will be recalled that these incorporate into small blocks of material the various circuit functions traditionally requiring many separate components. Appropriately, they are sometimes called micro-circuits, and their effect on audio equipment is likely to make it clear that size is determined more by convenience in the handling of controls than by the quantity of 'bits' behind the panel. Power amplifiers are unlikely to shrink to invisibility because of heat dissipation problems and the need to transform the mains supply voltage, but they will certainly get smaller. Tuners and control units will necessarily retain a certain frontage to facilitate adjustment of knobs, pushing of buttons and tuning-in of stations, but as the sheer bulk of circuitry declines we may see the emergence of flat picture-frame units. However, hi-fi equipment is not packed into space probe rockets and does not employ the many thousands of stages needed in computers, so there is no desperate pressure on space or weight; and as in most installations, turntables, pickups, tape recorders and record collections are already more bulky than the tuner/amplifier complex, there is no point in pursuing miniaturisation indefinitely. Integrated circuits will be used in hi-fi units for reasons of economy, consistency and convenience, though the Ad-man's natural reaction is to make the most of any new development: the first FM tuners with I.C. front-ends, for instance, were announced in a manner that seemed to herald the millennium!

Pickups will undoubtedly continue to improve, mainly in the direction of better tracking performance at lower playing weights, thus eliminating actual groove damage that can still occur at high frequencies due to the effective mass at the stylus needing accelerations which the groove cannot impart without suffering some permanent deformation. The distortions are minute as judged subjectively, but they are still there. Also, new types of transducer will be evolved, using various electro-mechanical properties of new materials to translate movement into electrical signals. Some solid-state cartridges using controlled semi-conductive properties are already here, and these will be joined by yet more sophisticated designs, perhaps using electrolytic action or – an old dream – light beams to scan the groove. Arms may give up their bearings and just float, perhaps disappear completely, the cartridge managing to move across the record almost without support.

A component with genuine room for improvement in terms of sound quality is the loudspeaker. As new diaphragm materials are evolved, speakers are likely to become more consistent in performance from model to model, with less coloration and a smoother frequency response. On the most expensive types more manufacturers will follow an existing trend by paying attention to electrical equalisation of response by means of complex circuits built into the cabinet with the crossovers. Following the example of some professional monitoring loudspeakers, domestic models may appear with built-in power amplifiers, the whole amplifier/speaker assembly being designed as an entity for the best overall performance instead of the speaker having to provide a flat acoustic response simply because the amplifier system has a flat electrical response. This might be borne in mind when choosing equipment in the higher price brackets, where purchase of separated control unit and power amplifier could make good sense in terms of future developments even though it might seem a little *infra dig* at a time when the trend is to integrated units.

Incorporation of power amplifiers into speakers will ease the design of full-range electrostatic types by obviating the need to offer for general use an electrical impedance characteristic that is not really natural for this type of transducer. The combined speaker-amplifier might even lead to types in which the sort of negative feedback at present applied within amplifier circuits is extended to cover the acoustic output of the whole system. It could even happen that physicists investigating the properties of matter with other ends in view will hit upon some means of producing mechanical vibrations in the surface molecules of materials by simple application of electrical AC voltages, thus interposing less obstructions between signals and sound than arise even with electrostatic speakers. The recalcitrant problem of geometrical radiation characteristics in relation to accurate stereophony is more likely to find a solution via the ELS principle than any other known at present. This may come about by commercial production of suitably shaped mid-range and treble units,* or even by some ingenious system of progressively delayed signals across the width of very large diaphragms. The latter would be for the more distant future, when stereo systems might be built into complete walls. There will also be some automatic control of, or compensation for, listening room acoustics – a scheme already on offer in the U.S.A.†

Another persistent irritant in music reproduction is background noise. With good equipment and the best disc records or radio transmissions, today's techniques are very satisfactory; but sound sources are not always of the best, and discs particularly are subject to variations arising from dust and manufacturing limitations. Even when background noise is not such as to mask the final subtleties of decaying reverberation, our *impressions* of dynamic range still seem to be affected by any noise present. In the concert hall, the loudness

* See 'Whither Stereo?' by the author, *Audio Annual*, 1966.

† Acoustette sound balancer by Altec.

level of ambient audience noise can on occasion fall to 30 phons, whereas in an average home the combined effects of local and reproduced noise usually place the background in a 'quiet' room at 40 phons or more. Thus music with *ppp* passages also at this level, although heard clearly in both situations, will in the concert hall be 10 dB above the noise and at home almost in it. Only the listener can improve the local acoustic background noise – not always practical without moving house! – but the contribution of reproduced noise can be significant even when playing records completely free from dust and imbedded imperfections. Such noise usually originates at the tape recording stage and is particularly noticeable on some 'reissues' or in cases where recording rights have changed hands between companies, with corresponding extra tape copying.

As mentioned before, various techniques are employed to minimise recorded background noise, one particularly effective one being the Dolby system. Evolved for use by the recording companies, it has now been applied to some recorders and cassette players, and is in principle applicable to the total recording/reproducing system.* In the latter case the dynamic range of music signals would be compressed before going on to the master tape, giving a 'stretched' signal-to-noise ratio on disc (or radio transmission) with corresponding expansion in the replay amplifier. Noise introduced at *any* stage would be attenuated, including that from radio interference or dust on records, with the original full dynamics eventually reproduced without blemish. This is a theoretical possibility, though it would require international agreement for satisfactory implementation, as exactly similar compression and expansion circuits would have to be used by all broadcasting authorities, recording companies and amplifier manufacturers – a real case here for a mass-produced plug-in integrated circuit module. Use of the facility could be optional in domestic replay equipment, and if the BBC joined in, the AM/FM dynamic range problem would be solved by simple omission of a dynamic expander from AM receivers.

On discs, a combination of such compressed dynamics and the pre-distortion processes discussed earlier could give the gramophone record an extremely strong position from the hi-fi point of view, particularly for any music lover wishing to compile his own programmes rather than rely on (much improved) radio quality. It is possible that such developments would lead to a micro-microgroove disc record of small diameter, perhaps for rotation at a lower speed – perhaps even with more signal channels (a double groove?) for extra stereo information. This of course is very conjectural and would depend

* 'Reducing Unwanted Noise', by D. P. Robinson, *Audio Annual*, 1967.

greatly on developments in ultra-miniaturisation of pickup transducers, final mastering of the noise problems created by dust, and – very important – what happens to tape recording and other possible media. In the early 1970s tape is attempting a major commercial assault on disc via the pre-recorded cassette medium, with an eye on the classical market following great successes in the popular field. There are a number of technical advances* currently tending to overcome the quality limitations of low-speed narrow tape, though mechanical problems still demand close attention if the popular compact cassette player is to satisfy hi-fi criteria for wow, flutter and drop-out. Some cassette recordings are now issued with partially 'Dolbyed' signals, the replay machine introducing an appropriate dynamic expansion at high frequencies by means of a Dolby 'B' circuit – to reduce tape hiss. Things are certainly on the move, and tape does have an enormous advantage when it comes to extra stereo channels, which can be added and reproduced without particular difficulty.

Ideas thrown out by research in various normally unrelated fields will play their part in the hi-fi future. There are magneto-optical phenomena to be explored, perhaps using microscopic light beams to implant magnetic patterns in manganese bismuthide or similar substances. Recordings may one day take the form of small plug-in cards rather like 35 mm colour transparencies, with a minute pencil of light scanning the card rather as a reader's eyes run back and forth across this page, 'reading' off the musical waveforms for amplification electrically. Possibly the demand for recorded music at high technical quality will make it practicable to provide a complete repertoire via wire as part of the telephone system, with appropriate charges for wide-range, noise-free signals to couple straight into the home amplifier. It's anyone's guess.

Now, to round off this speculative last chapter, we shall move from guesses about the future of hi-fi to some opinions and impressions concerning the place of reproduction in the world of music as a whole, with particular reference to the use and abuse of technique in the service of art.

Abuse can only arise in a reproductive process when departures from perfection are large enough for it to be a matter of opinion whether departure A is better or worse than departure B. The very best reproductions of paintings are for practical purposes indistinguishable from the originals, and if the exact tone of one small detail is not quite correct when examined in a particular light, no one says that the craftsman is abusing his skill. However, if the whole range of red dyes needed for picture reproduction was not available, and if

* 'Tape Prospects and Problems', by Graham Balmain, *Audio Annual*, 1966. 'Casettes and Coatings', by Graham Balmain, *Audio Annual*, 1968.

for some technical reason the relevant colours could not be simulated with other mixtures, alternatives would have to be used and someone would have to choose them. On the basis of their knowledge of the painter's style, their assessment of probable expert and public reaction, and their own aesthetic judgements of what is preferable, craftsmen A and B might use turquoise and lemon-yellow respectively for the normally red areas of picture. At a public showing the reaction is mixed: of those without prior prejudice, some like it and some don't; of those who feel more competent to pass judgement, some philosophise, some say that A is abusing his skill, some say B; and at least one 'expert' will claim that both men are either incompetent or irresponsible for failing to use red, despite a clear explanation of the situation in the catalogue.

Music reproduction has its analogies, worse confounded by the absence of an original 'picture' against which sound quality may be judged. A group of musical performers will seem to produce different sound qualities when heard from various points in a studio, and unless the recording producer decides to be conveyed around in a suspended seat on a crane in order to find the best listening position and layout of musicians from the 'live' point of view, with facilities for instant transfer to a monitoring room for A–B comparison with the sound as heard in domestic conditions, it cannot be said that there is any one original sound picture to reproduce. Also, even if this extraordinary procedure were adopted, that 'best' listening position would be a personal opinion not necessarily shared by all potential consumers. There is a common assumption in the hi-fi fraternity that our goal is the re-creation at the listener's ears of a sound pattern identical to that obtained at a good seat in a concert hall or at the opera. This may seem reasonable, but it does not bear close examination in terms of what can actually be done in practical recording with two-channel stereophony. For good or ill we are in the hands of the recording producer, and the most we can hope of our domestic reproduction is to create the sort of balance – in all the senses of that word – determined by the producer during recording. If we are lucky our loudspeakers may reveal more detail and offer better transients or greater depth of bass than those used for monitoring, but the general pattern and perspective of the sound we hear will be that chosen by the producer. This need not cause alarm, for since there must be a large element of choice in aural 'stage management', it is good that this is in the hands (ears!) of someone well accustomed to the sounds of live music and the influence of studio acoustics. Better he in charge than the conductor, for conductors are used to hearing orchestras at rather too close a range for relaxed listening and often insist on musicologically pedantic points when it comes to balancing, sometimes to the detriment of natural sound.

Conductors tend to listen to the score, while good producers also listen to the music, which includes sound quality as well as notes.

This is not to say that there are no mistakes. Multi-microphone stereo recordings do sometimes reveal themselves a little at the aural seams, with slight changes of reverberant perspective between orchestral sections or soloists and orchestra. Concerto recordings may be found in which a cadenza has clearly been recorded either in a different studio or with changed microphone balance, the soloist apparently moving into a drier acoustic for this and reverting to a more spacious setting when the orchestra returns. Off-stage effects in opera are sometimes too obviously cooked up, with reverberation associated with an extreme left- or right-hand sound extending back behind one loudspeaker only, when in terms of the supposed scene the sounds in question should excite ambience within the main hall also. On one operatic recital record both orchestra and singer are set in reverberant acoustics, but the orchestra's ambience is stereophonic and the singer's monophonic – the latter central only, with and beyond rather than around the voice. Of course, one can easily be pedantic on such points by carping, for instance, about the Decca recording of 'Götterdämmerung' because in the last Act the Gibichungs' Hall appears to collapse in double-mono only, or by complaining of an underground train passing beneath the Kingsway Hall during an EMI disc of Bloch's Violin Concerto. Both recordings are in other respects magnificent.

In opera recording it is an enormous advantage that the orchestra is liberated from the confines of the theatre pit and is able to expand in a proper concert hall acoustic. This has revealed a lot of fine musical detail not normally presented with very great care. Sometimes the deliberate employment of changing acoustics is most telling, as in Decca's use of artificial reverberation for the dungeon scene in 'Fidelio'. In the opera house Leonora and Rocco may be given a little acoustic atmosphere by entering at the far back of the stage, but they are no longer in a reverberant acoustic once they have moved to the foreground, despite the fact that they are still in the dungeon. Recording here does something that has not (yet) been tried in the theatre. For engrossing comment on the aesthetics of recording technique as applied to opera see John Culshaw's fascinating book on the whole Decca 'Ring' enterprise.*

With chamber works the correct amount of reverberation is more arguable and depends very much on taste regarding replay level, as is perhaps appropriate for this rather more personal, intimate type of music. The less evident the recorded reverberation, the less natural is the reproduction at any

* *Ring Resounding*, by John Culshaw.

volume below an 'in the room' level, and as preferences may vary in this respect from one composer to another, a listener might judge Brahms to be abused and Beethoven well served, or vice versa, by a similar recording style. These composers are chosen deliberately for a personal example: I prefer Brahms' chamber music at a relaxed, fairly distant sort of level, and Beethoven, if anything, rather close. But at one time my recording of the Brahms Piano Quintet had a dry, very immediate balance which insisted on a high replay level for naturalness – it sounded rather flat and lost if turned down. The Beethoven Op. 18 Quartets, however, although recorded by the same company, were given a studio setting which called for a lower replay loudness not fully to my taste. This brings us back to psycho-acoustics and the quest for perfection: who is to say which was the more natural recording?

A nice problem in the ethics of hi-fi which may loom more significantly as the years pass concerns the manner in which electronically produced musical sounds are recorded. A new world of tone-colours is being evolved by and for a new type of composer, whose workbench is littered with signal generators, mixers and tape recorders and who seldom sits down at the piano keyboard. Some traditional composers like Messiaen have called upon electronic resources for particular unusual effects, but many of the 'new men' have washed their hands of the acoustic past and synthesise *all* their sounds as waveforms heard via loudspeakers. Should these sounds be played in a hall to be picked up by microphones, or would it be more appropriate to feed the electrical signals straight into the disc cutter? If the latter, to what 'original' would our fidelity apply, the sounds produced by the composer's private speakers or the hypothetical sound that would arise from perfect reproduction of the electrical wave patterns? This is a technical problem very much for the future, perhaps insignificant compared with the musical problems thrown up by many of the creative minds working in this field, as a distressing number of those who might be exploiting new tonal possibilities in the cause of music are proponents of plink-plonk serialism.

In less rarefied musical regions there are tricks of the recording trade which extend rather beyond ordinary tinkering with microphone placings, without descending (if that is the word) to open gimmickry. Sometimes attempts are made to alter the tonal and/or reverberant balance of a recording dynamically as the music proceeds. RCA Victor's 'Dynagroove' records are said to do this, the object being to present the music in the most favourable possible light when heard at low loudness levels via cheap record players, without producing an obviously unnatural effect on better equipment. Subjective results are variable from the hi-fi point of view, not always bad, occasionally very good indeed: the Shostakovich 5th Symphony recording mentioned in the previous chapter is a Dynagroove issue.

At a more popular level attempts have been made to exploit recording techniques for the sake of sheer sonic impact, with unashamed use of directional effects – sometimes with movement – and variations in acoustic of a rather lush and overblown kind. Decca 'Phase 4' and EMI 'Studio Two' recordings come into this category, and though I personally dislike some of the results it must be admitted that when applied to the light Classics they probably attract the initial musical interest of many who might otherwise miss a whole world of aesthetic experience. Leopold Stokowski has been an advocate of this view, following in the popularising tradition of his collaboration with Walt Disney in making the film 'Fantasia'. Stokowski has also maintained that for home listening we should, ideally, have four-channel stereo, not using the extra channels for re-creation of a more natural ambience, but placing the speakers around the room to envelop the listener within music coming from all lateral directions. Some of the first 4-channel 'pop' recordings have employed this approach very successfully.

The important thing in this nether region suspended between electronic technique, psychology, acoustics and musical aesthetics is to keep a sense of proportion, to avoid believing that the whole tradition of Western music is threatened simply because of a passing craze for artificial reverberation or general acceptance of gutless background music. We may be a long way from the musical Utopia envisaged by Berlioz,* with the whole economic system geared to musico-aesthetic ends, and there is more than a grain of truth in the notion that 'Mozart for everyone reads more like Muzak for everyplace',† but I rather support Culshaw:‡ 'Mass communication by means of records and television and radio does not, as some Jeremiahs predict, necessarily mean a lowering of standards; on the contrary, over a long enough period, standards are bound to rise. The better is still the enemy of the good, and the general accessibility of art in all forms is a fine thing, even in its unconscious effect on those who profess to have no interest in it'.

There are those to whom the very idea of music being heard primarily from gramophone records – and by large numbers at that – is anathema. It is claimed, not simply that home listening is different, or not so exciting, or the music less spontaneous, but that it is somehow not musical at all, simply

* *Evenings in the Orchestra*, by Hector Berlioz; 25th evening, 'Euphonia, a tale of the future'.

† *Electronic Music Review*, January 1967.

‡ *Ring Resounding*, p. 259.

because one has the freedom to go back and play a movement or passage again! Such a view has been offered seriously in the gramophone press* and makes one wonder for how long people can reside in ivory towers before realising that the world has moved away, leaving them suspended in open space. In fact, of course, radio and the gramophone (and 'Fantasia') have promoted a more widespread appreciation of good music, have increased the attendance at concert halls and opera houses, and contributed to a general enlivening of amateur musical activity. A survey conducted by the Musical Instrument Association in 1964 revealed, in Britain, a quite phenomenal number of musical societies of every conceivable complexion, with more than half a million people involved annually in 350 separate music festivals, and opera alone occupying over 50,000. Paralleled with this are record sales of musical instruments – due partly, no doubt, to a great expansion of banging, scraping and blowing in schools – and whilst the purchase of over ten million recorders (flute, not tape) in recent years does not mean that the music of Scarlatti or Telemann is about to make the 'top ten', we may yet find that the cheap upright pianos cluttering our junk shops signify a change of instrumental taste rather than an abandonment of home playing. Even the 'pop' groups with their electric guitars seem to have caused a minor boom in the playing of more serious music on acoustic instruments, many players moving on to better things after the first flush of electronically assisted excitement. But whether or not concert-going flourishes and however many or few play instruments, the important consideration is the experience and enjoyment of music.

The present musical public is too large to be accommodated without a great deal of reproduced music, and if all the people who now listen to music with any regularity ceased to use reproductive media and attempted to gain entrance to live concerts for the same total listening time, the facts of architecture, transport and social life would produce nothing but a hopeless musical traffic jam. Extrapolating, the number of people appreciative of and desirous of hearing good music will continue to grow, but so will the cost of building concert halls and maintaining orchestras and opera/ballet companies. Inevitably, then, a vast preponderance of listening to music will in future be done in domestic surroundings. This situation, and various misgivings about lack of spontaneity and absence of genuine continuity in music-making during recording sessions due to the splicing together of various 'takes' on tape, have

* 'The Case Against Recordings', by Tibor Kozma, *The Stereophile*, Autumn 1966; and 'Are Records Musical?' by Hans Keller, *Audio Record Review*, February 1966, followed by an opposing view from Yehudi Menuhin in the March issue.

provoked debates about the nature of musical experience when listening to recordings, some critics – as noted earlier – believing it to be almost non-existent. However, we may perhaps ignore this extreme view, as it is perfectly obvious that while home listening may not give quite the same sort of experience as that obtained in the concert hall, it can and does involve one musically, intensely, and at times to the point of tears or dancing joy that would be inhibited in a public place.

On these lines, pianist Glenn Gould has declared the concert hall to be be doomed* and extolled the musical virtues of recordings because he expects before long to see the evolution of technical reproducing facilities to the point where the home listener has so much control over what he hears – interpretively as well as sonically – that listening will become a far more *creative* and musicologically interesting activity than it can be today in the concert hall. In accordance with this belief, Gould announced his decision to become a 'microphone-only' musician and adopted an almost moral-seeing attitude of superiority towards the old-fashioned way of presenting music. Gould may, in fact, be right in a broad sense about what will happen in the future, though I feel that if his prediction about the decline of the concert hall turns out to be correct it will not be because the home listener will have more control over reproduced performance qualities, but because such control will become *less necessary* as it becomes possible to reproduce ever more accurately the sound picture and musical performance designated by producer and conductor. That designation may well evolve away from a 'natural' sound, it being an increasingly common view among recording producers that music presented for home listening may to some extent make its own aural rules and offer an experience to be judged in its own right, rather than attempt to mimic a visit to concert hall or opera house.

There is still the question of spontaneity, and some risk that over-rehearsal and the endeavour to avoid *any* mistakes in recording will lead to a deadening of style. This can work both ways, as neatly explained by Ivan March:† 'Spontaneity is much easier to achieve at a live concert than in the recording studio, although its presence is not certain even with the advantage of a live audience. I remember once playing at a performance of Respighis *Pines of Rome*. In the course of rehearsals we went through the complete score four times. The most lively account (but also probably – though not necessarily, with an English orchestra – the least accurate) was the first. The second was

* 'The Prospects of Recording', by Glenn Gould, *High Fidelity Magazine*, April 1966.

† *The Great Records*, edited by Ivan March.

a commendable amalgam of polish and vivacity, but from then on each consecutive play-through deteriorated and the final concert performance was the poorest of all (both musically and technically), although the audience, which had not heard the others, responded vociferously.'

A vociferous response comes, not only because of performance quality in the technical sense, but from the nature of the occasion. There clearly is a certain excitement and uniqueness about music 'in the flesh', a psychological 'something' arising from seeing the performers actually creating beautiful sounds with skin, gut, wood, brass and sweat. Also, it is sometimes claimed that despite the perfect intonation, etc, available on records, the artists lack some indefinable power attained when an audience is present. This is probably a feeling rather than a fact, arising from the psychological situation, as can sometimes be shown by playing recordings of broadcast concerts thought to be wonderful by those present at the time, but later not having the lasting musical value of a performance recorded for home consumption. Anyway, there are compensations for these real or imaginary deficiencies as described by Peter Gammond:* 'The conditions for listening to music should definitely not resemble those for watching a football match. There is no real pleasure in sitting amongst a fidgeting and coughing multitude to listen to the intimate delights of Schubert, nor in having to get there, and away from the place again, by all the tortures and indignities known to our transport system. A cup of tea with Mahler at home is infinitely preferable to a warm beer in a crowded buffet. My private image of hell is being condemned for ever to stand listening to Brahms in the central arena of the Albert Hall during a Prom. I have only done it once!' I do not subscribe to quite such an extreme view – there are some very happy Prom memories, and I am still a fairly regular concert-goer – but for any particular type of music there are undoubtedly a number of social and practical difficulties for large numbers of people that are avoided with recordings heard in the home.

The other big compensation for home listening is the traditional one: performances on record are more nearly perfect technically, and the collections of top talent available, for instance, in major operatic works are seldom found in one place outside the recording studio. Top talent may also be juxtaposed in another musically valuable way. Music in the home permits a comparison of styles and phrasing of a sort otherwise quite impossible. For instance, there are sometimes radio programmes devoted to analyses of several performances of particular musical works, and one I particularly remember surveyed the various recordings of Schubert's great *Winterreise* song cycle, offering an

* *Audio Record Review*, Editorial, February 1966.

insight into interpretative differences – and their validity – that will certainly increase the reward from future hearings, whoever the singer.

The most serious criticism of music-at-the-touch-of-a-button has come from Benjamin Britten, a composer who, oddly enough, has clearly benefited enormously from recorded performances of his work. In his much-quoted Aspen address* he stated that reproduced music 'is not part of true musical experience', part of the reason being that music 'demands some preparation, some effort, a journey to a special place, saving up for a ticket, some homework on the programme perhaps, some clarification of the ears and sharpening of the instincts'. There have been various expressions of disagreement with this, though my feeling is that the psychology of the viewpoint is correct, the mistake being in Britten's implication that the *state of mind* which is desirable for listening to music cannot, for some reason, be induced when the music is recorded. If we equate saving up for a ticket with saving to buy a record, then all the prerequisites except 'a journey to a special place' may be met when the music is to be reproduced at home. There is also the point that the social paraphernalia of going out can actually detract from concern for the music by diverting attention to such matters as transport, timing, dress, etc.

I feel that there is a certain value in listening to music as a small group rather than alone, as may have been apparent from the previous chapter. This does have the advantage that one can organise successful recorded concerts which satisfy *all* of Britten's desiderata – including that journey (for friends and neighbours) and a small financial sacrifice (for a suitable charity). I am sure this helps to bridge the gap that so worries Benjamin Britten by making music in the home more nearly part of true musical experience, and for those genuinely intent on musical enjoyment there is really no serious barrier between composer and lone listener.

A chapter called *Music in the Home* concludes a book about *Hi-Fi in the Home*. Ideally the two are synonymous – indeed the former title was originally proposed for the whole book, though as it seemed possible that confusion might arise with piano practise, ballad singing and a well-known London hi-fi shop,† we opted for hi-fi. But in essence, music in the home it is, and if these pages have contributed in some small way to increased musical enjoyment and better understanding of the techniques whereby it may be achieved, they will have served their purpose.

* *On Receiving the First Aspen Award*, by Benjamin Britten.
† Thomas Heinitz, *Music in the Home*.

GLOSSARY

OF AUDIO TERMS AND ABBREVIATIONS

THE FOLLOWING list includes all the specialised terms and expressions employed in this book, plus many others in common use, and should therefore be useful for general reference when studying audio literature. That part of hi-fi language concerned with circuit theory and the detailed functioning of valves and transistors is excluded, as definitions to cover these would require long explanations or assume prior electronic knowledge. Generally, any special words used in definitions are themselves defined, thus making the glossary self-consistent. A particular style used for an abbreviation does not mean that it is an approved standard; in some literature lower-case versions may be found where capitals are given, and vice versa.

A Ampere.

A–B Designates type of audio comparison test in which changes of sound quality may be assessed by direct and immediate switch-over. An A–B comparison between items A and B. Left and right stereo channels are sometimes designated A and B.

ABR Auxiliary bass radiator.

AC Alternating current.

Acoustics Science or study of sound as an objective phenomenon. In popular parlance, applied particularly to acoustical character of halls and rooms.

Acoustic feedback Unwanted acoustic interaction between output and input of an audio system, usually between loudspeaker and microphone or pickup. Can lead to continuous oscillation or a tendency thereto.

Acoustic resistance unit Device comprising absorbent material clamped in an openwork frame for fitting over normal port or over an additional vent on reflex loudspeaker enclosures. Helps to control Q–factor of the speaker/enclosure system.

Acoustic suspension Principle used in loudspeaker systems employing a sealed cabinet, whereby acoustic stiffness of the enclosed air volume provides main restoring force for the speaker diaphragm.

Aerial Device for capturing radio signals to feed input of receiver or tuner. May be a wire, rod or tuned dipole (the latter possibly elaborated with extra elements), or an internal ferrite rod. External device also known as antenna.

AES Audio Engineering Society (U.S.A.).

AF Audio frequency.

AFC Automatic frequency control (in tuner).

AGC Automatic gain control (in tuner or tape recorder).

Alignment Process of adjusting tuned circuits in a radio receiver for optimum performance; accuracy of such adjustment as set. Also, sometimes refers to setting-up of pickup arm.

Alignment protractor Device for measuring or indicating errors in lateral alignment of a pickup arm. Locates on turntable centre spindle, pickup stylus fitting in a small hole.

Alternating current To-and-fro movement of electricity (AC).

AM Amplitude modulation.

Ambience Acoustic coloration added by concert hall or listening room.

Ambi-stereophony Experimental technique for producing an impression of extra reverberation from above and to the sides and rear when reproducing two-channel stereo from the front.

Amp Abbreviation of ampere (A).

Amperage Magnitude of electrical current flow. Number of amperes (A).

Ampere Unit of electrical current flow (A).

Amplification Increase in signal magnitude achieved in an amplifying circuit.

Amplification factor Degree of amplification in a circuit; output divided by input.

Amplifier Circuit unit providing amplification or increase in magnitude of signal, from low to high voltage or current (preamplifier) or to power output (power amplifier).

Amplitude Magnitude, size. Peak deviation in a complex waveform.

Amplitude distortion Alteration of sensitivity or amplification with change of signal level. Arises from non-linearity of transfer characteristic and produces harmonic distortion and intermodulation distortion.

Amplitude modulation Type of radio transmission (AM) in which the audio modulating signal varies the amplitude of the radio carrier.

Amplitude pickup Pickup with voltage proportional to amplitude of stylus displacement. Employs a piezo-electric or strain-gauge type of transducer and compensates approximately for recording characteristic when used with appropriate load impedance.

AM suppression Extent to which an FM receiver suppresses changes of amplitude in received signals, thereby improving both S/N and the ability to reject unwanted stations even when they are near to or at the frequency of the wanted one.

Anechoic Without reverberation. Special type of room or chamber with acoustically absorbent boundaries.

Aneroid barometer Device for measuring changes of atmospheric pressure by allowing external air to deform a sealed but compliant metal chamber.

Antenna *see* aerial.

Anti-phase Situation in which two similar or identical signals are disposed in phase opposition. Waveform patterns displaced by 180 degrees and therefore tending to cancel.

Antiphonal Strictly, refers to traditional plainsong antiphons of the Roman Catholic Church, where alternating passages are sung by separate choirs. Also applied more generally to music in which one section responds to, complements or opposes another, using spatially separated instruments or voices.

Anti-skating device Mechanism providing outward force on a pickup arm to counteract inward bias caused by groove/stylus friction and offset geometry.

Antistatic Device or substance for dissipating or preventing build-up of static electric charges. Usually associated with gramophone records.

Arm Commonly, pickup arm. Often applied to the whole pickup assembly though strictly excludes the cartridge.

Armature Vibrating metal part in pickup or some other transducers.

Arrival-time Time at which sound information arrives at the ears. Usually applied to stereo perception, where the exact relative arrival-times at the two ears of particular acoustic wave patterns transmitted at the same instant play a part in determining the apparent directions of sound sources.

ARU Acoustic resistance unit.

Atmospheric pressure Force per unit area exerted on all objects from all directions due to weight of the atmosphere. Normal figure at sea level is 14·7 lb. per square inch (1·034 Kgm/sq.cm).

Atom Fundamental unit or matter. Made up from electrical charges (normally balanced within any one atom in the absence of ionization) and different for each chemical element.

Atonal Having no key centre. Music without key or obvious tune and harmony because of equal value attached to all twelve semitones in the octave. Opposite of diatonic.

Attack Percussive or transient quality of some musical sounds. Sometimes used of loudspeaker with good transient performance.

Attenuator Device or circuit for reducing signal amplitude.

Audio Science and art of sound recording and reproduction. Sound frequencies or signals.

Aural Connected with hearing.

Auto-changer Record playing mechanism with facility to change from disc to disc automatically at ends of sides.

Auto-couplings Arrangement of sequence on a recording occupying several disc sides to facilitate use on an auto-changer.

Automatic frequency control Circuit function in a radio receiver (AFC) designed to keep the system tuned accurately to the desired station by counteracting any tendency to drift.

Automatic gain control Circuit function in a radio receiver (AGC) designed to maintain amplitude of IF signal fed to demodulator at a more-or-less constant level over a wide range of aerial input voltages. Sometimes applied to signal compression system used in 'automatic' tape recorders. Also called automatic volume control (AVC).

Automatic volume control See automatic gain control.

Aux Auxiliary. Extra input on amplifier, etc.

Auxiliary bass radiator Additional speaker cone system used in place of port in flapping baffle type of enclosure (ABR).

AVC Automatic volume control (in receiver).

BAF Bonded acetate fibre; material used for damping in some loudspeaker enclosures.

Baffle Structure for isolating front and rear of loudspeaker diaphragm.

Bal Balance (of stereo channels).

Balance Four meanings: (1) equality of gain or sensitivity between left and right stereo channels; (2) tonal balance between bass, middle and treble; (3) balance of loudness between instrumental and/or vocal parts; (4) ratio of direct to reverberant sound.

Balance control Potentiometer used to adjust differential gain of left and right stereo channels. Enables user to compensate for unbalanced signals or transducers.

Balanced (line or circuit) System of connections in which the two signal-carrying conductors are equally 'live' with respect to earth. Signals conveyed by twin-feeder.

Balun Device for converting a balanced circuit or line to an unbalanced condition, or vice versa. Usually a transformer with appropriate windings and connections.

Band II VHF radio band covering 87·5–108 MHz. Used in Britain for FM broadcasting.

Bandwidth Width of a frequency band between lower and upper limits (normally taken at −3 dB or half-power points, but sometimes −6 dB for radio work). With audio, terminating frequencies are given, but for FM and IF purposes it is more usual to quote the total width in kHz or MHz.

Bass Low frequency end of audio spectrum, below approximately 150 Hz. In musical notation, top line of the bass clef is A_1 (220 Hz).

Bass reflex *see* reflex.

Beat Difference tone formed when two signals are mixed via a non-linear device – including the ear. Process of 'beating' to produce such a tone or tones. Also, audible fluctuation of amplitude when two tones are very close together in frequency, moving in and out of phase.

Bias High frequency signal used to linearise magnetisation process in tape recording. Fixed voltage or current to set valves or transistors at optimum operating point. Side-thrust on pickup arm.

Bias compensation Provision of outward force on a pickup arm to counteract inward bias caused by groove/stylus friction and offset geometry.

Binaural Heard with two ears.

Bi-radial stylus See elliptical stylus.

BKSTS British Kinematograph, Sound and Television Society.

Bottoming Pickup stylus reaching rounded bottom of record groove due to smaller than optimum tip radius as viewed from front or rear.

Braiding Commonly refers to outer conductor of a flexible screened cable. Made up from 'braided' fine bare wires.

Bulk eraser Device for erasing recording from a whole tape at once on its spool by application of an intense alternating magnetic field.

C Symbol for capacitance.

Cantilever Rigid bar supported at one end. Commonly refers to stylus arm in pickup.

Capacitance Reactive electrostatic property, measured in microfarads (μF).

Capacitor Circuit component with specific capacitance value.

Capacitor microphone Microphone employing a diaphragm so positioned in relation to a fixed member that movement caused by sound-waves produces a change of capacitance, this being translated into an audio voltage.

Capstan Accurate spindle which, in conjunction with pinch-wheel, drives tape in a tape recorder.

Capture effect Peculiarity of FM reception whereby an unwanted station on a similar or even identical frequency, and at a level not far below that of the wanted station, remains inaudible – even though it may itself be frequency modulated.

Capture ratio Necessary difference in received level of two FM stations on the same frequency for the capture effect to operate.

Cardioid Microphone with 'heart-shaped' polar response, making it most sensitive in one direction.

Carrier Continuous high frequency signal used to 'carry' audio information in the form of modulation. May be radio wave, or IF in a receiver.

Carrier–wave Radio wave used to 'carry' audio signals as modulation.

Cartridge Detachable transducer-plus-stylus part of pickup head. Also tape cassette of continuous loop type, usually with eight tracks and a speed of $3\frac{3}{4}$ i/s.

Cassette Preloaded container with tape and spools for use on cassette tape recorder, usually with four tracks and a speed of $1\frac{7}{8}$ i/s.

Cathode-follower Valve circuit with especially low output impedance suitable for feeding signals via long cables.

CCIR International Radio Consultative Committee (CCIR commonly refers to tape replay characteristics).

Ceramic Type of man-made piezo-electric or permanent magnet material used in some pickups and loudspeakers respectively.

Changer Auto-changer.

Channel Sequence of circuits or components handling one specific signal. In domestic equipment, usually one of the two stereo channels.

Channel separation Degree to which left and right stereo signals are isolated in pickup, amplifier, etc.

Charge Excess or deficit of electrons or protons, imparting positive or negative polarity to the charged object.

Chassis Metalwork on which circuit components are assembled. Usually connected to 'earthy' side of circuit.

Choke Alternative name for an inductor, especially when used in a smoothing or filtering circuit.

Chromatic Use of notes not forming part of the prevailing major or minor keys.

Circuit An arrangement of interconnected electrical and/or electronic components to perform some particular task. Theoretical diagram of such an arrangement.

Class–A Type of power amplifier in which all the valves or transistors operate on linear portions of their transfer characteristics all the time. Circuit draws constant current from power supply irrespective of signal level.

Class–AB Type of power amplifier in which valves or transistors operate on linear portions of their transfer characteristics at low amplitudes (Class–A), changing to a regime in which positive and negative portions of the signal waveform are shared between push-pull components at high levels (Class–B).

Class–B Type of power amplifier in which some valves or transistors (usually the output pair) split positive and negative portions of the signal waveform between them, each operating from a low initial current point. Current drawn by circuit from power supply rises with signal level.

Clipping Form of distortion due to severe overloading, tips of the audio waveform becoming clipped as the system is incapable of providing any more output despite increasing input.

cm/sec Centimetres per second. Velocity expressed in these units and referring either to linear speed (of disc surface or tape) or alternating velocity of record groove, loudspeaker cone, etc.

c/o Cut-off or crossover.

Coaxial cable Type of screened cable with central conductor surrounded by an outer screen. Dimensions and materials are controlled to give a particular RF impedance value (usually 70–80 ohms) for coupling aerial to tuners with unbalanced input circuits.

Coil Winding of wire to achieve a certain inductance value or to couple to other windings. May be with or without a magnetic core, depending on frequency or function.

Coincident *see* crossed-pair.

Coloration Alteration of sound quality according to resonances or other peculiarities in an audio system. A 'character' imparted to reproduce sound, often and mainly by loudspeakers.

Column Type of loudspeaker enclosure in which rear of drive unit receives resonant bass loading from a column of air. Speaker may be mounted at one end or, in sophisticated versions, some distance from the narrow end of a tapered column (possibly folded).

Comparator Device with elaborate switching arrangements for interconnecting any combination of pickup, turntable, tuner, tape recorder, amplifier and speakers for demonstration purposes.

Compatible Four meanings: (1) stereo radio signal receivable as mono on a single-channel receiver; (2) stereo disc record claimed to be playable with a mono pickup; (3) mono pickup claimed to be suitable for playing stereo records; (4) stylus in stereo pickup of size suitable for both types of record.

Compliance Yielding quality due to springiness or elasticity, measured in

Compliance Units (cu.). Reciprocal of stiffness. Static compliance is measure of yield produced by a fixed force, dynamic compliance represents yield with alternating force; latter often less than former due to resistive behaviour of compliant material.

Compliance Unit Equals 10^{-6} cm/dyne.

Component board Rigid board carrying components in electronic equipment. Usually with printed circuit on one side and components on the other.

Component System Domestic audio system employing separate components such as tuner, amplifier and speakers rather than an integrated console.

Compression Reduction of a signal's dynamic range by raising the level of quiet passages and/or lowering the level of loud passages. Necessary when S/N ratio of proposed transmission or recording system is smaller than dynamic range of material.

Condenser *see* capacitor.

Condenser microphone *see* capacitor microphone.

Conduction Process of electrical current flow through a conductor.

Conductor Material which permits flow of electrical current due to low or medium resistance. Opposite of insulator. Also used in reference to signal-carrying wires.

Cone Diaphragm of conventional moving-coil speaker.

Cone breakup Effect when loudspeaker cone ceases to move as a whole. Usually applies at the higher frequencies within operating band of any drive unit, manifested as a multitude of minor response irregularities and degraded transient performance.

Cone surround Strip or roll of compliant material fitted to periphery of loudspeaker cone to seal it to the frame yet permit axial movement.

Console Generally, a type of equipment cabinet; but also refers to one-piece radiogram or stereogram as distinct from component audio system, particularly in the U.S.A.

Contrapuntal Using counterpoint.

Control unit Preamplifier part of audio system designed for use displaced from power amplifier. Accepts signals from pickup, microphone, tuner, etc, applying any necessary equalisation and incorporating volume and tone controls.

Convection Upward movement of air or water as it becomes less dense when heated.

Conversion efficiency *see* efficiency.

Core Usually, ferro-magnetic centre, or structure around and within which the windings of a transformer or inductor are housed.

Correction Compensation for some response peculiarity. Equalisation of a recording or transmission characteristic during reproduction.

Counterpoint Combination of simultaneous parts in music, each of separate significance but contributing to a coherent texture.

Counterweight Weight fitted at rear of pickup arm behind the pivot to 'counter' the weight of the arm/head assembly and permit adjustment of stylus force to an appropriate value.

Coupling Circuit, component or connection involved in feeding a signal from one stage, transducer or instrument to another.

c.p.s. Cycles per second (Hz preferred).

Crolyn Name used by the Dupont Corporation (U.S.A.) for recording tape coated with chromium dioxide.

Cross-field bias System of tape recording using a separate head for application of high frequency bias. Extra head is positioned in relation to record head so that its HF magnetic field has less demagnetising effect at high audio frequencies than in the conventional arrangement using a common head. Effect is to extend overall audio response at the top end.

Crossed-pair Two directional microphones arranged one above the other, or otherwise very close together, with axes diverging at approximately 90 degrees. Used for stereo recording in the manner of Blumlein.

Crossover Circuit for dividing output of an amplifier into various frequency bands to feed appropriate loudspeaker units.

Crossover distortion Can arise in Class–B amplifier circuits due to discontinuity as signal waveform swings across 'zero' line. Experienced at low rather than high output levels.

Crossover frequency Frequency at which a loudspeaker crossover divides the signal or 'crosses over' from one section to another.

Crosstalk Breakthrough of signal between two stereo channels. Expressed as level of unwanted signal in relation to signal in wanted channel. Measured in decibels (dB).

Crystal Natural piezo-electrical transducer used in some pickups and microphones. Solid-state diode. Also, resonant quartz crystal sometimes used for tuning purposes in radio reception.

c/s Cycles per second (Hz preferred).

c.u. Compliance unit.

Current Electrical flow, measured in amperes (A).

Curve-bending Reshaping of frequency response curve to serve some particular purpose.

Cut-off frequency Effective limiting point in a frequency response. Often taken to be point at which level has fallen by 3 dB.

Cutter *see* disc-cutter.

Cycle Ultimate repetitive pattern in any vibrating system, mechanical, electrical or acoustic. One complete cycle or period comprises change of pressure, velocity, voltage or current from 'zero', up to a maximum in one direction, down through 'zero' to a maximum in the other direction, then back to 'zero'.

Damped Employing electrical, mechanical or acoustic damping to reduce resonance.

Damping Process of reducing resonant effects by use of resistance or its mechanical and acoustic equivalents. Material used for this purpose. Absorbent material applied to a surface.

Damping factor Ratio of loudspeaker impedance to amplifier source impedance. Large ratio improves speaker damping.

dB Decibel.

DC Direct current.

Decibel Logarithmic unit representing ratios and used for expressing wide-ranging quantities on a simple linear scale. Decibel figure equals logarithm of a voltage ratio multiplied by 20 and logarithm of a power ratio multiplied by 10.

Decoder *see* multiplex decoder.

Decoupling Use of capacitors and other components to 'decouple' separate stages in an amplifier or tuner – and to isolate such stages from the HT supply – at audio and sub-sonic frequencies. Also, mechanical decoupling by means of compliance in some pickups and loudspeakers, or between cabinets and rooms to overcome acoustic feedback.

De-emphasis Reduction of high frequency level in reproduction to compensate for corresponding pre-emphasis applied to recorded or broadcast material. Improves overall S/N ratio.

Defluxer *see* demagnetiser.

Demagnetiser Device for removing residual magnetism from tape-heads and tape-deck fittings, or for erasing recording from short stretches of tape. Works by applying an alternating magnetic field. Also known as defluxer.

Demodulator Part of tuner circuit for extracting audio modulation from radio frequency or intermediate frequency signal.

Derived centre channel Signal for feeding a third, central loudspeaker in an otherwise conventional two-channel stereo system. Obtained by addition of left and right via a circuit which, ideally, does not degrade separation of signals fed to outer speakers.

Detector *see* demodulator.

Deviation In FM transmission and reception, amount by which carrier frequency is shifted by audio modulation. Agreed maximum modulation corresponds to a deviation of ± 75 kHz.

Diaphragm Sound generating element in a loudspeaker. May be a cone driven at its apex or dome driven at its periphery in moving-coil units, or a stretched sheet driven over its whole area in electrostatic units.

Diatonic Use only of notes forming the major or minor keys. Temporary departure from this is chromaticism, total departure is atonalism or serialism.

Difference signal *see* sum-and-difference.

Difference tone Formed when two signals beat together. Equal in frequency to difference between the generating tones.

Diffraction Bending of sound-waves around an obstacle. Takes place readily when object is small in comparison with sound wave-length, but object casts an acoustic 'shadow' when it is greater than a wavelength in diameter.

DIN Deutscher Industrie Normen (German Industrial Standards). Commonly refers to standard plugs, sockets and tape equalisation characteristics, but includes some general hi-fi specifications.

Diode Crystal or valve with two elements having high back-to-front resistance ratio.

Dipole (aerial) Commonly, main resonant limb in a VHF aerial system from which RF signal is taken by feeder to the receiver.

Direct Current Movement of electricity in one direction only (DC).

Direct disc Disc record produced as a one-off recording by direct cutting rather than indirectly by a plating, copying and pressing process. In commercial mass production, a disc so cut is used to imprint raised groove patterns on the Master.

Directional Applied to microphone, aerial or loudspeaker: having greater sensitivity or output in a particular direction.

Director Element in a VHF aerial system placed in front of main dipole to improve sensitivity in the forward direction.

Disc-cutter Mechanism used for making direct-cut disc recordings; comprises recording lathe and a cutter-head fed from special amplifier.

Discriminator Demodulator in frequency modulation (FM) receiver.

Dished Symmetrical departure from flatness of disc record, giving saucer-like shape with central region displaced from plane of the periphery.

Dissonance Inharmonious and consequently harsh combination of tones.

Distortion Strictly, any deviation from the original in reproduction. Usually refers to harmonic distortion or intermodulation, both resulting from amplitude distortion.

Dolby system Noise reducing process which compresses the dynamics of a signal before recording on to tape and expands them again on replay for disc recording. Net result is a reduction in level of any background noise introduced between compression and expansion stages. Processed signal sometimes known as 'stretched' because of increased gap between signal and noise.

Doppler effect Change of frequency or pitch when sound source is moving relative to observer. Applies to loudspeaker cone carrying high frequency signals when subjected to large low frequency movements; hence Doppler distortion.

Double-mono An identical signal heard via the two channels of a stereo system. Ideally, should seem to come from a point midway between the two loudspeakers.

Doublet Type of transducer system, particularly loudspeaker, in which energy is radiated equally and freely from both sides of the diaphragm.

DP Double play (tape thickness).

Drift Tendency of tuner circuits to shift away from optimum adjustment as components warm up. Counteracted by AFC.

Driver Stage Penultimate stage in a power amplifier designed to feed or 'drive' the output stage.

Driver transformer Transformer incorporated in some power amplifiers for feeding signal from driver stage to output stage.

Drive cord In a tuner or receiver, fine cord used in conjunction with pulleys to drive a pointer across tuning dial.

Drive unit Loudspeaker transducer unit as distinct from enclosure or cabinet.

Drop-out Momentary reduction or disappearance of signal due to inconsistent tape coating or mechanical defect.

Dual concentric Type of loudspeaker unit in which separately driven bass and treble cones or diaphragms are mounted coaxially.

Dual cone Type of loudspeaker drive unit with two cones driven from one moving-coil. Inner cone of smaller diameter and designed for greater efficiency at high frequencies.

Dubbing Copying of a recording by direct transfer or superimposition.

Dynagroove Recording and processing system evolved by RCA Victor (U.S.A.). Said to compensate automatically for technical limitations in reproduction arising from pickups and domestic replay preferences.

Dynamic Moving-coil (m.c.).

Dynamic range Range of signal amplitudes from highest to lowest found in acoustic programme material (in phons), or the range which a device will handle (in dB). In equipment, upper limit is set by overload point and lower limit by background noise.

Dyne Unit of force equal to approximately one milligram weight.

Earth That part of a circuit or system of interconnections used as 'zero' reference for signals; normally common to chassis or screening elements in audio equipment. Sometimes connected to a genuine external earth point or earth pin of mains socket. Same as 'ground'.

Earthed Connected to that part of a circuit or system of interconnections used as 'zero' reference for signals, or to a true external earth.

Earth-loop Arrangement of interconnections between circuit units and/or input sources resulting in more than one path for the 'earth' side of signal-carrying cables. Resulting loop can lead to hum pick-up.

Earthy Connected to or associated with the earthed, grounded or screened part of a circuit, component or signal.

Echo Box Popular name for reverberation unit. Device for producing artificial 'echo' or reverberation.

Editing Process of coupling and arranging lengths of tape recording to suit a particular programme purpose.

Efficiency Ratio of output to input power in a transducer. With loudspeakers, percentage of electrical input available as acoustic output.

EIA Electronic Industries Association (U.S.A.).

Eigentone Room resonance produced by parallel walls.

Electro-acoustic Concerned with transduction of energy between electrical and acoustical forms.

Electrode Strictly, positive or negative pole of an electric cell (in battery); commonly, any of the functional elements in a valve or transistor.

Electrolytic Commonly, a type of capacitor; also an electrochemical action.

Electromagnetic Type of energy radiated as radio-waves, heat and light. Inter-relationship of electrical and magnetic processes. Device employing electromagnetism. Induction of signals by electromagnetic fields.

Electro-mechanical Devices or equipment employing a mixture of electrical and mechanical components (i.e. tape recorder). Item or process involving interaction of mechanical and electrical forces.

Electro-motive force Electrical force or pressure, measured in volts (V).

Electron Fundamental negative electrical particle.

Electronic Involving the use of 'active' components such as valves, transistors and diodes, though also often applied to any complex circuit employing resistors, capacitors, inductors, transformers, etc.

Electrostatic Concerned with forces and fields associated with electric charges; strictly, fixed or 'static' charges, but now also commonly applied to alternating electric fields. Loudspeaker employing electrostatic forces to actuate radiating diaphragm(s).

Element (aerial) Resonant limb in a VHF or other aerial system, possibly one of several.

Elliptical stylus Pickup stylus designed to minimise tracing distortion by placing a small radius in contact with the groove walls as viewed from the sides and a larger radius across the groove as viewed from front or rear.

ELS Electrostatic loudspeaker (also ESL).

EMF Electro-motive force.

Emitter-follower Transistor circuit with especially low output impedance suitable for feeding signals via long cables.

Encoding Commonly, processing a pair of stereo signals to produce the composite modulation for a multiplex radio transmitter.

EP Extra play (disc record).

Equalisation Electrical correction for a recording characteristic or compensation for a frequency-sensitive component.

Equal temperament Musical scale in which each semitone is given an equal interval corresponding to the twelfth root of two (1·059463), or just under 6 per cent away from its neighbours.

Erase head Tape-head designed to apply a strong high frequency magnetic field across a fairly wide gap for erasure of earlier material as tape passes through a recorder switched to record mode.

Erasure Removal of tape recording with strong magnetic field.

ESL Electrostatic loudspeaker (also ELS).

Fading Drift up and down in level of signals from distant radio stations due to changes in the upper atmosphere. Sometimes signals 'fade' into the background noise.

FBA Federation of British Audio (manufacturers' body).

FCC Federal Communications Commission (U.S.A.).

Feedback Signal from output of amplifier or electronic network applied to input in anti-phase (hence negative feedback) to reduce distortion and noise, and to flatten or shape frequency response. Also, unwanted acoustic feedback (positive).

Feeder Cable used to feed an RF signal from aerial system to receiver.

Feed spool Reel on a tape recorder from which the tape is drawn during recording or replay. Left-hand spool on a conventional machine.

Ferrite Iron based composite material with controlled magnetic properties. Various uses include rod aerials in portable radios.

Ferro-magnetic Magnetic substance based on iron.

fff Molto fortissimo, extremely loud.

ffrr Full frequency range recording (Decca trade name).

ffss Full frequency stereophonic sound (Decca trade name).

Field Lines of force surrounding or extending from some magnetic, electrostatic or electromagnetic device. Some magnetic field is present whenever an electric current is flowing, and some electrostatic field is present whenever an EMF exists between two points.

Fifth Pitch or frequency ratio of two to three. Span of four diatonic intervals covered by five notes, hence fifth.

Figure-of-eight Microphone with polar response shaped like a figure eight, making it most sensitive to front and rear and insensitive to the sides.

Filter Circuit to attentuate above, below, or at a particular frequency.

Flapping baffle Type of loudspeaker enclosure based on the reflex principle but using a compliantly mounted panel, baffle or additional drive unit cone in place of the port.

Flutter Waver of pitch caused by spurious fluctuations of speed in the recording medium at rates above about 10 Hz. Heard as a sort of bubbly roughness.

Flux density Strength of magnetic field, usually given in gauss (G).

Flywheel Symmetrical wheel, often with a concentration of mass at the periphery, which rotates to provide momentum for speed stabilisation.

FM Frequency modulation.

Formant Favoured frequency, or group of frequencies, contributing to instrumental tone-colour.

Fourier Scientist who devised classic system of harmonic waveform analysis.

Four-track Tape recorder designed to record four tracks on magnetic tape. A tape recorded in this fashion.

f.r. Frequency range or response.

Free-air resonance Natural basic diaphragm resonance of a loudspeaker unit as measured with the unit unmounted in an enclosure or baffle.

Free-field Unrestricted surroundings in which performance of a radiating device (acoustical or electromagnetic) is not complicated by restrictions or reflections. Also, type of induced-magnet pickup cartridge.

Frequency Rate of vibration or oscillation. Number of complete cycles in one second.

Frequency-changer Stage in a superhet receiver used to change the incoming RF signal, of whatever frequency, into an IF signal of fixed-frequency.

Frequency discriminator *see* discriminator.

Frequency doubling Production by loudspeaker of an acoustic output at second harmonic of the signal input frequency. Can occur at low frequencies due to inadequate loading and other factors.

Frequency modulation Type of radio transmission (FM) permitting reception at high quality and with low background noise. Only employed at very high radio frequencies (VHF) and with audio modulating signal varying frequency rather than amplitude of the radio carrier.

Frequency range Effective or operating limits of equipment.

Frequency response Frequency range covered by equipment within stated decibel limits.

Fringe area Region at the extremities of a transmitter's normal service area in which received signal strength is very low, requiring extra care with aerials.

Front-end Input stages of tuner or pre-amplifier.

Full-track Tape recorder designed to record a single track across the full width of magnetic tape. A tape recorded in this fashion.

Fundamental Lowest frequency component in a complex waveform. Basic resonance of loudspeaker.

Fuse Circuit protection device in which a fine wire breaks continuity by melting when the safe current value is exceeded.

G Gauss. Unit of magnetic flux density.

g Acceleration due to gravity.

Gain Voltage amplification factor.

Ganged Controls linked for operation by a single knob; particularly in stereo preamplifiers where most controls work in pairs.

Gap Vertical slit in the magnetic material of a tape-head, forming poles across which a magnetising field occurs during recording, and into which a magnetic signal is induced during replay. Magnetic gap in moving-coil speaker or microphone.

Gauss Measure of magnetic flux density. One gauss equals one maxwell, or line, per square centimetre (G).

Generator Test instrument producing signals of known frequency and waveform, etc. Also sometimes applied to signal-producing element in a transducer.

Glissando Sliding from one note to another, sounding the intermediate pitches on the way.

gm Gram (approximately $\frac{1}{28}$ of an ounce).

Gram Gramophone = Phonograph. Used on amplifiers to label pickup sockets and pickup position on input selector switch.

Ground *see* earth.

Grounded *see* earthed.

H Henry.

H and N Hum and noise.

Haas effect Characteristic of hearing whereby an identical sound emitted from two or more points seems to come from the nearest point due to the latter's precedence in time.

Half-power bandwidth *see* power bandwidth.

Half section Type of filter circuit, commonly used in loudspeaker cross-overs.

Half-track *see* twin-track.

Hangover Effect of poor loudspeaker transient response whereby a tendency to ring adds a blurbed 'hangover' to the sound, particularly at low frequencies.

Harmonic Frequency multiple of a fundamental tone. Twice the fundamental frequency is the second harmonic, three times is the third and so on. On some musical instruments harmonics may be played in isolation from the natural fundamental tone.

Harmonic distortion Product of amplitude distortion whereby harmonics are added to original signal (HD).

Harmonic series Complete succession of harmonics in ascending order. Musical intervals between successive harmonics depend on their positions in the series.

Harmony Strictly, clothing the main melodic line in music with subsidiary melodies and chords; or relationships between lines of more nearly equal importance in polyphonic music. Popularly, degree of consonance or pleasantness between concurrent musical lines, dependent on harmonic relationship.

HC Handling capacity.

HD Harmonic distortion.

Head Tape-head or pickup head.

Headphones Small transducers, usually mounted in pairs on a frame to fit over the head and designed to make intimate contact with the ears, facilitating private listening for pleasure and useful for monitoring during recording of live programme material.

Head shell Detachable part of pickup arm designed to carry the cartridge, with connections for conveying signals from the transducer to contact pins.

Heater Filament in a valve, energised by the low tension (LT) supply (normally 50 Hz AC).

Heat-sink Metal structure (sometimes part of chassis) used to conduct heat away from transistors and other components. Prevents excessive temperature rise in these components.

Henry Unit of inductance (H).

Hertz Unit of frequency, equals one cycle per second (Hz).

HF High frequency.

HFDA High Fidelity Dealers' Association (U.K.).

Hi-fi Abbreviation of high fidelity.

High-definition Clearly defined, with full detail. Applied to reproduction of pictures in which every facet is visible and accurate.

High fidelity High degree of truthfulness in sound recording and reproduction. Equipment or sound sources exhibiting this characteristic.

High-flux Refers to loudspeaker drive unit employing a magnet system with high flux density, usually over about 12,000 gauss.

High-pass Applied to filter or circuit which attenuates low frequencies.

High tension Relatively high DC voltage (HT) in a valve circuit needed for amplification purposes; so called to distinguish it from low tension (LT) used for valve heaters. By habit, often also applied to the rather low 'HT' used in transistor circuits.

Hole-in-the-middle Weakness or absence of apparent sound sources in the region midway between loudspeakers when reproducing a stereo signal.

Horn Acoustic device with cross-sectional area expanding or flaring according to a particular mathematical law.

Horn-loading Acoustic load offered to a loudspeaker drive unit by horn type enclosure.

Howlback Unwanted positive acoustic feedback producing continuous oscillation or a tendency thereto.

HT High tension (voltage).

Hum Unwanted low frequency tone in reproduction. Of fixed frequency and usually due to 50 Hz mains and its harmonics.

Hum-field Alternating electrical or magnetic field surrounding equipment or wiring carrying voltages or currents derived from the AC mains.

Hum-loop Arrangement of leads or interconnections causing AC mains hum to be added to audio signals. Occurs when wanted signals are made to share conducting paths with hum currents, or when signal-carrying cables form a 'loop' into which hum is induced.

Hz Hertz

IB Infinite baffle (loudspeaker mounting).

IC Integrated circuit, in which circuit junctions and components are contained on a microscopic scale within a solid block of material.

Idler Wheel used in tape recorder or turntable mechanism for transmitting drive between rotating components, usually rubber or plastic tyred.

IEC International Electrotechnical Commission.

IF Intermediate frequency.

IF transformer Component in tuner or radio receiver used to couple or feed IF signal between successive amplifying transistors or valves. Windings usually tuned with capacitors to resonate at the fixed IF frequency.

IHF Institute of High Fidelity (U.S.A.). Was IHFM.

IM Intermodulation.

Impedance Opposition to alternating current flow, may be reactive and/or resistive. Measured in ohms (Ω). Also, analogous mechanical and acoustic impedance.

Induced Produced by induction across a space due to electric or magnetic fields.

Induced-magnet Type of pickup transducer.

Inductance Reactive electro-magnetic property; measured in Henrys (H).

Induction motor Type of AC electric motor where current in a stator winding produces a rotating magnetic flux which induces current in a rotor winding. Speed of rotor lags behind natural synchronous speed according to mechanical load.

Inductor Circuit component with specific inductance value.

Inertia Opposition to change of position or velocity due to mass or effective mass (moment of inertia).

Infinite Baffle Type of loudspeaker mounting (IB) where there is no air path between front and rear of speaker diaphragm. Ideally, a very extensive plane surface, but term also often applied to acoustic suspension type of enclosure.

Inharmonic Not having a harmonic relationship. Dissonant.

In-line *see* crossed-pair.

Inner groove distortion Increase of tracing distortion at small record radii due to reduced linear groove speed and correspondingly shorter recorded wavelengths. High frequency undulations just traceable with a given stylus radius at the outer grooves will be untraceable, and therefore give rise to distortion, at the inner grooves.

Input impedance Effective impedance at input terminals of a circuit or device. The load 'seen' by whatever signal source is connected to the input.

Instability Oscillation or tendency to oscillation in an amplifier or other circuit; variation of voltage from a power supply as current consumption changes.

Insulator Material which prevents flow of electrical current due to extremely high resistance. Opposite of conductor.

Integrated amplifier Circuit unit combining the functions of preamplifier and power amplifier, usually stereo.

Integrated circuit Circuit in which conventional components are replaced by microscopic pieces of material, transistor junctions, minute interconnections, etc, within a solid circuit block (IC).

Intensity In acoustics, the level of sound power (per unit of area). Sometimes mistakenly used to denote pressure or velocity.

Interaural Pertaining to differences between sounds at the two ears.

Intermediate frequency Fixed frequency (IF) at which most of the amplification takes place before demodulation in a superhet receiver. All incoming RF signals are changed to this frequency (modulation remaining intact) at an early stage.

Intermodulation Product of amplitude distortion whereby harmonically unrelated components are added to the original signals.

Ionic Pertaining to ionization or type of loudspeaker using this (ionophone).

Ionization Creation of charged electrical state in matter (often a gas) by addition to or subtraction from natural complement of electrons in the atoms.

Ionophone Type of high frequency speaker unit in which acoustic pressures are generated directly in ionized air by audio modulation of an electric spark discharge.

I/P Input

i.p.s. Inches per second (also i/s).

i/s Inches per second (also i.p.s.).

J.O.K.E. System Jointly operated kompression expansion system. Scheme for encoding 4-channel signals into two, devised by Michael Gerzon.

K Kilohm (when used after figures).

Kc/s Kilocycles per second (kHz preferred).

kHz Kilohertz.

Kilocycle One thousand cycles.

Kilohertz One thousand cycles per second (kHz)

Kilohm One thousand ohms (K or KΩ).

Kilowatt One thousand watts (kW).

kW Kilowatt. One thousand watts.

L Symbol for inductance.

Labyrinth Type of loudspeaker cabinet with partitions producing a long, convoluted path between rear of drive unit and outlet point. Lined with damping material to confine resonant effects to very lowest frequencies.

Landline Cable system used for conveying audio and other messages over long distances. Often employed by broadcasting authorities to send programmes from studios to transmitters.

Larynx-tone Basic speech or singing tone generated by vocal folds in the larynx. Harmonic content is modified by other elements in vocal mechanism to produce vowel sounds and voiced consonants.

Leader tape Coloured plastic tape for attaching to and indicating beginnings and ends of tape recordings, or for splicing into such recordings.

Leakage current Small current that 'leaks' through an imperfect capacitor when a steady DC voltage is applied across it. More generally, any stray currents due to poor insulation or odd circuit arrangements.

Legato Bound together, smooth transition between notes. Opposite of staccato.

Level Strength of a signal for particular descriptive or test purpose.

LF Low frequency.

LH Left-hand.

Limiting Generally, any process whereby amplitude becomes limited by a signal clipping process. In an FM receiver, deliberate use of this to remove all amplitude changes – including unwanted AM components – before IF signal passes to demodulator.

Line input Amplifier or tape recorder input point designed to take some standard nominal signal level at a particular impedance (professionally

1mW at 600 ohms). Also frequently applied to any amplifier or recorder input devoid of particular equalisations or especially high sensitivity.

Line output Output point on an amplifier (often preamplifier), tape recorder or tape unit providing a signal of useful voltage level for monitoring, tape recording or passing on to an associated power amplifier. On professional equipment, at an impedance of 600 ohms.

Linear When applied to transfer characteristic of amplifier, transducer, etc, indicates straightness of line on corresponding graph – necessary to avoid amplitude distortion. Also refers to forward velocity of record groove past stylus, or tape past tape-head.

Linearity Generally applies to transfer characteristic of an amplifier or other audio device. Good linearity indicates constancy of amplification factor with change of signal level, and hence low distortion.

Live Connected to or associated with the signal-carrying or electrically sensitive part of a circuit or component. Opposite of earthed or grounded. Sometimes refers to point in a circuit carrying dangerously high voltages or to that side of the AC mains supply not at earth or neutral potential.

Load Impedance of circuit or component connected to output of amplifier or transducer. Also, acoustic load offered to loudspeaker by enclosure system.

Load stability Extent to which an amplifier's stability depends on the impedance and reactive make-up of the load.

Localising faculty Psycho-physiological mechanism which translates acoustic differences at the two ears into aural awareness of the directions of sound sources.

Logarithmic Using a scale or law with linear units proportional to ratios of some function rather than proportional to equal finite bits of that function. Thus on the logarithmic frequency scales used for many graphs in this book a distance of about $\frac{5}{16}$ in. equals a ratio of one octave regardless of the fact that this is a span of 20 Hz (20–40 Hz) at one end of the scale and 10 kHz (10–20 kHz) at the other. The decibel is a logarithmic unit.

Long throw Applies to moving-coil in a loudspeaker unit designed for large cone movements. Coil cylinder longer than magnetic gap.

Long wave Radio band extending (approximately) from 150 kHz to 260 kHz (2000–1154m).

Loudness Subjective aural sensation related logarithmically to objective sound intensity. Generally, loudness levels are measured in phons with zero at the hearing threshold, though a unit of actual loudness magnitude is the sone, one sone corresponding to a level of 40 phons.

Loudness control Potentiometer in preamplifier used to adjust audio signal level according to misapplied theories of hearing. Output falls less rapidly at low frequencies as level is reduced, i.e. response is flat at high settings but bass (and sometimes also extreme treble) rises at low settings. Facility is often switchable.

Loudspeaker Transducer system for converting electrical energy into sound energy. Fed by a power amplifier to reproduce music and speech.

Low-pass Applied to filter or circuit which attenuates high frequencies.

Low tension Relatively low voltage (LT) in a valve circuit needed to supply valve heaters; usually but not necessarily AC.

LP Long Play (tape thickness) or long-playing (disc record).

LS Loudspeaker.

LT Low tension (voltage).

LW Long wave (radio band).

L + R The 'sum' element in a stereo sum-and-difference signal complex.

L − R The 'difference' element in a stereo sum-and-difference signal complex.

M Megohm (when used after figures).

m Metre.

mA Milliamp.

Maestoso Majestic, dignified.

Mag Magnetic.

Magic-eye Small electronic indicator sometimes used to display signal level in tape recorders, tuning in radio receivers, etc.

Magnetic field Lines of force surrounding or extending from a magnetic device. Fixed with a permanent magnet and alternating with transformers, motors, etc.

Magnetic flux Magnetic field or strength thereof, latter usually measured in gauss (G).

Magnetic pickup Gramophone pickup employing an electromagnetic transducer (moving-magnet, moving-iron, variable reluctance, induced magnet, moving-coil). Output voltage is proportional to rate of alternating stylus motion or velocity.

Magnetic tape Plastic strip coated with a special iron oxide film permitting storage and replay of signals by controlled magnetisation.

Magnetisation Implanting a magnetic field. In tape recording, re-orientation of minute magnetic particles in tape coating according to applied signal.

Magneto-optical Interaction of magnetic fields and light.

Main amplifier Commonly, power amplifier as distinct from preamplifier or control unit.

Mains Domestic electrical AC power supply. Approximately 240 V at 50 Hz in Europe, and 110 V at 60 Hz in America.

Mains transformer Transformer used in power supply section of equipment for converting mains voltage to various other voltages needed by the circuit.

Masking Hearing effect whereby a sound normally within the range of audibility is pushed below the hearing threshold (masked) by another sound.

Master Original recording from which all commercial copies ultimately derive. Basically, the master tape; but for gramophone record reproduction a durable negative imprint called the Master is made from the original positive direct-cut lacquer.

Mass Quantity of matter. Inertia thereof measured in grams or milligrams.

Matching Efficient coupling between electrical or acoustic components achieved by using similar impedances.

Matching impedance Impedance effecting maximum power transfer across a coupling. With a power amplifier, load impedance specified for stated power output.

Matrix Shell or stamper carrying a negative imprint of record groove and used for pressing final commercial discs. Derived from a positive Mother which came from a negative Master made from original direct-cut lacquer. Word also applied to some types of circuit.

m.c. Moving-coil.

Mechanical impedance Opposition to change of position or velocity due to inertia (effective mass), stiffness and mechanical resistance (friction), or any combination of these three.

Medium wave Radio band extending (approximately) from 520 kHz to 1650 kHz (577–182m).

Megahertz One million cycles per second (MHz).

Megohm One million ohms (M or MΩ).

Meter Indicating device employing a needle pointer and responding to current or voltage.

mgm Milligram.

mH Millihenry.

MHz Megahertz.

Mic Microphone.

Micro-circuit A tiny integrated circuit.

Microfarad Unit of capacitance (μF).

Microgroove Groove of an LP or EP disc record, as distinct from coarse groove of 78 r.p.m. discs.

Micron One-thousandth of a millimetre (μ).

Microphone Transducer with sensitive diaphragm for converting sound energy into electrical energy. Feeds tape recorder or sound amplifying system.

Microphony Sensitivity of circuit components, usually valves, to acoustic or mechanical disturbance. Produces microphone-like behaviour as electrodes vibrate at their resonant frequencies, thus modulating current flow and causing corresponding sounds to come from loudspeaker.

Microsecond One-millionth of a second (μS).

Microvolt One-millionth of a volt (μV).

Mid-range Frequency band extending from about 500 Hz to a few kilohertz.

Mil Thousandth of an inch (also thou.).

Milliamp One thousandth of an ampere (mA).

Milligram One thousandth of a gram.

Millihenry One thousandth of a Henry (mH).

Millisecond One thousandth of a second (mS).

Millivolt One thousandth of a volt (mV).

Milliwatt One thousandth of a watt (mW).

Mixer Circuit facilitating electrical mixing, in desired proportion, of a number of audio signals.

Mode In a particular fashion or manner. Applied to resonant system, refers to one or more of the possible frequencies or types of resonance.

Modulation Audio signal as carried by storage or transmission medium. Geometrical deviation of record groove, alignment of magnetic particles on tape, deviation of amplitude or frequency in radio wave. Also, move from one key to another in music.

Molecule Smallest possible particle of any chemical compound or element. In the case of some elements, one atom.

Molto fortissimo Extremely loud.

Molto pianissimo Extremely soft (quiet).

Momentum Quantity of motion of a moving body, proportional to both velocity and mass; impetus gained by movement. May be regarded as a sort of dynamic inertia. A rotating body is said to have angular momentum.

Monaural One-eared. Often wrongly used in place of monophonic (mono).

Monitor Head Extra tape-head on a tape recorder used for replay purposes only. Enables user to monitor recorded signal an instant after recording.

Monitoring 'Listening in' during recording or transmission of a programme to judge and/or control quality. Watching such a programme as registered on a meter with a view to restricting dynamic range.

Mono Monophony or monophonic.

Monophonic Single channelled audio device or system.

Mother Positive shell used in disc record manufacture. A few are produced from a negative Master and are used in turn to produce the negative stampers or matrices employed for actual disc pressing.

Motor Commonly refers to complete gramophone turntable mechanism, including driving motor.

Mouth Large end of a horn, remote from drive unit and positioned or shaped to radiate sound towards the listener.

Moving-coil Type of loudspeaker, microphone or pickup transducer in which a coil of wire moves in a magnetic field.

Moving-iron Type of magnetic pickup or headphone transducer.

Moving-magnet Type of magnetic pickup transducer.

MPX Multiplex.

mS Millisecond (also mSec.).

Multi-path reception Arrival of VHF radio signal via several paths due to obstructions, reflecting objects, etc. Results in differing times of arrival from various directions.

Multi-path distortion Audio waveform distortion (sometimes very severe) resulting from particular VHF/FM signal arriving at different times by various paths.

Multiplex Frequency modulation stereo radio system (MPX).

Multiplex decoder Circuit to derive stereo information from multiplex signal in radio tuner.

Mu-metal An alloy of iron and nickel with properties making it particularly suitable for magnetic screening.

Music power Power rating based on non-sustained tones.

Muting Facility for silencing a signal temporarily, especially on tuners.

mV Millivolt.

mV/cm/sec Millivolt per centimetre per second (pickup sensitivity).

MW Medium wave (radio band).

mW Milliwatt.

NAB National Association of Broadcasters (U.S.A.). Commonly refers to various tape standards (Also NARTB).

Negative Electrical polarity of a DC voltage source (battery or power supply), or of one point in a circuit in relation to another. Point from which electrons flow; opposite of positive; an electrical charge. Also applied to half cycles on appropriate side of 'zero' line in an AC waveform.

Negative feedback *see* feedback.

Neutral That side of the AC mains supply which sits, electrically, at approximately earth potential. Not to be confused with true earth connected to third pin of power sockets.

NFB Negative feedback.

Noise Unwanted background signal comprising a mixture of random electrical agitations and specific interfering tones such as hum or whistles.

Noise-level Amplitude of unwanted background noise.

Non-linearity Curvature of input-to-output transfer characteristic of audio system. Results in amplitude distortion: alteration of sensitivity or amplification factor with change of signal level.

Objective Measurable, physical, amenable to scientific investigation. Opposite of subjective.

Octave Pitch or frequency interval of two-to-one. A span of seven diatonic intervals covered by eight notes, hence octave.

Offset angle Angle by which centre-line of a pickup head or cartridge is offset, in a conventional arm, from a line joining pivot and stylus. Combines with overhang to minimise lateral tracking error.

Ohm Unit of electrical resistance, reactance or impedance (Ω).

Omnidirectional Having equal sensitivity in all directions (microphone) or equal output in all directions (loudspeaker). Polar response is spherical.

One note bass Effect of excessive LF resonance in a loudspeaker, whereby all bass notes in music tend to have a similar character or pitch.

O/P Output.

Open circuit Absence of electrical continuity. An open circuit component has suffered a break where there would normally be a continuous conducting path. Opposite of short circuit.

Oscillation Vibration or movement in regular sequence – of mechanical, electrical or acoustical system. Resonance is manifested by a tendency to oscillate at particular frequencies.

Oscillator Circuit for producing continuous electrical oscillation at a chosen frequencey.

Output impedance Effective impedance at output terminals of a circuit or device as 'seen' by the load. Source impedance at that point. Often applied wrongly to power amplifiers, where matching impedance for maximum power transfer is usually much greater than the output impedance.

Output stage Final stage of a power amplifier designed to feed power to a loudspeaker load.

Output transformer Transformer in valve power amplifier to couple output valves to loudspeaker load.

Overhang Amount by which pickup stylus overhangs turntable spindle when a conventional arm is moved to record centre. Combines with offset angle to minimise lateral tracking error.

Overload Condition in which equipment or recording medium has been driven beyond it signal handling capacity, and hence into distortion.

Overload margin Amount by which input signal to a device or circuit (commonly a preamplifier) may be raised above the nominal sensitivity figure before there is overloading of stages preceding the volume control. Given in decibels (dB).

Overtone A tone accompanying the fundamental in a musical note. May or may not be a harmonic.

PA Power amplifier or public address.

Parallel Electrical connection in which similar or equivalent points (on separate devices or components) are connected together. Any current flowing is thus split between the various paths. Opposite of series.

Parallel tracking Movement of pickup cartridge exactly parallel to record radius, with stylus always positioned on that radius for zero lateral tracking error. Unconventional pickup arm designed to achieve this.

Partial Any one component in a complex sound. In musical sounds, applied to overtones or harmonics.

Passive radiator *see* auxiliary bass radiator.

Peak programme meter Circuit and associated meter designed to indicate real amplitude peaks in an audio signal (PPM).

Peak-to-peak Magnitude of voltage or current as measured between extreme positive and negative excursions (p–p).

Perfect pitch Capacity to remember exact musical pitch, with consequent awareness when reproduced music is off-key.

Period Time taken for one cycle of an alternating quantity or waveform. Periodic time.

pF Picofarad.

Phase Point in a cyclic pattern. Two similar repetitive signals may be in phase (aiding), out of phase (opposing), or anywhere in between.

Phase-angle Measurement of phase difference between two waveforms Given in degrees, one complete cycle containing 360 degrees.

Phase invertor *see* reflex.

Phasing Arranging the connections in a stereo system so that signals representing a central sound source are in phase on arrival at the two loudspeakers. Usually applied just to speaker connections, where reversal of the leads to one only puts the signals in or out of phase.

Phon Unit used to express loudness level above standard threshold. Equal to decibel scale at 1 kHz, but takes account of variation of subjective impression with intensity at other frequencies.

Phono Phonograph = Gramophone. *see* Gram.

Phono plug Type of connector for use in signal circuits, having single central pin ('live') and an outer shield ('earth'). Used on screened cables and with equivalent phono sockets.

Pickup Device for producing electrical signals from gramophone records.

Pickup adaptor On some amplifiers or control units, small plug-in device which automatically adjusts sensitivity and frequency response of input circuit to suit a particular type of pickup.

Pickup arm Mechanical assembly of arm, pivots, counterweights, etc, for carrying pickup cartridge across a disc record.

Pickup head Part of a pickup containing the transducer and situated at free end of the arm. Often removable from the arm and usually in the form of a head 'shell' carrying a cartridge.

Picofarad One millionth of a microfarad (pF).

Piezo-electric Effect, used in some transducers, whereby certain substances generate voltages when subjected to mechanical stress.

Pilot tone In stereo multiplex broadcasting, a 19 kHz tone at the equivalent of 10 per cent FM modulation ($\pm 7\frac{1}{2}$ kHz) transmitted with the sum-and-difference signals for use in receiver when re-constituting stereo signal.

Pinch-effect An aspect of tracing distortion in disc reproduction whereby the pickup stylus undergoes a slight vertical movement at twice the recorded lateral frequency. Happens with both mono and stereo records, though no corresponding vertical output components are produced if the pickup is insensitive to vertical modulation.

Pinch-wheel Presses tape against capstan to obtain drive in tape recorder.

Pitch Quality of a sound which determines its position in the musical scale. Subjective equivalent of frequency.

Player-unit Combination of turntable and pickup designed as an entity. Often mounted on a plinth with lid for free-standing use.

Playing weight Downward force applied at pickup stylus. See also tracking weight.

PM Permanent magnet. Often applied to m.c. loudspeakers, some of which at one time used electrically energised magnets.

Point-source Applied to a sound source, usually a loudspeaker, in which the effective area from which sound is radiated (or seems to be radiated) is very small.

Polarity Positive/negative (+ or −) terminals of battery or power supply, or North/South of a magnet. Sometimes refers to phase in circuits or transducers.

Polar response Plotted shape (polar diagram) showing variation of microphone sensitivity, aerial sensitivity or loudspeaker output in various directions.

Pole-pieces End-pieces in magnetic system, across which a field is applied or induced.

Port Opening or vent in bass reflex loudspeaker cabinet.

Positive Electrical polarity of a DC voltage source (battery or power supply) or of one point in a circuit in relation to another. Point towards which electrons flow; opposite of negative; an electrical charge. Also applied to half-cycles on appropriate side of 'zero' line in an AC waveform.

Pot Potentiometer.

Potentiometer Resistor with movable tapping point, or slider. Used for volume and tone controls.

Power Rate at which energy is converted or dissipated. Measured in watts (W).

Power amplifier Circuit unit designed to supply audio power to a loudspeaker load. Driven by voltage or current signal from preamplifier.

Power bandwidth Frequency band over which a power amplifier will deliver not less than half its normal full rated power. Sometimes called half-power bandwidth.

Power handling capacity Literally, the amount of electrical AC power that a device or circuit is capable of handling at audio frequencies. Maximum power output of amplifier; maximum power that can be safely fed into loudspeaker.

Power pack Assembly of circuit components comprising the power supply in an audio or other electronic system. Sometimes housed on a separate chassis.

Power response Variation with frequency of maximum continuous wattage output from a power amplifier for stated nominal distortion. Represents upper power boundary of system or variation of overload point with frequency, and should not be confused with normal low-power frequency response.

Power supply That part of equipment circuitry used to supply HT and LT, usually working from the mains supply via a transformer.

p-p Peak to peak (of waveform).

p.p. Push-pull.

PPM Peak programme meter.

ppp Molto pianissimo, extremely soft (quiet).

Pre-amp Preamplifier.

Preamplifier Circuit unit designed to accept signals from pickup, microphone, tape-head, etc, applying any necessary equalisation and incorporating volume and tone controls. Output feeds power amplifier(s).

Precedence effect *see* Haas effect.

Pre-distortion Modification of recorded musical waveforms on disc to compensate for tracing distortion introduced by replay stylus with a finite tip radius.

Pre-echo Faint impression of oncoming loud passage due to slight deformation of adjacent groove on disc records.

Pre-emphasis Boosting of high frequencies in recording or broadcasting.

Pre-recorded Generally applies to a magnetic tape carrying a commercial recording; a tape record as opposed to blank tape.

Presence Quality of immediacy or vividness in reproduced sound, obtained with close microphone techniques and/or a boost in the upper-middle frequency band.

Pre-set control Control requiring adjustment to suit particular conditions of use or to match interconnected units when equipment is first installed. Often positioned at rear of units for screwdriver adjustment.

Pressure Sound pressure level due to passing or generated sound-waves. Instantaneous deviation from atmospheric pressure due to sound.

Pressure-changes Alternating change of air pressure at a given point, above and below atmospheric pressure, due to passing sound-waves.

Pressure-pad Applies gentle force to hold moving tape against heads in tape recorder.

Pressure unit Small moving-coil loudspeaker drive unit, usually with a dome of metal or plastic as its diaphragm, designed for use at high acoustic pressures in the throat of a horn. Term sometimes applied to similar units used as tweeters without horns.

Primary Input winding on a transformer.

Printed circuit Lay-out of circuit interconnections as conducting lines 'printed' on a rigid board by an etching process. Components are usually mounted on far side of board, with terminating wires passing through to be soldered to the 'circuit' where appropriate.

Print-through Unwanted transfer of signal by magnetic printing between adjacent tape layers.

Proton Fundamental positive electrical particle.

p.s.n. Power supply needed.

Psycho-acoustic Concerned with problems and phenomena in the psychology of hearing.

PU Pickup.

Push-pull Type of circuit (commonly power amplifier output stage) in which balanced components work in phase opposition for greater efficiency and lower distortion (p.p.).

p.w. Playing weight.

Q-factor Selectivity of resonant device or circuit.

QP Quadruple play (tape thickness).

Quadraphony Type of stereophony employing four signal channels to convey a 360 degree sound-field.

Quarter section Type of filter circuit, used in loudspeaker crossovers.

Quarter-track *see* four-track.

Quietening Term sometimes used for circuit facility in receiver which suppresses background noise when tuning between stations. Not to be confused with quieting.

Quieting Amount (in dB) by which level of background noise is reduced when radio tuner is fed with RF signal of specified voltage. Not to be confused with quietening.

R Symbol for resistance.

Radio frequency Strictly, can be any frequency from the lowest normally used for radio purposes (about 70 kHz) up to the infra-red band. Term is used in practice to distinguish incoming radio signals (RF) from other signals that might be met in a receiver.

Radiogram Combined record player and radio receiver housed in a cabinet.

Radio link Radio transmitter and receiver positioned for efficient conveyance of audio or video information between two specific points.

Radio tuner *see* tuner.

Radio wave Transmitted electromagnetic radiation at any frequency above about 70 kHz.

Rake angle Angle that pickup stylus shank makes to the record when viewed from the side. Not to be confused with vertical tracking angle.

Random noise Signal generated in circuits by molecular and other agitations, and in tape replay by random distribution of magnetic particles. Sounds like a rushing or hissing noise when fed to a loudspeaker. Called 'white' noise if its components are distributed evenly in frequency.

Rarefy To make less dense; opposite of compress. A rarefaction is the part of a sound-wave where pressure is below that due to the atmosphere.

Reactance Opposition to alternating current flow offered by capacitance or inductance. Measured in ohms, but differing from resistance due to energy storage rather than dissipation. Also, analogous mechanical and acoustic reactance offered by stiffness or mass.

Rear suspension In moving-coil speaker units, compliant support near apex of cone. Helps to keep coil central in magnetic air gap.

Receiver Tuner-amplifier. Also, a radio complete with built-in loudspeaker.

Record changer *see* auto-changer.

Recording characteristic Agreed frequency response shape applied to commercial recordings, whereby a flat overall response is obtained by use of an appropriate inverse equalisation when replaying.

Recording level indicator Device for registering amplitude of signals fed to recording head in a tape recorder. May be a meter or magic-eye, and normally calibrated so that a particular deflection corresponds to acceptable maximum tape modulation.

Record/Play Name of tape-head combining record and replay functions. Label on tape recorder function switch (R/P).

Record Player Simple unit, usually portable, for playing disc records without undue attention to quality of reproduction. Incorporates turntable, pickup, amplifier and speakers.

Rectifier Type of diode designed specifically for producing direct current (DC) supply from alternating current (AC) source.

Reflector (aerial) Element in a VHF aerial system placed behind the main dipole to improve sensitivity in the forward direction and reduce sensitivity to the rear.

Reflex Type of loudspeaker cabinet with an outlet (vent or port) permitting enclosed air to be tuned for a coupled resonance effect with drive unit cone. Results in improved efficiency at low frequencies due to inversion of phase within enclosure so that outlet from port aids radiation from front of cone.

Resistance Non-reactive opposition to current flow, measured in ohms (Ω)

Also, mechanical property (friction) and acoustical property (radiation resistance).

Resistive reflex Modification of reflex type loudspeaker enclosure in which internal space is filled with fluffy damping material to reduce resonant effects.

Resistor Circuit component with specific resistance value.

Resonance Natural tendency of a mechanical, acoustical or electrical system to respond, vibrate, oscillate or 'ring' at particular frequencies.

Response Normally refers to frequency response if used without qualification.

Response curve Graphical representation of frequency response, with frequency (in Hz) on a horizontal logarithmic scale and signal level, usually in dB, on a vertical scale.

Restoring force Elastic force tending to return a suspended object to its neutral position when it is deflected against a compliance.

Reverberation Repetitive reflection of sound in an enclosed space. Contributing to ambience.

Reverberation time Time taken for sound intensity in an enclosed space to fall by 60 dB from an initial steady state (r.t.).

Reverberation unit Device used for processing audio signals to simulate added reverberation.

RF Radio frequency.

RH Right-hand.

RIAA Record Industry Association of America (commonly refers to disc replay characteristic).

Ribbon Electro-acoustic transducer employing a thin ribbon of aluminium alloy suspended in a magnetic field. Used for some microphones and high frequency speaker units.

Ringing Resonant reaction of a device or circuit when actuated by a transient signal. Latter triggers a short-term local oscillation at frequency of self-resonance.

Ring-main Type of domestic AC mains wiring system in which a number of socket outlets, each rated at 13 amps maximum consumption, are connected in parallel on a 'ring' running from and returning to a main supply point fused at 30 amps.

Ripple Residual alternating voltage 'sitting on top' of a nominal DC supply; one cause of hum in amplifiers. Also, sometimes applied to small surface undulations on disc records.

Rise-time Measure of one aspect of transient response, i.e. the finite time taken for an amplifier to respond to an instantaneous voltage step.

RMS Root mean square.

Roll-off The frequency at which a filter, equaliser or tone control begins to attenuate. Rate of slope.

Root mean square Effective amplitude of an AC voltage waveform (RMS); equal to the DC voltage or current that would produce the same power dissipation in a given load. Always less than peak amplitude except with a square-wave.

R/P Record/Play.

r.p.m. Revolutions per minute.

r.t. Reverberation time.

Rumble Unwanted low frequency vibrations transmitted to pickup by turntable.

Run-in Applied to groove at beginning of side on a disc record. 'Runs in' to close-grooved section carrying the recording.

Scheiber System Scheme for encoding 4-channel signals into two, devised by Peter Scheiber.

Screen Metallic shield or wrapping to provide electrostatic screening between, within or around circuits, components or cables.

Screened Protected from surrounding undesirable electrostatic fields by a screen. By use of special ferrous materials, screening may also exclude magnetic fields.

Screened lead Signal-carrying cable with surrounding sheath of braided conductive material to provide electrostatic screening of inner conductor.

Secondary Output winding of a transformer.

Selectivity Sharpness of tuning in mechanical, acoustic or circuit devices. In radio receiver, narrowness of frequency band accepted either side of wanted station.

Self-powered Not requiring external high tension (HT) and low tension (LT) supplies. Has own mains power circuit.

Semi-conductor Component such as transistor or crystal diode dependent on semi-conducting properties of certain junctions.

Semitone Interval between each of twelve notes in a chromatic octave.

Sensitivity Measure of signal level needed to actuate tuner, amplifier or transducer for a stated output. Lower the input, higher the sensitivity.

Separation Degree to which left and right stereo signals are isolated in pickup, amplifier, etc. Measured in decibels (dB).

Serialism Twelve-tone or dodecaphonic music which, by attaching equal importance to all twelve semitones in the octave, shows no regard for key centres or normal harmony. Totally atonal music; opposite of diatonic.

Series Electrical connection in which devices or components are placed 'end to end'. Any current flowing thus has to pass through each element in turn.

Service area Area normally covered by a radio or television transmitter, giving a good received RF signal level.

Shell Commonly, pickup head shell designed to carry the cartridge and usually detachable from the arm.

SHF Super high frequency. Applies to some radio links feeding audio signal to transmitters.

Short circuit Direct conducting path of negligible resistance placed across two points in a circuit or device. Opposite of open circuit.

Short wave Radio band extending (approximately) from 4 MHz to 25 MHz (75-11.5m).

Sidebands Frequencies other than that of the pure radio carrier wave necessarily added to the carrier when modulation is present.

Side-thrust Unwanted lateral force on pickup arm.

Signal Audio 'information' handled by equipment as alternating voltages or currents, whether directly as audio or as modulation on RF and IF carriers.

Signal generator Electronic unit, usually a piece of test equipment, which uses an internal oscillator to generate signals at calibrated voltage levels over a range of frequencies.

Signal-to-noise ratio Ratio of desired or reference signal voltage to unwanted random noise and hum. Measured in decibels (dB).

Sine-wave Waveform of a pure single-frequency tone.

Skating force Inward bias on pickup arm caused by groove/stylus friction and offset geometry.

Slope Steepness of sloping portion on a frequency response curve. Usually stated in decibels per octave (dB/octave).

Smoothing Removal of alternating 'ripple' from a DC supply derived via rectifiers from AC.

Smoothing capacitor Component used in a power supply circuit for filtering away residual AC ripple, thus 'smoothing' the DC supply.

S/N Signal-to-noise, measured as a ratio in decibels (dB).

Solid-state Circuit or component employing transistors rather than valves.

Sone Unit of subjective loudness magnitude. Takes a level of 40 phons as equal to one sone, two sones corresponding to twice the loudness, and so on.

Sonic To do with sound.

Sound-field The region of space occupied by sounds radiating from a given source, or within which a listener might hear sounds.

Sound-in-the-round Popular term for quadraphony.

Sound-on-sound Dubbing from one track to another while adding a fresh signal to a tape recording.

Sound-stage Space between two loudspeakers in a stereo system across which subjective sound images are spread, as on a 'stage'.

Sound-wave Sound as propagated through the air in wave motion.

Source impedance Effective impedance of a circuit or device as 'seen' when looking back from the next stage.

SP Standard play (tape thickness).

Spark discharge Flow of electric current through air made possible by ionization of the atoms due to high voltage applied across a gap.

Speaker Abbreviation of loudspeaker (LS).

Spectrum Range of frequencies for a particular purpose.

Speech-coil Coil in moving-coil loudspeaker.

Splice Accurate join in a length of magnetic tape. Strictly diagonal, but term also applied to right-angled join.

Square-wave AC waveform in which voltage or current leaps instantaneously to a peak value, stays there for half the cyclic period, then falls (instantaneously) to the reverse peak value, remaining there for the next half cycle, completing the period by returning (again instantaneously) to zero. Such a waveform is very rich in harmonics and demands good transient and frequency response for accurate reproduction.

Squawker Loudspeaker unit for use at mid frequencies only.

Stability Freedom from oscillation or tendency to oscillate in an amplifier or other circuit; constancy of voltage from a power supply despite variations in current consumption.

Staccato Detached, broken-up. Opposite of legato.

Stage Part of an electronic circuit that can be regarded as a functional entity. Valve or transistor and its associated components.

Stamper *see* matrix.

Standing-wave Resonant effect in an enclosed space, often a room, whereby at particular frequencies a fixed pattern of sound intensity is created due to coincident phases of reflected sounds.

Static Normally refers to electrostatic charges, often generated by rubbing motions on objects made from insulating materials – such as records.

Steep-cut Applied to filter which attenuates very rapidly above or below its cut-off frequency.

Stereo Stereophony or stereophonic.

Stereogram Traditional one-piece radiogram employing two channels for nominal stereo reproduction, with loudspeakers mounted at the cabinet's extremities.

Stereophony System of sound recording and reproduction carrying information about spatial disposition of sound sources. Usually two-channel.

Stiffness Opposition to motion due to rigidity of moving parts. Reciprocal of compliance.

Strain-gauge Device for measuring mechanical stress or displacement.

Straight-through amplifier Facility provided on some tape recorders whereby the internal amplifier circuits may be switched for use as a complete preamplifier and power amplifier combination in conjunction with external signal sources and loudspeakers.

Strobe Stroboscope.

Stroboscope Pattern of dots or bands which appears stationary when moving at a certain velocity if illuminated by AC lighting at a particular frequency. Used on turntables to indicate correct rotational speed with light from 50 Hz or 60 Hz mains.

Stylus In a pickup cartridge, specially shaped piece of hard material (usually sapphire or diamond) used to trace record groove.

Stylus arm Stylus bar.

Stylus bar Cantilever carrying pickup stylus.

Sub-carrier In stereo multiplex radio, a 38 kHz carrier which is AM modulated with the stereo difference signal (L − R) and then suppressed before the resulting modulation side-bands are added to the sum signal (L + R) and a 19 kHz pilot-tone to form a composite multiplex signal for normal FM modulation of the VHF carrier.

Subjective Mental, internal, existing as consciousness, sensations, feelings, etc. May correlate with external stimuli but cannot be measured in any physical sense. Opposite to objective.

Sub-sonic Below the bottom frequency limit of human hearing; also, below the speed of sound.

Sum-and-difference Two types of signal derived from a normal pair of stereo signals, one comprising the sum of left and right and the other the difference. On disc records these two quantities are represented by lateral and vertical modulation components respectively. Term sometimes applied to type of moving-iron pickup whose primary response is to these parameters.

Superhet Superheterodyne, or strictly, supersonic heterodyne, meaning production of a beat tone at a supersonic frequency. Applies to type of radio receiver (the vast majority) in which the RF input, of whatever frequency, is converted by a heterodyne or 'beating' process to a fixed IF for ease of amplification.

Superimpose Recording one signal over another on tape without erasing the original.

Supersonic Above the upper frequency limit of human hearing; also, above the speed of sound.

Surround Strip or roll of compliant material fitted to periphery of loudspeaker cone to seal it to the frame yet permit axial movement.

SW Short wave (radio band).

Swinger Disc record with 'centre' hole positioned eccentrically in relation to groove spiral, causing pickup to oscillate laterally as disc rotates.

Switch Device for making or breaking contact between electrical circuits, or for selecting various functions.

Synchronous motor Type of AC electric motor in which rotor speed is related directly to frequency of power supply (hence synchronism) and remains thus up to full mechanical load.

Tag Strip Length of insulating material such as Paxolin fitted with a row of pressed-on soldering tags. Used for mounting components or terminating leads with soldered joints.

Take-up spool Reel on a tape recorder to which the tape is fed during recording or replay. Right-hand spool on a conventional machine.

Tape deck Purely mechanical part of a tape recorder, comprising complete tape transport system, with motors, drive pulleys, linkages, etc. Normally all mounted on a rigid deck plate.

Tape-head Component scanned by tape for recording or replay. Applies magnetising field when recording and produces electrical output when replaying.

Tape hiss Sound of the noise signal produced by tiny random agglomerations of magnetic particles in tape coating.

Tape recorder Device for recording and replaying audio signals via the medium of oxide-coated magnetic tape. Incorporates tape transport mechanism and necessary electronic circuitry.

Tape transport That part of a tape recorder's mechanism concerned with transporting the tape from spool to spool, at constant velocity and with proper control, past the tape-heads.

Tape unit Tape recorder without power amplifiers or loudspeakers, intended for use with external amplifiers as part of a hi-fi component system. Recording circuitry and tape-head replay pre-amplifier stages are normally included.

Tempered scale *see* equal temperament.

Template Flat sheet or strip of material, often paper or cardboard, used for marking out the position, shape or fixing holes for some device, unit or component.

Tenor Part of audio spectrum extending approximately from 150 Hz to 500 Hz. In musical notation, tenor clef covers D_1 to E (146·83–329·63 Hz).

Terminal board Strip or block of insulating material fitted with screw-terminals. Suitable for mounting components or terminating leads when soldering is not possible or appropriate.

Test-tape Magnetic tape recording carrying special tones or other material such as selected bands of white noise. Used for measuring replay frequency response, wow and flutter, etc., of tape recorders.

THD Total harmonic distortion.

Thou' Thousandth of an inch (also mil).

Threshold of hearing Lowest perceptible level of sound intensity as measured in otherwise silent surroundings. Varies with frequency, but for young people (by statistical sample) corresponds to a sound pressure of 0·0002 dyne per square centimetre at around 2 kHz.

Threshold of pain Level of sound intensity at which sensation becomes painful. Usually in the region of 120–130 dB above a standard lower threshold of 0·0002 dyne per square centimetre.

Throat Small end of a horn, usually coupled to speaker drive unit.

Timbre Characteristic tonal quality of a musical instrument.

Time-constant Product of resistance and capacitance in an R/C circuit. Used to specify response curve shapes achievable by such circuitry. Given in microseconds (μS).

Tip Mass Effective inertia at tip of pickup stylus as seen by record groove at high frequencies. Given in milligrams (mgm).

Tonal Quality of audio signal or musical sounds in relation to balance between bass, middle and treble frequency bands.

Tone-arm Antique name for pickup arm (still used in the U.S.A.) derived from acoustical days when sound actually travelled along the arm.

Tone-colour Characteristic tonal quality of a musical instrument.

Tone control Potentiometer or switch in preamplifier used to adjust tonal balance of audio signal. Simplest type of control merely attenuates treble, while sophisticated circuits have a range of boost and cut for both bass and treble.

Top Treble.

Top-cut Reduction of the level of treble or high frequencies in sound reproduction.

Topping Pickup stylus reaching 'shoulders' or top edges of record groove due to larger than optimum tip radius as viewed from front or rear.

Torque Turning force applied to a rotating device.

TP Triple play (tape thickness).

t.p.i. Tape position indicator.

Tracing Accuracy with which a pickup stylus tip follows the geometry of recorded modulations.

Tracing distortion Inability of pickup stylus to follow recorded modulations due to finite tip radius.

Track Recorded path along the length of a magnetic tape. Also, sometimes used for a recorded band on disc or as a verb in connection with pickup tracking.

Trackability Ease with which a pickup cartridge will track recorded material of high amplitude and velocity. Ability to track such passages at a low playing weight indicates high trackability – i.e. low mechanical impedance at stylus tip.

Tracking Accuracy with which a pickup stylus follows the dynamics of recorded modulations.

Tracking distortion Inability of pickup stylus to follow recorded modulations due to high mechanical impedance or low playing weight.

Tracking error Angular difference between centre-line of pickup cartridge and a line drawn at right-angles to the record radius at point of stylus contact; this is lateral tracking error. Vertical tracking error applies to stereo pickups only and is the angular difference between path taken by stylus when deflected upwards and the effective recorded 'vertical' component on stereo records, the latter now standardised at 15 degrees forward from true vertical. Tracking error also sometimes applied to ganged controls in stereo preamplifier, indicating degree of left/right difference over a stated range of control.

Tracking force *see* tracking weight.

Tracking weight Downward force applied at pickup stylus to ensure proper tracking of recorded groove modulations.

Transcription Strictly, process of copying or re-recording. Word commonly applied to better quality electro-mechanical devices – particularly turntables – implying a professional standard of performance.

Transducer Device for converting from one form of energy to another; i.e. loudspeaker converts from electrical to acoustic.

Transfer characteristic Overall input-to-output amplitude response. Ideally, a straight line.

Transformer An arrangement of magnetically coupled windings, usually on a ferro-magnetic core. AC voltages are induced from one winding to another in required ratios.

Transient Sudden change of state. Sharp wavefront of percussive sound, plucked string, etc.

Transient response Behaviour of amplifier or transducer when fed with transient signals. Transients tend to excite resonances, which indicate poor performance.

Transistor Semi-conductor device with three or more elements, having current amplifying properties. Also, popular name for a small radio using transistors.

Transistor sound Characteristic sound quality attributed to some power amplifiers employing transistor circuitry. Possibly due to improved transient response (desirable) and/or crossover distortion arising from Class–B output stages (undesirable).

Transistorised Electronic circuit employing transistors rather than valves.

Transmission line Term from electronics, applied in audio to type of loudspeaker enclosure resembling a labyrinth but filled with absorbent material.

Treble High frequency end of audio spectrum, above 2–3 kHz. In musical notation, bottom line on the treble clef is E (329·63 Hz).

Tremolo Rapid iteration of a note, or variation of amplitude.

Trimmer Small adjustable capacitor, usually pre-set to 'trim' RF circuits within a radio receiver for accurate tuning.

TT Turntable.

Tube Valve.

Tuned Adjusted to resonate, respond or reject at a particular frequency.

Tuner Circuit unit for converting received radio transmissions into audio.

Tuner-amplifier Circuit unit combining the functions of radio tuner, preamplifier and power amplifier. Sometimes called a receiver.

Tuning Process of adjusting a resonant device or circuit to react at a particular frequency. In tuner, adjusting to respond to wanted transmission.

Tuning indicator Meter or electronic device used to indicate accuracy of tuning in a receiver.

Turnover Frequency at which an audio system undergoes a change in mode of operation, or a response curve changes its shape or slope.

Turntable Accurate circular platter with a central bearing together with associated driving mechanism, for rotation of gramophone records.

Tutti All. A passage employing the full orchestra.

Tweeter Loudspeaker unit for use at high frequencies only.

Twin-feeder Type of flat twin insulated cable in which the two conductors are spaced to give a particular RF impedance value (often 300 ohms) for coupling aerials to tuners with balanced input circuits.

Twin-track Tape recorder designed to record two tracks only on magnetic tape. A tape recorded in this fashion.

Two-track Twin-track.

Ultrasonic Mechanical vibrations or waves at frequencies above the upper limit of human hearing.

Unbalanced (line or circuit). System of connections in which one side of the circuit is earthed. Signals conveyed by coaxial cable or other screened lead.

Unilinear Comprising a single line. A musical waveform, no matter how complex or however many instruments may be playing together, is essentially a single undulating line of sound pressure.

Upper-middle Frequency band around 2–5 kHz. Emphasis in this region adds vividness, brightness or 'presence' to reproduced music.

V Volt.

Valve Vacuum tube device used for amplifying or rectifying. Has internal heater and various other electrodes.

Variable reluctance Magnetic transducer principle used in some pickups. A species of moving-iron transducer.

Varigroove System of variable groove spacing on microgroove records in accordance with musical dynamics. Permits accommodation of more recording time per radial inch of record.

Velocity Rate of change. Often applied to stylus motion generated by record groove, i.e. recorded velocity. Also linear velocity of record groove or tape.

Velocity pickup Pickup with output voltage proportional to rate of alternating stylus motion – hence velocity. Usually employs a magnetic type of transducer. Must be used with equalisation for recording characteristic.

Vent Opening or port in bass reflex loudspeaker cabinet; hence vented enclosure.

Vertical tracking angle In stereo pickup cartridge, 'vertical' stylus motion path in relation to true vertical; should be 15 degrees forward. Not to be confused with stylus rake angle.

VHF Very high frequency (radio band).

Vibration-curve Visual display of a vibration or sound covering many cycles.

Vibrato Rapid undulation of pitch.

Video Visual information as coded for recording or transmission by radio, tape or other non-photographic means. Techniques, circuits and equipment for handling visual pictures in electrical form.

Virgin tape Untouched magnetic recording tape straight from the manufacturer.

Voice-coil Coil in moving-coil loudspeaker.

Vol. Volume.

Volt Unit of electrical force or potential (V).

Voltage Magnitude of electrical pressure or potential. Number of volts (V).

Volume control Potentiometer used to adjust level of signal fed to power amplifier section of an audio system, thus controlling electrical power delivered to loudspeaker and consequently the sound volume.

Volume unit Type of audio signal level meter which does not respond accurately to short-term transient peaks, but which is useful for monitoring programme material with limited dynamics.

VU Volume unit.

W and F Wow and Flutter.

W Watt.

Warp Applied to disc record, refers to an irregular departure from flatness causing pickup to rise and fall as disc rotates. Very rapid undulations are sometimes referred to as ripples.

Watt Unit of electrical power (W); volts multiplied by amps.

Wattage Magnitude of electrical power. Number of watts (W).

Waveband Band of frequencies or wavelength for a particular purpose, usually radio, i.e. LW, MW, SW, VHF, etc.

Waveform Shape obtained when a signal is displayed graphically as a plot of amplitude against time. Mental picture thereof.

Waveform compensation *see* pre-distortion.

Wavefront Leading edge or surface of a sound (or other) wave. Applicable also to the continuous succession of fresh wave-fronts generated in music with every minor change of sound quality.

Wavelength Linear distance between points of equal phase in a propagated wave. Distance traversed during one cycle of oscillation.

Wave motion Mode of energy propagation in which the medium undergoes a sequence of pressures, tensions or movements corresponding to frequency or waveform of passing 'information'.

Weighted Applied to a noise-level figure or a S/N ratio which takes account of the ear's decreasing acuity at low frequencies. A figure so adjusted will be noticeably better if hum or other low frequency noise plays a disproportionately large part in the total.

White noise Random noise covering all audible frequencies without any particular coloration or emphasis. Sounds like rushing or hissing.

Wobble Collective term for wow and flutter, speed fluctuations in a recording medium.

Woofer Loudspeaker unit for use at low frequencies only.

Wow Waver of pitch caused by spurious fluctuations of speed in the recording medium at rates below about 10 Hz.

Zenith-GE Applied to multiplex stereo radio system combining major features proposed by these two companies (U.S.A.) and now generally adopted throughout the world.

λ Symbol for wavelength.

μ Micron.

μf Microfarad.

μS Microsecond (also μSec).

μV Microvolt.

Ω Ohm.

♭ Musical flat.

Musical sharp.

BIBLIOGRAPHY

Books and pamphlets referred to in the text and a selection of others for the beginner.

MUSICAL AND GENERAL

Beranek, Leo L.	*Music, Acoustics and Architecture*	John Wiley
Berlioz, Hector	*Evenings in the Orchestra* (trans. Fortescue)	Penguin
Briggs, G. A.	*Musical Instruments and Audio*	Rank Wharfedale
Britten, Benjamin	*On Receiving the First Aspen Award*	Faber & Faber
Culshaw, John	*Ring Resounding*	Secker & Warburg
Gammond, Peter	*Terms Used in Music*	Phoenix House
Jeans, Sir James	*Science and Music*	Cambridge University Press
March, Ivan	*The Great Records*	Long Playing Record Library
Scholes, P. A.	*Concise Oxford Dictionary of Music*	Oxford University Press
Williamson, H. S.	*Introducing the Orchestra*	Faber & Faber
Wilson, W. J.	*The Stereo Index*	Wilson Stereo Library

AUDIO AND HI-FI

Baldock, Crabbe and West	*Five Speakers – How to Make Them*	Henslow Year Books
Borwick, John	*Hi-Fi for Beginners*	Henslow Year Books
Borwick, John	*Know Your Gramophone*	General Gramophone Publications
Briggs, G. A.	*A to Z in Audio*	Rank Wharfedale
Briggs, G. A.	*Audio and Acoustics*	Rank Wharfedale
Briggs, G. A.	*Cabinet Handbook*	Rank Wharfedale
Briggs, G. A.	*More About Loudspeakers*	Rank Wharfedale
Brown, Clement	*Introduction to Hi-Fi*	Newnes
Brown, Clement	*Questions and Answers on Audio*	Newnes
Brown, R. Douglas	*Tape Recording and Hi-Fi*	Arco
Cooke, Raymond E.	*How to Choose a Loudspeaker*	KEF Electronics
Cooke, Fincham and Jones	*You and Your Loudspeaker*	KEF Electronics

Goodmans	*Goodmans High Fidelity Manual*	Goodmans Loudspeakers
Hadden, H. Burrell	*Practical Stereophony*	Iliffe
KEF	*Loudspeaker Enclosure Design*	KEF Electronics
King, Gordon J.	*The Hi-Fi and Tape Recorder Handbook*	Butterworth
Link	*Audio Talk* (audio terms defined)	Link House Publications
Matthews, C. N. G.	*Introduction to Electronics*	Museum Press
Slot, G.	*Audio Quality*	Iliffe
Slot, G.	*From Microphone to Ear*	Centrex (Macmillan in U.K.)
Smith, Wheater	*Cabinet Making for Beginners*	Henslow Year Books
Villchur, Edgar	*Reproduction of Sound*	Acoustic Research (U.S.A.)
Walton, J.	*Pickups, the Key to Hi-Fi*	Pitman
Webb, B. J.	*Stereo for Beginners*	Henslow Year Books
West, Ralph	*Loudspeakers in Your Home*	Henslow Year Books
Wharfedale	*Cabinet Construction Sheet*	Rank Wharfedale

SOME MORE ADVANCED BOOKS

Crowhurst, N. H.	*High Fidelity Sound Engineering*	Newnes
Guy, P. J.	*Disc Reproduction*	Focal Press
Jordan, E. J.	*Loudspeakers*	Focal Press
Mackenzie, G. W.	*Acoustics*	Focal Press
Moir, James	*High Quality Sound Reproduction*	Chapman & Hall
Olson, Harry F.	*Acoustical Engineering*	D. Van Nostrand
Taylor, C. A.	*The Physics of Musical Sounds*	English Universities Press

MAGAZINES AND ANNUALS

Art of Record Buying, The	U.K. yearly	Synopsis of classical record review recommendations (EMG Handmade Gramophones)
Audio	U.S.A. monthly	Audio articles, news, equipment reports

Audio Annual	U.K. yearly	Audio articles, reprints of equipment reports from *Hi-Fi News* and *Studio Sound* (Link House Publications)
Audio Record Review	Renamed *Record Review*	——
Gramophone, The	U.K. monthly	Record reviews, musical articles, news, equipment reports
Hi-Fi News (incorporating *Record Review*)	U.K. monthly	Audio and musical articles, news, record reviews equipment reports
Hi-Fi Sound	U.K. monthly	Audio articles, news, some record reviews, equipment reports
Hi-Fi Year Book	U.K. yearly	Comprehensive equipment catalogue with specifications. (IPC Electrical-Electronic Year Books)
High Fidelity Magazine	U.S.A. monthly	Audio and musical articles, news, record reviews, equipment reports
Journal of the Audio Engineering Society	U.S.A. quarterly	Audio articles, generally at a high technical level
Monthly Letter, The	U.K. monthly	Record reviews. By subscription only from EMG Handmade Gramophones
Record Review	See *Hi-Fi News*	——
Records and Recording	U.K. monthly	Record reviews, musical articles, news, equipment reports
Stereo Review	U.S.A. monthly	Audio articles, news, equipment reports
Stereophile, The	U.S.A. monthly	Audio articles, equipment reports. By subscription only from Box 49, Elwyn, Pa. 19063, U.S.A.

Studio Sound (incorporating *Tape Recorder*)	U.K. monthly	Professional and advanced-amateur recording, studio affairs, tape recorders, equipment reports
Tape Recorder	See *Studio Sound*	——
Tape Recording Magazine	U.K. monthly	Tape recording articles, equipment reports
Wireless World	U.K. monthly	General electronics, occasional audio articles

TEST RECORDS

An Audio Obstacle Course (TTR–101)	Shure Electronics
Blank Disc	Wilson Stereo Library
Hi-Fi Sound Stereo Test Record (HFS 69)	Haymarket Press Group
How to Give Yourself a Stereo Checkout (SKL 4861)	Decca
Seven Steps to Better Listening (STR–101)	CBS (U.S.A. only)
The Enjoyment of Stereo (SEOM 6)	EMI

RECORD LIBRARIES AND OTHER BODIES

Audio Engineering Society, 124 East 40th Street, New York, N.Y. 10016, U.S.A.

Federation of British Audio, 49 Russell Square, London, W.C.1.

High Fidelity Dealers' Association, 19 Conway Street, Fitzroy Square, London, W1P 6DY.

Institute of High Fidelity Inc., 516 Fifth Avenue, New York, N.Y. 10036, U.S.A.

Long Playing Record Library Ltd., Squires Gate Station Approach, Blackpool, Lancs.

National Federation of Gramophone Societies, 3 Mosley Avenue, Holcombe Brook, Bury, Lancs.

Performing Right Society, 29 Berners Street, London, W.1.

Phonographic Performance Ltd., 62 Oxford Street, London, W.1.

Wilson Stereo Library Ltd., 104 Norwood High Street, London, S.E.27

MANUFACTURERS AND RETAILERS MENTIONED IN THE BOOK (British or European addresses)

Acoustec *see* Koss

Acoustical Manufacturing Company Limited, Huntingdon.

Acoustic Research *see* Bell & Howell
ADC *see* KEF
AEG (Great Britain) Limited, Lonsdale Chambers, 27 Chancery Lane, London, W.C.2.
Akai *see* Pullin
Altec Lansing *see* Carston
Amplivox Limited, Beresford Avenue, Wembley, Middlesex.
Armstrong Audio Limited, Warlters Road, London, N.7.
Audio Technica *see* Shriro
Bang & Olufsen U.K. Division, Eastbrook Road, Gloucester.
Barnet Factors Limited, 4 Lisle Street, London, W.C.2.
Bell & Howell Limited, Audio Products Division, Alperton House, Bridgewater Road, Wembley, Middlesex.
BFF *see* Bonded Fibre.
Bonded Fibre Fabrics Limited, Regal House, London Road, Twickenham, Middlesex.
Brush Clevite Company Limited, Thornhill, Southampton, SO9 1QX.
Carston Electronics Limited, 71 Oakley Road, Chinnor, Oxon.
Celestion Limited, Ferry Works, Thames Ditton, Surrey.
Chilton *see* Magnetic Tapes
Connoisseur *see* Sugden
A. Davies & Company, 56 Wellesley Road, London, N.W.5.
Decca Special Products, Ingate Place, Queenstown Road, London, S.W.8.
Dolby Laboratories Inc., 346 Clapham Road, London, S.W.9.
EMG Handmade Gramophones Limited, 26 Soho Square, London, W.1.
Fane Acoustics Limited, Hick Lane, Batley, Yorks.
Farnell-Tandberg Limited, Hereford House, North Court, Vicar Lane, Leeds 2, Yorkshire.
Ferrograph Company Limited, Mercury House, 195 Knightsbridge, London, S.W.7.
Fisher *see* Getz
Garrard Engineering Limited, Newcastle Street, Swindon, Wilts.
Getz Bros. & Co. Inc., 2 Harwood Place, London, W.1
Goldring Manufacturing Company Limited, 486/488 High Road, Leytonstone, London, E.11.
Goodman Loudspeakers Limited, Axiom Works, Wembley, Middlesex.
Grundig (GB) Limited, Newlands Park, London, S.E.26.
Hampstead High Fidelity, 91 Heath Street, London, N.W.3.
Heathkit *see* Heath

Heath (Gloucester) Limited, Gloucester, GL2 6EE.
P. F. & A. R. Helme, Summerbridge, Harrogate, Yorks.
Horns, 6 South Parade, Oxford.
Howland-West Limited, 2 Park End, South Hill Park, London, N.W.3.
J-Beam Engineering Limited, Rothersthorpe Crescent, Northampton.
KEF Electronics Limited, Tovil, Maidstone, Kent.
Koss Electron F.R.L., via Bellini 7, 20054 Nova, Milanaise, Italy.
Lafayette *see* Barnet
Largs of Holborn, 76/77 High Holborn, London, WC1V 6NA.
H. J. Leak & Company Limited, Bradford Road, Idle, Bradford, Yorks.
Lowther Manufacturing Company, Lowther House, St. Mark's Road, Bromley, Kent.
Magnetic Tapes Limited, Chilton Works, Garden Road, Richmond, Surrey.
Metrosound (Sales) Limited, Audio Works, Cartersfield Road, Waltham Abbey, Essex.
B. H. Morris & Company (Radio) Limited, 84–88 Nelson Street, London, E.1.
Music in the Home *see* Thomas Heinitz.
Neat *see* Howland-West
Ortofon *see* Metrosound
Peak Sound (Harrow) Limited, 32 St. Judes Road, Englefield Green, Egham, Surrey.
Peerless *see* Helme
Philips Electrical Limited, Century House, Shaftesbury Avenue, London, W.C.2.
Phoenix Bookcases, 36a St. Martin's Lane, London, W.C.2.
Pioneer *see* Shriro.
Power Judd & Company Limited, 94 East Hill, London, S.W.16.
Protecta Systems Limited, 30 Parker Street, London, WIM 2D5.
Pullin Photographic, P.O. Box 70, Great West Road, Brentford, Middlesex.
Quad *see* Acoustical
Radford Electronics Limited, Ashton Vale Estate, Ashton Vale Road, Bristol, 3.
Rank Wharfedale Limited, Bradford Road, Idle, Bradford, Yorks.
Recordaway *see* Protecta
Record Housing, Brook Road, London, N.22.
Rectavox Company, Central Buildings, Wallsend, Northumberland.
Revox, 90 High Street, Eton, Windsor, Berkshire.
Richard Allen Radio Limited, Bradford Road, Gomersal, Cleckheaton, Yorks.
Rogers Developments (Electronics) Limited, 4–14 Barmeston Road, London, S.E.6.

Sansui *see* Brush Clevite.

Shriro (*U.K.*) *Limited*, Electronics Division, Lynwood House, 24/32 Kilburn High Road, London, N.W.6.

Shure Electronics Limited, 84 Blackfriars Road, London, S.E.1.

S.M.E. Limited, Steyning, Sussex.

Sony (*U.K.*) *Limited*, Pyrene House, Sunbury Grove, Sunbury-on-Thames, Middlesex.

A. R. Sugden & Co. (*Engineers*) *Limited*, Market Street, Brighouse, Yorks.

Tandberg *see* Farnell.

Tape Recorder Spares Limited, Harmsworth House, 9 Harmsworth Street, London, S.E.17.

Telefunken *see* A.E.G.

Thomas Heinitz, Music in the Home, 35 Moscow Road, Queensway, London, W.2.

Thorens *see* Metrosound

Trio *see* Morris.

C. E. Watts Limited, Darby House, Sunbury-on-Thames, Middlesex.

Wharfedale *see* Rank Wharfedale.

INDEX

A

Abbreviations 272
Acoustic feedback 156, 223
  suspension speaker 128
Ader, Clément 146
Aerials 172, 213
  illustrated 215
Aerial matching 214
Alignment protractor 207
Ambience 16, 33, 42, 136, 152, 242
Ambi-stereophony 256
Amplifiers 48, 109, 118, 175
  illustrated 181
Amplifier prices 180
  stability 119
Amplitude distortion 42, 69
Anscomb, J. H. 157
Antiphony 34, 153
Atmospheric pressure 33, 36
Audio books 317
  Fair 186
  magazines 318
  terms 272
Autochangers 81
Automatic frequency control (AFC) 103
  gain control (AGC) 100

B

Bach 34, 249
Baldock, R. N. 57, 178, 230
Balmain, G. C. 263
Barbirolli, Sir John 152
Bartok 153
Bass drum 18
  response (loudspeaker limits) 130, 166
Bassoon 25, 37
Beethoven 10, 34, 249, 266
Bell Telephone Labs 29, 147
Beranek, L. L. 255
Berlioz 35, 153, 237, 267
Bias in tape recording 64
Blank disc 205
Bloch 265
Block diagram 52
Blumlein, A. D. 146, 256
B-Minor Mass (Bach) 34
Bottoming 71
Brahms 249, 266, 270
Brass bands 29
Briggs, G. A. 26
Brittain, F. H. 144
Britten, Benjamin 271
Budget systems 187

C

Canby, E. T. 256
Cantilever 57
Cartridges 56, 76, 86, 169
  illustrated 171
Cartridge alignment 206
  fitting 202
  prices 169
Cassettes (tape) 69, 90, 96, 232, 263
Cello 20, 28
Ceramic pickups 76, 111, 169
Channel separation 75
Clarinet 10, 24, 163
  Quintet and Concerto (Mozart) 10
Class-A and Class-B 122, 175
Coaxial feeder 214
Coincident stereo microphones 146, 256
Coleridge, S. T. 252
Collecting records 238
Coloration 40, 180
Column speakers 128
Complete hi-fi systems 187
Compliance 74
Component hi-fi systems 53
Concert programmes 249
Cone breakup 130
Connecting equipment 212
  pickups 208
Contra-bassoon 19, 20
Control units 109
Cor anglais 20, 24
Corelli 249
Corner speaker mounting 223
Crossover distortion 122, 175
Crosstalk 75, 77, 142
Crystal pickups 58, 76, 111
Culshaw, John 265, 267
Curtain of sound 147
Cymbals 19, 25, 44

D
Damping factor 133
Decibels 29
Decoders (multiplex) 106
Demodulation 49
Derived centre channel 118
Diminishing returns in sound quality 159
Directional hearing 139
Direct recording 161
Disc record quality 67, 241
records: the future 253, 262
replay curve 76, 111
Disney, Walt 267
Distortion 42, 44, 116, 120, 122, 244
Divided strings 153
Dixieland 35
Dolby 'B' system 263
noise reducing system 262
Double-bass 18, 20, 32, 37, 44
Drop-out (tape) 89
Dvořák 10
Dynagroove (RCA-Victor) 73, 266
Dynamic range 29, 40, 67, 68, 253, 256, 262

E
Ear drum 33, 36
Early stereo experiments 146
Earthing 205, 210, 213, 220, 222
Earth-loops 220
Eccentric records 233
Eigentones 184, 193
Electronic components 53
music 266
Electrostatic speakers 131, 225, 261
Elliptical styli 72, 253
Enock, Joseph 145
Equipment cabinets 198

F
Fantasia 267, 268
Fidelio (Beethoven) 265
Filters 114, 176, 244
Flute 21, 23, 27, 28, 163
Flutter 43, 44, 83, 91, 163, 166
Flux density (speakers) 133
FM pre-emphasis 101
reception 99, 172, 253
Formants 25, 28
Four-channel stereo 96, 153, 180, 226, 256, 267
Fourier analysis 24
Free standing equipment 157
French horn 20, 25, 44
Frequency 14, 17, 18, 37
response 37, 44
Furniture 198
illustrated 199
Future recording media 263

G
Gabrieli 34
Gammond, Peter 270
George, Dr. W. H. 27
Gerzon, M. 256
G-Major Symphony (Dvořák) 10
Götterdämmerung (Wagner) 151, 259, 265
Gould, Glenn 269
Gramophone Societies 246, 320
Grande Messe des Morts (Berlioz Requiem) 153
Gregorian Chant 33
Guitar 26, 239, 268

H
Haddy, Arthur 69
Hafler, D. 256
Hall, Bishop Joseph 259
Handling records 235
Harmonic distortion 42, 121
Harmonics 19, 23, 25, 36
Harold in Italy (Berlioz) 237
Harp 18, 21, 26, 28
Harpsichord 26
Haydn 35, 249
Headphones 185
illustrated 186
Hearing thresholds 29, 30, 33
Hi-fi and cars 159
components: the future 260
perfectionists 252, 259
systems: prices 187
Hole-in-the-middle 144
Holst 32, 43, 249
Home-made dipole 214
speaker enclosures 229
Horn loudspeakers 127
Hum 50, 85, 90, 167, 218, 222
Hum-loops 221
Human voice 28, 67

I

Ideal room ratios 195
Induced-magnet pickup 59
Inertia (pickups) 79
Infinite baffle speaker 128
Inner groove distortion 71, 245, 253
Integrated circuits 54, 260
Interconnections 211
Intermodulation 42, 116

J

Jazz 10, 35, 259
Jeans, Sir James 259
Judging overall sound quality 191
  loudspeakers 183

K

Keller, Hans 268
Kelly, S. 254
Kingsway Hall 150, 265
Kozma, Tibor 268

L

Labyrinth speaker enclosure 127
Leakey, D. M. 144
Limitations of 2-channel stereo 254
Listening room 184, 193, 224, 247
Live recording 161
London Symphony Orchestra 259
Loudness 16, 29, 31, 243, 248
  control 116
Loudspeakers 45, 124, 180, 260
  illustrated 182
  the future 260
Loudspeaker efficiency 177
  kits 229
  enclosures 125, 229
  height 225
  matching 179, 216
  phasing 226
  positioning 144, 184, 197, 223, 226
  prices 183
  quality 183
  resonance 125, 133

M

Magnetic pickups 59, 60, 76, 86
  recording tape 60
Mahler 31, 148, 152, 243
Mains supply 50, 212, 222
Manfred Symphony (Tchaikovsky), 124
Manufacturers' addresses 320
March, Ivan 269
Matching equipment 162, 179, 214, 216
Mechanical feedback 223
  impedance 74
  noise 86, 91, 163, 166
Menuhin, Yehudi 268
Messiaen 266
Microphony 156
Modulation (radio) 49
Monitoring loudspeakers 224, 261
Mono records 230
  reproduction 244
Moving-coil pickups 60
  speakers 45, 131
Moving-iron pickups 59
Moving-magnet pickups 59
Mozart 10, 31, 33, 249
Multi-directional ambience 256
Multi-path distortion 104
Multiplex stereo reception 105
Musical evenings 245
  Instrument Association 268
  instruments (pitch range) 17
Music for Strings, Percussion and Celeste (Bartok) 153
  power 123

N

Nabucco (Verdi) 251, 252, 259
Negative feedback 120
New Philharmonia Orchestra 152
Noise 41, 44, 89, 176, 261
Noise reduction 91, 262

O

Oboe 20, 24, 44
One-note bass 183
Opera recording 242, 258, 265
Orchestral bells 44
Organ 19, 26, 27, 44
Overloading 44, 109, 178
Overtones 21, 25
Overture, 1812 (Tchaikovsky) 249

P

Parry, Gordon 151
Pathétique Symphony (Tchaikovsky) 249
Phase 4 (Decca) 267
Philadelphia Orchestra 147
Phons 33, 262

Piano 18, 26, 28, 35, 268
backwards 26
concerto recording 242
Quintet (Brahms) 266
reproduction 163, 183
Piccolo 19, 20, 23, 37
Pickups 69, 202, 260
hi-fi specification 82
the future 260
Pickup arms 56, 78, 167
arm prices 168
cartridges 56, 76, 86, 169
cartridges illustrated 171
fitting 201
arms illustrated 170
matching 76, 218
response 77
Piezo-electric pickups 58, 76, 111, 169
Pilot-tone (multiplex) 107, 173
Pinch-effect 71
Pines of Rome (Respighi) 269
Pitch 16, 29
Planets (Holst) 32, 44
Playing weight 79
Plugs and sockets 211
Power amplifiers 48, 118, 261
bandwidth 122
(electrical) 48
handling capacity 42, 122
needed 177
output 121, 123, 177
supplies 50
Preamplifiers 109
Precedence effect 143
Pre-distortion (disc records) 73, 254, 262
Pre-echo (disc) 90
Pre-recorded tapes 68, 263
Presence control 114
Preset adjustments 217
Prices (complete installations) 159, 187
Print-through (tape) 89
Programme planning (for record concerts) 248
Prokofiev 31
Promenade Concerts 8, 248, 270
Pseudo-stereo 138

Q

Quadraphony 96, 153, 180, 226, 256, 267

R

Rachmaninov 249
Radio frequencies 49
quality 66, 99
the future 253
waves 48
Radiograms 156
Ravel 10
Record cabinets illustrated 234
cleaning 236
collection (basic list) 239
concerts 246
handling 235
libraries 238, 320
storage 233
Recording level indicators 88
Producers 264, 269
tape 60
Reflex speaker enclosures 126
Resonance 18, 40, 125, 133
Respighi 269
Response curves 38
Reverberation 34, 136
Ribbon speaker 131
Ring Cycle (Wagner) 150, 265
Ring-main 212
Rite of Spring (Stravinsky) 10
Robinson, D. P. 262
Rodgers and Hammerstein 35
Rodrigo 239
Room resonances 184, 193, 224
Royal Albert Hall 150, 270
Rumble 84, 115, 166

S

Sample concert programmes 249
Scarlatti 268
Scheiber coding system 256
Schubert 35, 270
Screening 219, 221
Shostakovich 35, 239, 267
Sibelius 10
Side-thrust (pickups) 81
Sight and sound 258
Signal-to-noise ratio 42, 89, 110, 164
Simulated concert hall 257
Sine-waves 21, 123
Snare drum 19, 25
Sofiensaal, Vienna 150
Solti, Georg 151
Sonex exhibition 187
Soprano 21, 171

Sound (definition) 14
Sound-waves 15, 22, 47
Speech reproduction 67
Square-wave response 119
Standing-waves 193
Stereophony 43, 135, 254
Stereo groove motions 74
  in small rooms 157
  misunderstandings 137
  radio 104
  studio techniques 148, 242, 265
  tape tracks 95
  the future 254
Stereograms 156
St. Matthew Passion (Bach) 34
Stokowski, Leopold 147, 267
Stravinsky 10, 249
String quartet 153, 242, 266
Stroboscope 84
Studio-Two (E.M.I.) 267
Stylus cleaning 237
  tip radius 71, 230, 254
  wear 231
Sum-and-difference signals 105
Superhet radio 49
Surface noise 230, 236
Swan of Tuonela (Sibelius) 10
Symphonie Fantastique (Berlioz) 249
Symphony of a Thousand (Mahler) 31, 148, 243

T

Tallis Fantasia (Vaughan Williams) 11
Tape bias 86
  distortion 87
  frequency response 65, 87, 92, 94
  heads 61
  hiss 89
  monitoring 97, 164
  recorders 60, 86, 161
  recorders illustrated 165
  recorder prices 164
  recording quality 68, 263
  records 68, 263
  replay curves 92, 111
  the future 263
  track positions 63
  transport mechanisms 62
  units 96
Tchaikovsky 124, 249
Technical terms 272
Telemann 268
Televised music 258
Test-records 230, 320
Timpani 18
Tip mass (pickup stylus) 74
Tonal balance 111, 242
Tone-colour 10, 16, 20, 39, 44, 266
Tone controls 112, 177, 243
Tracing distortion 70, 253
Tracking error 80
  (pickups) 74
Transient response 40, 44, 119
Transmission line loudspeaker 127
Transistor 48
  radio 48, 51
Triangle 19, 25
Trombone 10, 20, 25
Trumpet 21, 25, 44
Tuba 10, 19, 20, 25
Tuner prices 173
Tuner-amplifiers 51
  illustrated 174
Tuners 98, 172
  illustrated 174
Tuning fork 22
  indicators 103
Turnover cartridge 58
Turntables 55, 83, 166
Turntable fitting 201
  prices 167
Turntables illustrated 168
Tweeters 130
  illustrated 132
Two-channel stereo: limitations 254

V

Valves 48
Vaughan Williams 11
Verdi 34, 251
Vienna Philharmonic Orchestra 151
Viola 20
Violin 18, 20, 21, 22, 28, 29, 37
  concertos 258, 265
Vivaldi 249
Volume control 48, 116, 176, 243, 248

W

Wagner 10, 34, 150
Waveform 22, 23, 33, 36, 41, 48
Wavelength (sound) 15, 19
Webb, B. J. 137, 223
Weighted noise figures 85
West, R. L. 134
Winterreise (Schubert) 270
Woofers 131
Wow 43, 44, 83, 91, 163, 166, 234